NANORODS

Edited by **Orhan Yalçın**

Nanorods
Edited by Orhan Yalçın

Published by InTech
Janeza Trdine 9, 51000 Rijeka, Croatia

Publishing Process Manager Marko Rebrovic

Technical Editor Teodora Smiljanic

Cover Designer InTech Design Team

Additional hard copies can be obtained from orders@intechopen.com

Nanorods, Edited by Orhan Yalçın
p. cm.
ISBN 978-953-51-0209-0

For EU product safety concerns:
IN TECH d.o.o., Prolaz Marije Krucifikse Kozulić 3, 51000 Rijeka, Croatia,
info@intechopen.com or visit our website at intechopen.com.

Meet the editor

Orhan Yalçın was born in Niğde, Turkey and studied inorganic spin-Peierls systems. He received his PhD degree in Physics from Gebze Institute of Technology, Kocaeli, Turkey. He worked at Gaziosmanpaşa University, Gebze Institute of Technology, Erciyes University, and Bozok University. Since January 2011 Dr. Yalçın is an academic partner of the Niğde University. His current research interests include magnetic nanostructures (magnetic nanowires, magnetic nanoparticles and magnetic nano O-ring), ferromagnetic resonance, microwave technology, and spin-Peierls systems. Dr. Yalçın has published a number of papers in journals listed in the Science Citation Index and several invited chapters in books released by international publishers.

Contents

Preface XI

Chapter 1 **ZnO Nanorods Arrays and Heterostructures for the High Sensitive UV Photodetection 1**
Soumen Dhara and P.K. Giri

Chapter 2 **Structure, Morphology, and Optical Properties of the Compact, Vertically-Aligned ZnO Nanorod Thin Films by the Solution-Growth Technique 33**
Chu-Chi Ting

Chapter 3 **ZnO Nanorods: Synthesis by Catalyst-Free CVD and Thermal Growth from Salt Composites and Application to Nanodevices 51**
Oleg V. Kononenko, Arkady N. Redkin, Andrey N. Baranov, Gennady N. Panin, Artem A. Kovalenko and Anatoly A. Firsov

Chapter 4 **Collective Plasmonic States Emerged in Metallic Nanorod Array and Their Application 75**
Masanobu Iwanaga

Chapter 5 **ZnO Nanorod Arrays Synthesised Using Ultrasonic-Assisted Sol-Gel and Immersion Methods for Ultraviolet Photoconductive Sensor Applications 93**
Mohamad Hafiz Mamat, Zuraida Khusaimi, Musa Mohamed Zahidi and Mohamad Rusop Mahmood

Chapter 6 **Synthesis and Application of Nanorods 117**
Babak Sadeghi

Chapter 7 **Charge Transfer Within Multilayered Films of Gold Nanorods 129**
Mariana Chirea, Carlos M.Pereira and A.Fernando Silva

Chapter 8 **Preparation and Characterization of Gold Nanorods 159**
Qiaoling Li and Yahong Cao

Chapter 9 **The Controlled Growth of Long AlN Nanorods and *In Situ* Investigation on Their Field Emission Properties 179**
Fei Liu, Lifang Li, Zanjia Su, Shaozhi Deng, Jun Chen and Ningsheng Xu

Chapter 10 **Manipulation of Nanorods on Elastic Substrate, Modeling and Analysis 197**
A. H. Fereidoon, M. Moradi and S. Sadeghzadeh

Chapter 11 **Recent Developments in the Synthesis of Metal-Tipped Semiconductor Nanorods 217**
Sabyasachi Chakrabortty and Yinthai Chan

Preface

The book "Nanorods" is an overview of the fundamentals and applications of nanosciences and nanotechnologies. Nanoscience, nanotechnology and nanomaterials have become a central field of scientific and technical activity. Over the last years the interest in nanostructures and their applications in various electronic devices, effective optoelectronic devices, bio-sensors, photodetectors, solar cells, nanodevices, plasmonic structures has been increasing tremendously. This is caused by the unique properties of nanostructures and the outstanding performance of nanoscale devices. Nanorods is a new and interesting subject, which undergo revolutionary changes in the last years in the nanoscience and nanotechnology.

Dimension in the range of one to hundred nanometers, is called the nano regime. In recent years, nanorods, nanoparticles, quantum dots, nanocrystals etc. are in a class of nanostructures. As the dimensions of nano materials decrease down to the nanometer scale, the surface of nanostructures starts to exhibit new and interesting properties mainly due to quantum size effects.

There are eleven chapters in the book "Nanorods" and each chapter is about a different point of nanorods. This is a basic book for students, researchers and engineers. The chapters in this book are clear, to the point, and appealing. There are many methods for the fabrications of nanostructures (nanorods, nanoparticles, quantum dots, nanocrystals, nanowires, nanowhiskers, hybrid nanostructures and etc.). The methods described in this book are very powerful and have practical applications in the subjects of nanorods. The potential applications of nanorods are very attractive for bio-sensor, magneto-electronic, magneto-sensor, plasmonic state, nano-transistor, data storage media, microwave electronic devices in high frequency region, etc.

There have been increasing interests in nanocrystalline zinc oxide nanorods in recent years. Dhara and Giri presented a review of the recent achievements on the controlled growth of vertically aligned zinc oxide nanorod arrays and heterostructures in Chapter 1. In Chapter 2, Ting fabricated the dense and well-aligned zinc oxide nanorod thin films by the simple solution method. As for synthesis and application of zinc oxide nanorods, the catalyst-free chemical vapor deposition and the thermal growth from salt composites for nanodevices are represented by Kononenko et al. in

Chapter 3, collective plasmonic states are shown to emerge in metallic nanorod array by Iwanaga in Chapter 4 and the optical, structural, and electrical properties are reviewed by Mamat et al. in Chapter 5. A two-step growth method for nanorods developed to grow nanorods by changing the oxygen content in gas mixture during nucleation and growth steps in chapter 6 by Sadeghi. The hybrid nanostructures composed of gold nanorods as a molecular bridge by Chirea and the creation gold nanorods with three new methods by Li and Cao are presented in Chapter 7 and 8, respectively. In Chapter 9, Liu et al. studied the controlled growth of nanorods (nanocone, nanorod, nanocrater and ultra-nanorod) and in-situ investigation on their field emission properties. The applications of nanoparticle include medical and pharmaceutical sciences, electromechanical equipment, laboratory equipment for materials study and other applications have been observed by Moradi et al. in Chapter 10. In Chapter 11, Chakrabortty and Chan investigated the development of hybrid metal-semiconductor nanocrystals as a current research trend in the field of colloidal semiconductor nanocrystals.

This book is of interest to both fundamental research, such as the one conducted in Physics, Chemistry, Biology, Material Science, Medicine etc., and also to practicing scientists, students, researchers in applied material sciences and engineers.

I would like to thank all the authors for their efforts in preparing their chapters. The scientists and technicians have a duty to share their knowledge about their scieintific studies. I am very grateful to Ms. Mia Macek, Mr. Marko Rebrović and Ms. Silvia Vlase for their asistance, guidance, and support.

Dr. Orhan Yalçın
Department of Physics,
Niğde University,
Turkey

1

ZnO Nanorods Arrays and Heterostructures for the High Sensitive UV Photodetection

Soumen Dhara and P. K. Giri
Department of Physics, Indian Institute of Technology Guwahati, Guwahati, India

1. Introduction

In the field of semiconductor nanostructures, one–dimensional (1D) ZnO nanostructures (*e.g.* Nanowires, nanorods, nanobelts) are the most promising candidates due to their important physical properties and application prospects. Large surface–to–volume ratio and direct carrier conduction path of 1D ZnO nanostructures are the key factors for getting edge over other types of nanostructures. ZnO is a direct wide band gap materials having bandgap of ~3.37 eV and high excitonic binding energy, 60 meV at room temperature. 1D ZnO nanostructures are extensively studied for their applications in various electronic and optoelectronic devices, e.g., field effect transistors, ultra violet (UV) photodetectors, UV light emitting diodes, UV nanolaser, field emitter, solar cells etc.. (Huang et al., 2001a; Liao et al., 2007; Li et al., 2005; Kind et al., 2002; Soci et al., 2007; Alvi et al., 2010; Liu et al., 2009; Law et al., 2005; Law et al., 2006; Yeong et al., 2007; Xu et al., 2010; Gargas et al., 2009) Various types of ZnO nanorods (NRs) have been grown by several groups worldwide (Huang et al., 2001b; Wei et al., 2010; Ahn et al., 2004; Li et al., 2008; Dhara & Giri, 2011c; Chen et al., 2010; Giri et al., 2010) and they studied the effect of growth conditions on the morphology of the ZnO NRs. The surface of the nanostructures has crucial role in determining the electrical and optoelectronic properties of nano-devices. As the surface-to-volume ratio in NRs is very high, the surface states also play a key role on optical absorption, luminescence, photodetection and other properties. Thus, nanoscale electronic devices have the potential to achieve higher sensitivity and faster response than the bulk material.

Since the first report on UV photodetection from single ZnO nanowires by Kind et al. (Kind et al., 2002), many efforts have been made on 1D ZnO, including NRs to improve the photodection and photoresponse behaviours. It is known that, photodetection and photoresponse of the ZnO NRs depends on the surface condition, structural quality, methods of synthesis and rate of oxygen adsorption and photodesorption. Therefore, it is expected that arrays of NRs, surface modification or structural improvement can enhance the photosensitivity as well as photoresponse. In the steps towards this goal, various groups have put efforts to enhance the photoresponse and photosensitivity by using appropriate dopant, structural improvement, surface passivation, peizo-phototronic effect and making heterostructures with suitable organic or inorganic materials (Porter et al., 2005; Bera & Basak, 2010; Dhara & Giri, 2011a; Liu et al., 2010a; Yang et al., 2010; Chang et al., 2011). However, photosensitivity value and photoresponce time of the ZnO NWs based photodetectors will require significant improvements in order to meet future demands in variety of fields. At the same time it is also more important to understand the origin of improvement in the photodetection behaviours from ZnO NRs heterostructures in order to play with the photodetection properties to make the flexible photodetectors.

In this chapter we will present a review of the recent achievements on the controlled growth of vertically aligned ZnO NRs arrays and heterostructures by our group and other research groups. Then we will describe the basic properties of these arrays for the application of UV photodetection by means of crystal structures, optical absorption, emission, photoresponse, photosensitivity and photocurrent spectra. The effects of arrays and heterostructures on the mechanism of improved photodetection behavior are also discussed.

2. Growth of ZnO nanorods

ZnO is a II-VI group compound semiconductor whose ionic nature in between covalent and ionic semiconductor. Although the crystal structures shared by ZnO are wurtzite, zinc blende, and rocksalt, however at ambient conditions, only wurtzite phase is thermodynamically stable. The wurtzite structure has a hexagonal unit cell with two lattice parameters, a and c, in the ratio of c/a=1.633 and belongs to the space group of C_{6v}^4or $P6_3mc$. The hexagonal lattice of ZnO is characterized by two interconnecting sublattices of Zn^{2+} and O^{2-}, such that each Zn ion is surrounded by a tetrahedral of O ions, and vice-versa. The Zn terminated polar (0001) plane is the primary growth direction due to the lower surface energy of this plane.

ZnO NRs with controlled shape and order could be grown by thermal vapor deposition (TVD) (Huang et al., 2001b; Giri et al., 2010; Li et al., 2008; Yao et al., 2002), metal-organic chemical vapor deposition (Yuan & Zhang, 2004; Park et al., 2002; Kim et al., 2009), molecular beam epitaxy (Heo et al., 2002), hydrothermal/solvothermal methods (Breedon et al., 2009; Verges et al., 1990; Alvi et al., 2010; Tak & Yong, 2005; Pacholski et al., 2002; Song & Lim, 2007) and top down approach by etching (Wu et al., 2004). Among those techniques, vapor deposition and chemical methods are the widely used techniques for their versatility about controllability, repeatability, quality and mass production. MOCVD and MBE can give high quality ZnO NRs arrays, but use of these techniques are limited, due to the poor sample uniformity, low product yield, choices of substrate, and also the high experimental cost. In the vapor deposition method, the growth process follows either vapor–liquid–Solid (VLS) or vapor–solid (VS) mechanisms, depending on the growth conditions. On the other hand, the NWs are grown by chemical reaction with the seed layer in the hydrothermal/solvothermal methods with the assistance of cationic surfactant. In the growth of ZnO NRs, in general metal catalyst or ZnO seed layer are used to promote the one dimensional and vertical growth. In this case catalyst/seed layer act as a nucleation site and facilitate the one–dimensional growth.

2.1 Mechanosynthesis method

Mechanosynthesis method is generally used for the synthesis of binary metal oxide or complex oxide nanocrystals/quantum dots, however recently we successfully synthesized good quality ZnO NRs with varying sizes. Metal nanoparticles (Tsuzuki & McCormick, 2004; Ding et al., 1995), ZnO nanocrystals (Tsuzuki & McCormick, 2001; Ao et al., 2006), CdS quantum dots (Patra et al., 2011) and various complex oxide nanoparticles (Pullar et al., 2007; Mancheva et al., 2011) have been synthesized by several groups using mechanosynthesis technique. For the growth of the NRs by this method, a suitable surfactant should be chosen, which play a crucial role for the growth in one-direction. The important advantages of this method are NRs can be grown at room temperature and a very fast way, compared to any other chemical methods. In addition, size of NRs could be controlled by reaction time duration and ball to mass ratio.

We have synthesized ZnO NRs of various diameters by mechanosynthesis method at room temperature for reaction time as short as 30 minutes (Chakraborty et al., 2011; Dhara & Giri, 2011b). For the growth of ZnO NRs, mechanochemical reactions were carried out in a planetary ball-milling apparatus. Zinc acetate [$Zn(CH_3COO)_2$], N-cetyl, N, N, N-Trimethyl ammonium bromide (CTAB), a cationic surfactant and sodium hydroxide pellets were used as starting materials. The cationic surfactant plays a crucial role in this reaction and facilitate the growth along only one-direction. The reagents were first mixed together properly before starting milling process. Millings were performed at 300 rpm for the time durations 30 min, 2 and 5 h. After the mechanosynthesis reaction, the resultant product was washed several times by DI water and then with alcohol to remove the surfactant and other bi-products. In the next step, it was dried for 2 h at 100° C to remove the water moister and organic agents.

Figure 1 shows the field emission scanning electron microscope (FESEM) image of the ZnO NRs grown for 30 min reaction, which clearly shows a bundle of dense ZnO NRs. The inset shows the higher resolution isolated NRs of the same sample. The measured diameter and length of the NRs varies in the range 22-45 nm and 300-780 nm, respectively. The FESEM images of the ZnO NRs with reaction time 2 and 5h show similar morphology with smaller lengths in the range 200-600 nm.

Fig. 1. FESEM image of the ZnO NRs grown for 30 min reaction, agglomerated bundle of ZnO NRs are clearly visible. Inset shows the high resolution image of the isolated NRs.

2.2 Vapor-liquid-solid growth method

Vapor–phase synthesis method is the most extensivly explored method for the growth of one–dimensional nanostructructures. Among all vapor-based methods, the VLS mechanism seems to be the most successful in generating large quantities of nanowires with single crystalline structures. Wagner & Ellis (Wagner & Ellis, 1964) first reported this mechanism in the 1960s to produce micrometer-sized wires, later justified thermodynamically and kinetically by Givargizov in 1975 (Givargizov, 1975). In the early twenty–first century, this mechanism is extensively explored by several research groups worldwide to prepare nanowires and NRs from a rich variety of inorganic materials (Wu & Yang, 2000; Zhang et al., 2001; Wu & Yang, 2001; Gudiksen & Lieber, 2000; Wu et al., 2002b; Duan & Lieber, 2000; Pan et al., 2001; Gao et al., 2003; Chen et al., 2001; Wang et al., 2002b). The VLS growth mechanism is practically demonstrated by Yang group (Wu & Yang, 2001) with the help of in–situ transmission electron microscopy (TEM) techniques by monitoring the VLS growth mechanism in real time. In a typical VLS growth, the growth species is evaporated first, and then diffuses and dissolves into a liquid droplet (catalyst particle). The surface of the liquid has a large accommodation coefficient, and is therefore a preferred site for deposition. Saturated growth species in the liquid droplet will diffuse to and precipitate at the interface between the substrate and the liquid. The precipitation will first follow nucleation and then crystal growth. Continued precipitation or growth will separate the substrate and the liquid droplet, resulting in the growth of nanowires/NRs. Preferential 1D growth continues in the presence of reactant as long as the catalyst nanocluster remains in the liquid droplet state.

2.2.1 Self catalytic seed layer assisted growth

For VLS growth of the NWs, metal catalyst nanoisland/nanocluster is essential. However, undesired metal contamination is generally seen for the NRs grown at relatively lower temperature. For the binary compound, it is possible for one of these elements or the binary compound itself to serve as the VLS catalyst. The nanostructures grown by this process is named as self catalytic growth. The major advantage of a self-catalytic process is that it avoids undesired contamination from foreign metal atoms typically used as VLS catalysts. Different groups have reported the ZnO seed layer assisted catalyst free growth of ZnO NRs and studied its morphology and crystallinity by different methods (Li et al., 2006; Li et al., 2008; Li et al., 2009; Kim et al., 2009). Li et al. synthesized vertically aligned ZnO NRs with uniform length and diameter on silicon substrate by vapor-phase transport method and studied the structure, temperature dependent photoluminescence (PL) and field emission behaviours. In this case ZnO seed layer was prepared by pulsed laser deposition (PLD) technique. Kim et al. (Kim et al., 2009) obtained ZnO NRs by metal-organic chemical vapour deposition method with enhanced aspect ratio at relatively a low temperature (300 °C) by supplying additional Ar carrier gas at a high flow rates. In another work by Feng et al. (Feng et al., 2010), well-crystalline with excellent optical properties, flower-like zinc oxide NRs have been synthesized on Si(111) substrate using a PLD prepared Zn film as "self-catalyst" by the simple thermal evaporation oxidation of the metallic zinc powder at 800 °C. The crystalline quality of the ZnO seed layer strongly controlled the structural quality of the NRs. In most of the cases, synthesized NRs were not aligned, hence have limited applications in nanosize electronic and optoelectronic devices. The precise control over the NRs/nanowires lengths and diameters using a self-catalytic VLS technique is very difficult.

We have synthesized small diameter vertically grown ZnO NRs by self catalytic process using ZnO seed layer. First, high quality thin ZnO seed layer of thickness of 200 nm was deposited on the pre-cleaned, HF etched Si wafer by RF-magnetron sputtering. A mixture of high purity ZnO powder and high purity graphite powder at a weight ratio of 1:1 was used as a source. ZnO vapor was produced inside a horizontal quartz tube at 900°C, which was placed inside the muffle furnace. The ZnO vapor was deposited on the seed layer coated Si substrate in downstream direction at 800°C. The vapor deposition was carried out under the Argon gas flow for 30 min. After deposition the entire system was cooled down to room temperature and the synthesized product was characterized.

Figure 2 shows the ZnO seed layer assisted self catalytically grown ZnO NRs, which were grown vertically on the Si substrate. The diameter of the NRs varies in the range of 100-200 nm with a length up-to 1μm. Although the ZnO NRs are grown vertically but the growth orientation is random. From the FESEM image it is revealed that ZnO seed droplet is present on the top of the NRs. It was reported that the quality and diameters of the ZnO NRs depended on the crystallinity and particle size of the seed layer (Cui et al., 2005; Song & Lim, 2007; Zhao et al., 2005). In our case, the grown NRs are non uniform in diameter and length and also not well aligned due to the non uniform distribution of ZnO seed layer. As a result, these ZnO NRs are not suitable for further use in nanodevices. Then we move to the growth process of ZnO NRs by using a metal catalyst.

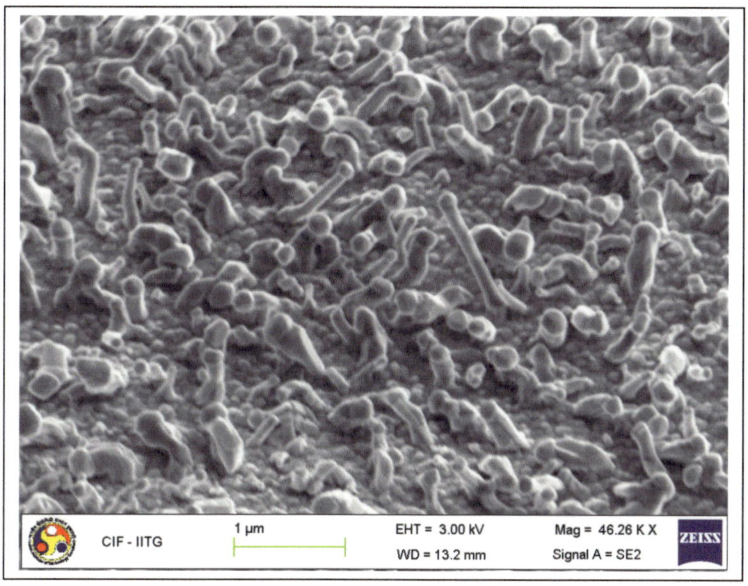

Fig. 2. 45° tilted FESEM image of the seed layer assisted self catalytically grown ZnO NRs.

2.2.2 Gold catalyst assisted growth

For the metal catalyst assisted growth of ZnO NRs, gold catalyst has got major popularity and extensively used due to its comparatively lower eutectic temperature (temperature require to form liquid droplet alloy of Au with the ZnO) and good solvent capability of forming liquid alloy with ZnO. Huang et al. (Huang et al., 2001b) first reported on the synthesis of highly crystalline ZnO nanowires via VLS growth mechanism using mono-dispersed Au colloid as catalyst. Diameter control of the nanowires was achieved by varying the Au layer thickness. They were also able to synthesized patterned nanowires network by patterning the Au catalyst on the substrate. Later, several groups have synthesized ZnO NRs with varieties of ordering using Au catalyst. He at al. (He et al., 2006), using AFM nanomachining technique together with catalytically activated vapor phase transport and the condensation deposition process, have grown a variety of patterned and featured ZnO NRs arrays. The grown pattern and feature are designed by the dotted catalyst prepared by using AFM tip indentation with controlled location, density, and geometrical shape. The vertical orientation of the NRs is achieved by the epitaxial growth on a single-crystal substrate. This technique allows a control over the location, shape, orientation, and density of the grown NRs arrays. Hejazi et al. (Hejazi & Hosseini, 2007) prepared Au-catalyzed ZnO NRs and studied the growth rate on lateral size of NRs, concentration and supersaturation of Zn atoms in the liquid droplet by a theoretical kinetic model, which is in good agreement with the experimental results. A general expression for the NR growth rate was obtained by materials' balance in the liquid droplet and growth front. Based on the derived formula, growth rate is inversely proportional to nanorod radius. A new understanding of the vapour-liquid-solid process of Au catalyzed ZnO NRs was presented by Kirkham et al. (Kirkham et al., 2007) by studying orientation relationships between the substrates, ZnO NRs and Au particles using x-ray texture analysis. From analysis, they claimed that the Au catalyst particles were solid during growth, and that growth proceeded by a surface diffusion process, rather than a bulk diffusion process. ZnO NRs are also grown successfully on the Si (100) or Al_2O_3 substrates by using Cu or NiO or tin as catalyst (Li et al., 2003; Lyu et al., 2002; Lyu et al., 2003; Wu et al., 2009; Gao et al., 2003).

ZnO NRs were synthesized by vapour deposition method on the Si substrate using Au as catalyst. ZnO vapour was prepared at 900°C from the mixture of commercial ZnO powder and graphite powder. ZnO NRs were grown at 800°C on the Au sputtered (of thickness ~5 nm) Si substrate.

Figure 3 shows the SEM image of the randomly oriented vertically grown ZnO NRs. The hexagonal facet of the ZnO NRs is clearly visible from the image. The diameters of the NRs vary from 100 nm to few hundreds of nm with length about few microns. Although the as-grown NRs are grown vertically but the diameter/length and orientation are not uniform. It is suggested that the non-uniformity is due to the non-uniform grain size of the Au in the sputtered film. It is also believed that lattice mismatch between Si and ZnO is responsible for non-uniform orientation.

2.2.3 Combined seed layer and gold catalyst assisted growth

As discussed before, the seed layer as well as the Au catalyst both failed to produces well-aligned ZnO NRs by vapor transport method. Zhao et al. (Zhao et al., 2005) first used the ZnO buffer layer along with Au catalyst and a well-aligned ZnO NRs is obtained. We also studied the effect of pre-depositing ZnO seed layer on the structure, morphology and optical properties of Au catalytic grown vertically aligned ZnO NRs arrays at different temperatures. (Giri et al., 2010) Based on the obtained results, it is understood that ZnO seed layer and Au layer together acting as the nucleation site and guide the NRs growth. So, for the Au/ZnO/Si substrate the nucleation sites of ZnO NRs have the same orientation as ZnO thin film by the effect of the seed layer. The catalyst layer transfers the orientation from seed layer to NRs leading to a vertically well–aligned growth (Zhao et al., 2005; Giri et al., 2010).

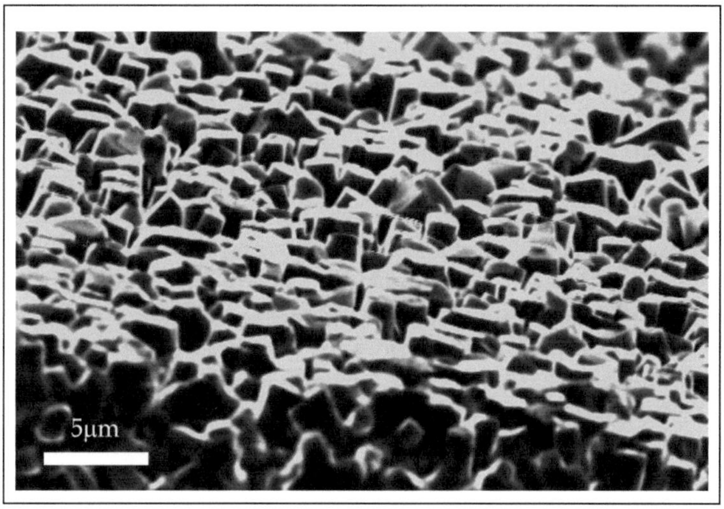

Fig. 3. SEM image of Au catalyst assisted randomly oriented ZnO NRs grown at 800°C.

In the first step of ZnO NRs growth, a ZnO seed layer was deposited by RF-magnetron sputtering followed by deposition of ultrathin Au layer by DC sputtering. Then ZnO NRs were grown in the temperature range of 700-900°C by vapour transport method, as described earlier.

Figure 4 shows typical SEM morphology of ZnO NRs grown on Au/ZnO/Si substrate at various growth temperatures. The NRs grew vertically on the substrate at 900°C, as seen from Fig. 2(a). The sizes of the NRs are in the range of few hundred nanometers and non-uniform diameters are due to variation in the local thickness of ZnO seed layer. ZnO seeds act as a nucleation sites for the NRs growth and importantly offers very negligible lattice mismatch or almost mismatch free interface between seed layer and NRs, which results in the high quality vertically aligned growth of ZnO NRs arrays. NRs grown at 900 and 850°C have larger diameter and highly aligned as comparable to the NRs grown at 700°C.

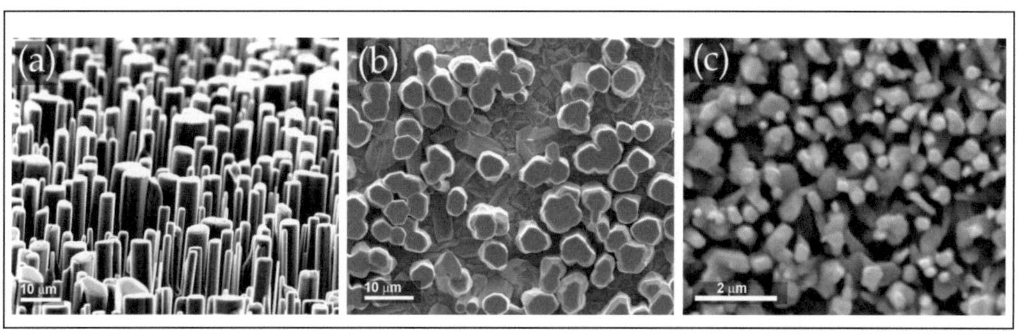

Fig. 4. SEM images of seeded layer and Au catalyst assisted grown aligned ZnO NRs grown at different substrate temperatures: (a) 900°C, (b) 850°C, (c) 700°C, respectively.

2.3 Aqueous chemical growth

Aqueous chemical growth methods are attractive for several reasons: low cost, less hazardous, and thus capable of easy scaling up; growth occurs at a relatively low temperature, compatible with flexible organic substrates; there is no need for the use of metal catalysts; in addition, there are a variety of parameters that can tuned to effectively control the morphologies and properties of the final products (Pearton et al., 2005; Xu et al., 2009; Guo et al., 2011b). The growth process ensures that a majority of the NRs in the array are in direct contact with the substrate and provide a continuous pathway for carrier transport, an important feature for future electronic devices based on these materials. Aqueous chemical methods have been demonstrated as a very powerful technique for the growth of 1D ZnO nanostructures via selective capping mechanisms. It is believed that molecular capping agents play a significant role in the kinetic control of the nanocrystal growth by preferentially adsorbing to specific crystal faces, thus inhibiting growth of that surface. Probably the most commonly used chemical agents in the existing literature for the hydrothermal synthesis of ZnO NRs are $Zn(NO_3)_2$ and hexamethylenetetramine (HMT) (Boyle et al., 2002; Vayssieres, 2003; Tak & Yong, 2005; Song & Lim, 2007). In this case, $Zn(NO_3)_2$ provides Zn^{2+} ions required for building up ZnO NRs. Using HMT as a structural director, Greene et al. (Greene et al., 2006)produced dense arrays of ZnO NRs in aqueous solution having controllable diameters of 30 – 100 nm and lengths of 2 – 5 μm. With addition of polyethylenimine (PEI) in the hydrothermal method, Qiu et al. able to synthesized well-aligned ZnO NRs arrays with a long length of more than 40 μm. However, without the additive PEI, the length of the NRs was not more than 5 μm. Guo et al. (Guo et al., 2011b) studied the factors influencing the size, morphology and orientation of the epitaxial ZnO NRs on the solution using hydrothermal method and discussed about tuning of the size and morphology.

The role of HMT in aqueous chemical method is still not clearly understood. HMT is a nonionic cyclic tertiary amine that can act as a Lewis base to metal ions and has been shown to be a bidentate ligand capable of bridging two zinc(II) ions in solution. In this case, HMT acts as a pH buffer by slowly decomposing to provide a gradual and controlled supply of ammonia, which can form ammonium hydroxide as well as complex zinc(II) to form $Zn(NH_3)_4^{2+}$ (Greene et al., 2006). Because dehydration of the zinc hydroxide intermediates controls the growth of ZnO, the slow release of hydroxide may have a profound effect on the kinetics of the reaction. Additionally, ligands such as HMT and ammonia can kinetically control species in solution by coordinating to zinc(II) and keeping the free zinc ion concentration low. HMT and ammonia can also coordinate to the ZnO crystal, hindering the growth of certain surfaces.

For the chemical growth of ZnO NRs, uniform distribution of ZnO nanocrystal seeds ware prepared on the Si substrate by thermal decomposition of a zinc acetate precursor. Then well-aligned ZnO NRs were synthesized by hydrolysis of zinc nitrate in water in the presence of HMT at 90°C. 25 mM equimolar concentration of zinc nitrate and HMT was used in the growth solution.

Figure 5 shows the well aligned ZnO NRs grown by aqueous chemical method with the help of ZnO seed layer. A high density ZnO NRs grew vertically on the substrate over a large area. The diameters of the NRs ranged from 30 to 40 nm and the length was about few microns. As a characteristic, hexagonal facet of the ZnO NRs are clearly seen from the top view image.

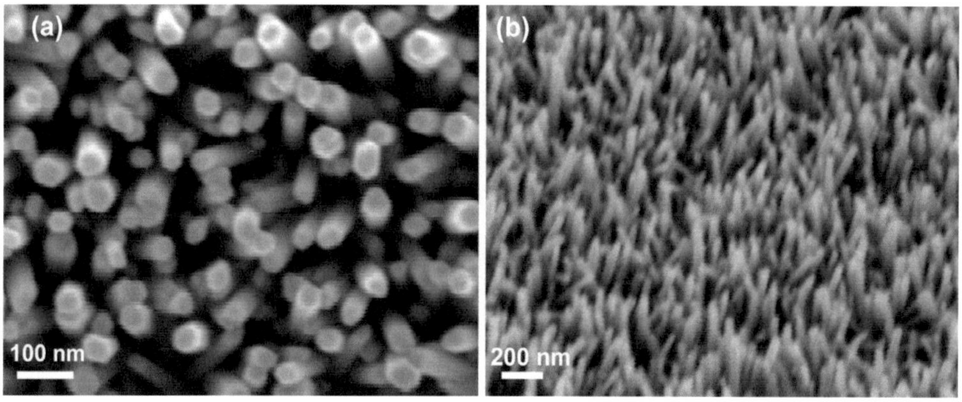

Fig. 5. FESEM images of ZnO nanorods grown on ZnO/Si substrate: (a) top view and (b) 45° tilted view.

3. Fabrication of ZnO nanorod heterostructures

It is considered that heterostructures are superior for the modulation of selective properties of that material. Using suitable external materials for the heterostructures, one can modify the properties of that material according to their requirements. In NRs structures, two types of heterostructures could be fabricated either longitudinal or radial/axial with suitable materials. Fabrication of planar semiconductor heterostructures for thin films is common, whereas the synthesis of one-dimensional heterostructures is difficult. Axial heterostructures, along the length of the NRs axis, have been reported for a few systems, such as InAs/InP, GaAs/GaP and Si/SiGe nanowires (Bjo¨rk et al., 2002; Gudiksen et al., 2002; Wu et al., 2002a). Recently there are reports on radial heterostructures of ZnO nanowires/NRs using several organic/inorganic materials (Bera & Basak, 2009a; Liu et al., 2010a; Bera & Basak, 2010; Liu et al., 2010b; Cheng et al., 2010a; Chang et al., 2011; Um et al., 2011). Bera et al. studied the radial heterostructure effect with poly(vinyl alcohol) on the photocarrier relaxation of the aqueous chemically grown ZnO nanowires. The photocurrent (PC) decay time during steady ultraviolet illumination has been reduced in the heterostructure, a decrease in the PC only by 12% of its maximum value under steady illumination for 15 min and a decrease in the PC by 49% of its maximum value during the same interval of time in the as-grown NWs. Three times enhancement in excitonic emission has been obtained by Liu et al. from the polymethyl methacrylate based ZnO NWs heterostructure. They explain this enhancement on the basis of surface states and energy band theory, due to the decrease in nonradiative process by surface modification. When ZnO NRs heterostructure was fabricated with another semiconducting material, ZnS, very high and faster photoconductivity and also enhanced UV PL intensity are obtained.

ZnO NRs covered with dense and uniform ultra small metal nanoparticles (NPs) is another form of heterostructures. Using suitable noble metal or low work function metal one could be able to achieve very high intense UV PL with significant reduction in visible emission, which is one of the most important requirements for the application in UV LED or laser. Earlier, Lin et al. (Lin et al., 2006) and later Cheng et al. (Cheng et al., 2010b) reported on the significant enhancement of UV PL intensity and subsequent reduction in the defect related visible emission from the ZnO NRs covered with ultra small Au NPs. It is also observed that, after certain size of the Au NPs, the UV PL intensity start decreasing. They proposed that the obtained enhancement is due to the defect loss along with the localized surface Plasmon assisted recombination. Whereas when the NRs surface is covered with Ag NPs, a significant improvement in the yellow-green light emission is obtained (Lin et al., 2011). Interestingly, it is also observed that NRs covered with some metal gives rise to the decrement of the PL intensity (Fang et al., 2011). Although a significant changes is obtained from the metal NPs covered NRs, however a general mechanism for all types of metal covering is yet to emerge.

Here we fabricated ZnO NRs heterostructure by capping the surface with thin layer of anthracene (Dhara & Giri, unpublished). Anthracene/ZnO NRs heterostructures was fabricated by dip coating of the NRs in the diluted anthracene solution. We also fabricated another two heterostructure systems one with decoration of Au NPs and other with Ti NPs (Dhara & Giri, unpublished). From these heterostructures we investigated the origin of the enhanced photoconduction and photoluminescence. Metal NPs decoration was done by directly depositing NPs on the surface of the NRs by sputtering in a controlled way. For systematic study we decorated the surface with different sizes of the NPs by varying the sputtering time. Transmission electron microscope (TEM) image (Fig. 6) of the Au sputtered ZnO NRs shows uniform distribution Au NPs with sizes 3-6 nm coated over the surface of the NWs. The NRs grown by combined ZnO seed layer and Au catalyst using vapor transport method is used for heterostructure fabrication. Due to the vertical alignment of the NRs, Au NPs density is more at the top surface.

4. Structural and optical properties of the ZnO NRs

After the synthesis of nanostructures, it is essential to characterize the as-grown sample to know the structure and related properties. Low-dimensional nanostructures, with possible quantum-confinement effects and large surface area, show distinct mechanical, electronic and optical properties, compared to the bulk materials counterpart. In this section, we will summarise the structural characteristics of the ZnO NRs by x-ray diffraction (XRD), and TEM imaging, followed by optical properties, in particular optical absorption and emission.

4.1 Structural characterization

The structural characterization of the mechanosynthesized NRs was done by XRD shows (Fig. 7) characteristic peaks of pure hexagonal wurtzite phase of ZnO. It is observed that full width at half maximum (FWHM) of the XRD peaks increase monotonically with increase in reaction time. It is primarily due to the reduction of size of the NRs with increase in milling time. With increasing reaction time, the size of NRs decreases and strain is induced during the milling process, resulting in broadening of the XRD peaks.

Figure 8 (a) shows the low magnification TEM image of the 30 min reacted ZnO NRs. Length and diameter of the NRs for ZNR-0.5h sample varies in the range of 300-800 nm and 25-40 nm, respectively. With increase in reaction time, both diameter and length of the ZnO NRs are decreased due to mechanical milling process. During milling, the strain is developed; however, for prolonged milling when the strain is high, the crystal breaks up and thus produces smaller sized NRs. After 5 h of milling, minimum diameter of ~15 nm is obtained (Fig. 8(b)). Fig. 8(b) also shows the high resolution lattice image of the NR with measured lattice spacing, 2.6 Å. The measured lattice spacing is in close agreement with the (002) plane of hexagonal structure. The selected area electron diffraction patterns (not shown) of the corresponding NR show the one-dimensional single-crystalline structures of the as-grown NRs.

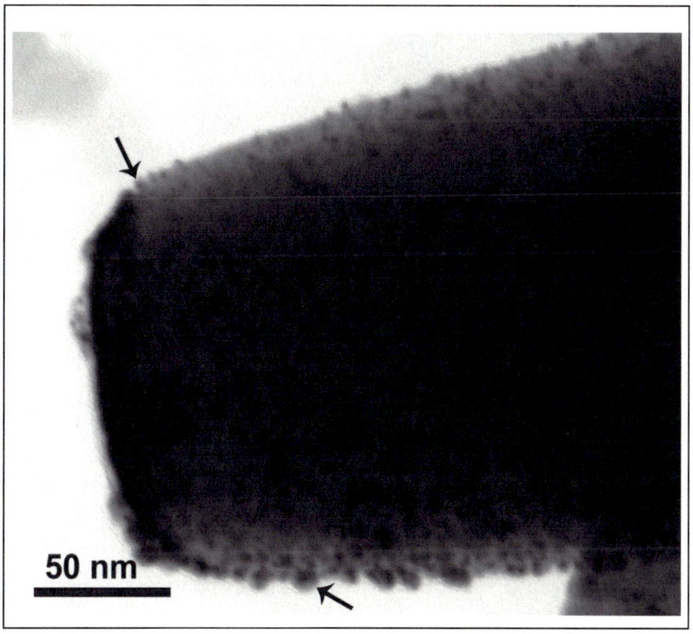

Fig. 6. TEM image of the Au NPs covered ZnO NRs, ultra small Au NPs on the surface of ZnO NRs are shown by solid arrows.

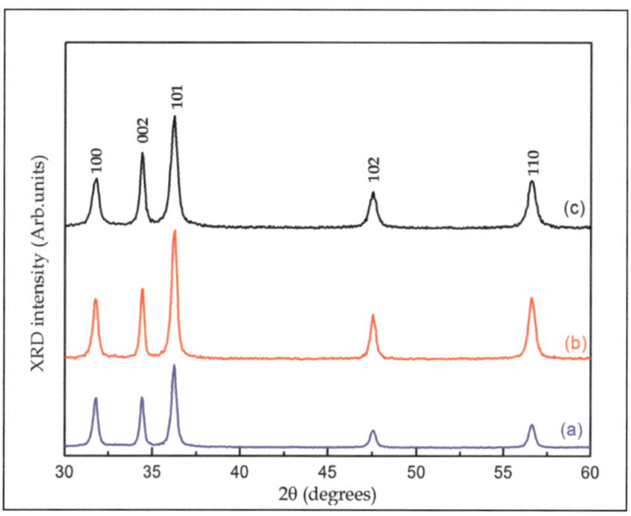

Fig. 7. XRD patterns of the mechanosynthesized ZnO NRs with reaction time: (a) 30 min, (b) 2 h, and (c) 5 h.

Fig. 8. TEM images of the mechanosynthesized ZnO NRs with reaction time; (a) 2 h, and (b) high-resolution lattice image of the 5h sample.

Figure 9 shows the XRD patterns of the ZnO NRs grown on Au coated ZnO seed layer at 900, 850 and 700°C respectively. The observed patterns shows only one strong diffraction peak indicates very high crystallinity. One strong (002) peak of hexagonal ZnO indicates the c-axis orientation of the single crystalline ZnO NRs, which are well aligned and the growth direction is perpendicular to the base surface. Relative intensities of the XRD peaks in Fig. 9 show that NRs grown at higher temperature have higher value of peak intensity, which confirms higher crystallinity. From XRD analysis, we have found that Au coating on ZnO seed layer induces a (111) orientation of the Au clusters at high temperature. Note that NRs grown without the seed layer does not show any preferred orientation and possess inferior crystallinity as compared to that grown with a seed layer. We have found that a substrate temperature below 800°C is not favourable for the growth of aligned NRs by VLS method.

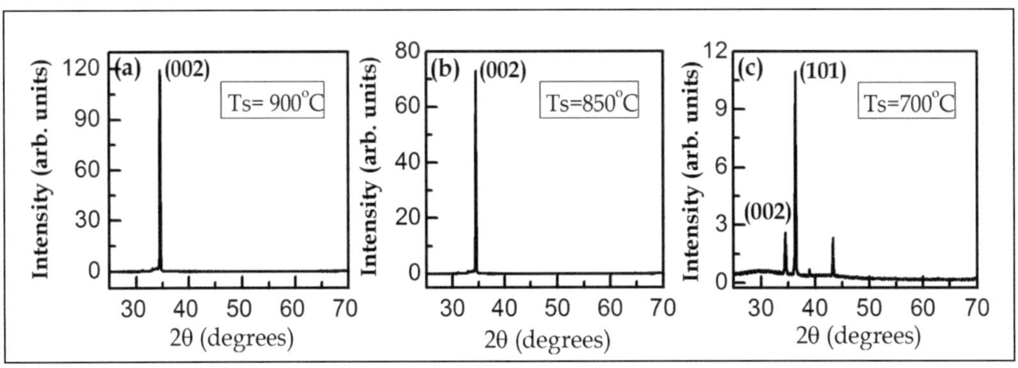

Fig. 9. XRD patterns of ZnO seed layer and Au catalyst assisted grown NRs array: grown at substrate temperature (a) 900, (b) 850 and (c) 700°C, respectively

4.2 Optical properties

As the energy band structure and bandgap reflects on the optical properties of the semiconductors, optical absorption spectroscopy is one of the important tool to probe the energy bandgap. UV-Vis absorption spectra of all the mechanosynthesis samples are shown in Fig. 10. Observed peaks in the UV region correspond to the excitonic absorption of ZnO. A clear blueshift in the absorption peak is observed from 369 nm to 365 nm, as the size reduces from 40 nm to 15 nm. The observed blueshift is indicative of the increase in bandgap with decrease in size of the NRs. This blueshift with size reduction cannot be attributed fully to quantum size effect in ZnO NRs as these NRs have diameters in the range 15-40 nm, which is much higher than excitonic-Bohr diameter in ZnO (~6.48 nm). Therefore, the change in bandgap is partly contributed by the strain induced band-widening. Rapid thermal annealing (RTA) is an effective and simple tool to reduce the strain as well as to improve structural quality. After RTA, a redshift in the excitonic absorption is observed from all the samples, with respect to as-synthesized sample. This redshift is an indication of the decrease in band gap energy as the result of recrystallization and strain relaxation of the NRs (Chakraborty et al., 2011).

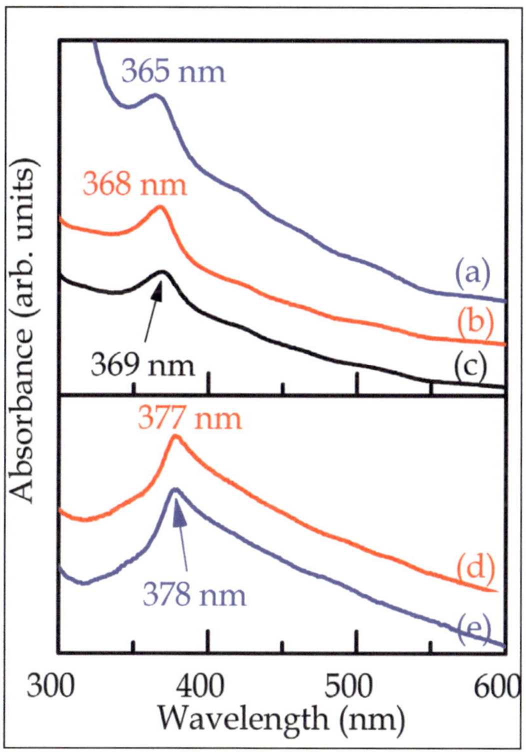

Fig. 10. UV-visible absorption spectra of (a) 5 h, (b) 2 h, and (c) 30 min mechanosynthesized ZnO NRs. Effect of RTA at (d) 500°C and (e) 700°C on the 2 h samples.

The room temperature PL spectra of the mechanosynthesized NRs show three distinct peaks (I-III) in the UV-blue region and one strong broad peak (IV) in the visible region. From 30 min to 5 h samples a blueshift in peak I is observed from 379 to 374 nm. This UV emission is due to the bound excitonic recombination. The peak II at ~390 nm is likely to be due to band-to-band transition between band tail states (Wang et al., 2002a). These band tail states are primarily caused by the presence of defects at the surface of the NRs. The peak III at

~409 nm is caused by the presence of zinc vacancy related defect states. The visible peak (IV) at 582 nm is very broad and it is likely to be related to the atomic disorder at the surface of the NRs caused by milling-induced lattice strain (Giri et al., 2007). An elegant review on presence of various defects in ZnO and corresponding emissions is presented by McCluskey et al. (McCluskey & Jokela, 2009). RTA-treated NRs show reduction in intensity of the peak IV as a result of strain relaxation, whereas intensity of the other three peaks is significantly enhanced. Interestingly, after RTA treatment, peaks II and III are shifted to higher wavelengths. The lattice strain may change the position of the intermediate defect-related states in the band structure of ZnO NRs. Recrystallization of NRs during RTA process is responsible for the change in the band gap and corresponding redshift in the PL spectra.

Figure 12 is corresponding to the PL spectra of the as-grown NRs grown at 900, 850, 700°C, respectively. VLS grown NRs shows two peaks in the PL spectra, one at UV region and other one at green region. The first one is the near band edge (NBE) related excitonic emission and latter one is the oxygen vacancy related defect emission, so called green emission band. The intensity of the UV PL gradually increases with the decrease in growth temperature. The lower intensity of NBE emission from vertically aligned NRs is primarily due to the lower area of absorption by the tip of the aligned NRs and corresponding emission. It is also possible that at higher temperature presence of oxygen vapour is relatively low compared to the low temperature region, which results in the formation of large no of oxygen vacancy states in the ZnO NRs. As a result strong green emission is observed from the NRs grown at higher temperature

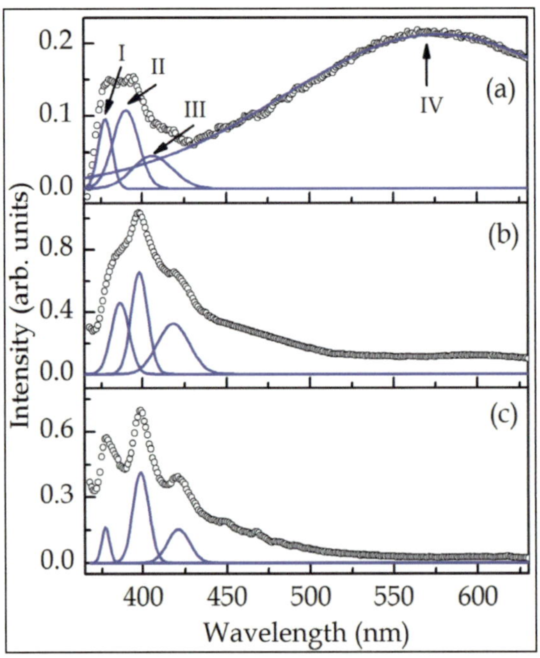

Fig. 11. PL spectra of 2 h mechanosynthesized ZnO NRs (a), after RTA at 500°C (b) and 700°C (c), respectively. Four peaks are fitted with Gaussian function (solid line) to the exp. data (symbol).

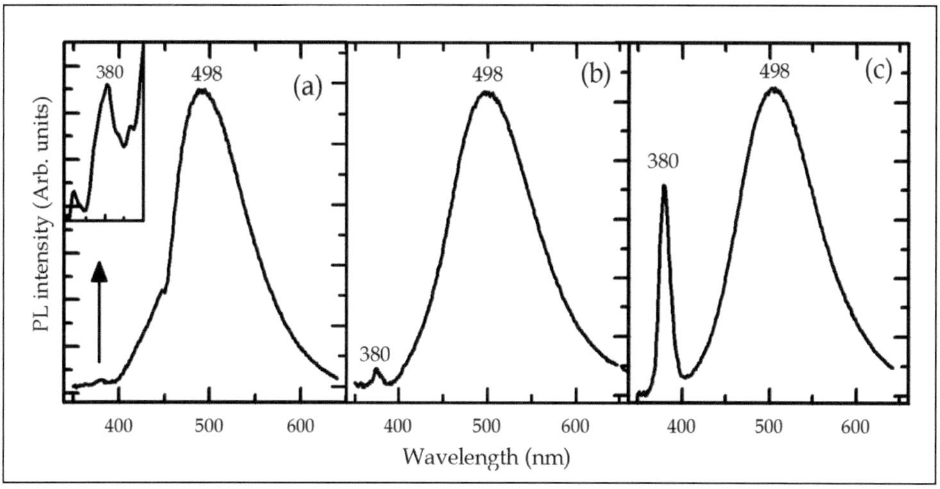

Fig. 12. PL spectra of combined seeded layer and Au catalyst grown aligned ZnO NRs at various substrate temperatures: (a) 900°C, (b) 850°C, (c) 700°C.

5. Photodetection behaviours of the ZnO NRs

Electronic conductivity of the ZnO NRs significantly enhanced when it is exposed to the light with wavelength below 380 nm. Using this property, ZnO NRs can be used for UV photodetectors. The dramatic change of conductance between dark and UV exposure suggest that the ZnO NRs photodetectors are also good candidates for optoelectronic switches, with the dark state as "OFF" and the UV exposed state as "ON" . In the step towards the efficient and faster photodetection from ZnO NRs/nanowires some important works have been done recently, which are summarize in Table 1. Several types of approaches have been reported e.g. structural improvement, efficient doping, and heterostructures formation with suitable external materials. Zhou et al. (Zhou et al., 2009) used nonsymmetrical Schottky-type (ST) contact devices and obtained higher sensitivity and faster reset time. Pt microelectrode arrays were first fabricated on a SiO_2 /Si substrate by UV lithography to make Schottky-type contact on one end of the nanowires and a focused-ion-beam (FIB) deposited Pt-Ga electrode on other end of the ZnO nanowire for a good Ohmic contact. He and coauthors utilized FIB technique to deposit Pt metal on ZnO nanowires to effectively reduce the contact resistance, and thus achieved high photoconductive gain as high as 10^8. Chang et al. report the synthesis of a ZnO NR/graphene heterostructure by a facile *in situ* solution growth method (Chang et al., 2011). By combining the attributes of photosensitive ZnO NRs and highly conductive graphene, they are able to fabricate a highly sensitive visible-blind ultra UV sensor. Recently, Park and coauthors obtained enhanced photoresponse from isopropyl alcohol treated ZnO nanowire devices by introducing surface roughness induced traps (Park et al., 2011). They propose that obtained enhancement is attributed to an increase in adsorbed oxygen on roughening induced surface traps.

Therefore it is very important to have detail understanding about current conduction mechanism and origin of enhancement from the heterostructures. It should be mentioned that, till now, the lack of well-established fabrication method and standard procedures make it difficult to compare the experimental results between different devices.

Morphology	Device Type	Light of Detection (nm)	Bias (V)	Maximum Photosensitivity	Photosensitivity enhancement factor from unmodified photodetector	Reference
Nanowire	Resistor	365	5	10^4-10^6	-	(Kind et al., 2002)
Nanorods	Resistor	325	2	19	-	(Ahn et al., 2004)
Nanowires film	Resistor	254	5	17.7	-	(Li et al., 2005)
Nanorod	FET	254	0.2	1000	-	(Park et al., 2005)
Nanowire	Resistor	390	5	10^4	-	(Soci et al., 2007)
Nanowires arrays	Resistor	325	3	18000	~2.6	(Bera & Basak, 2009a)
Nanowires arrays	Resistor	360	3	10^4	~2.8	(Bera & Basak, 2009b)
Nanowire	Resistor	365	1	1500	~4	(Zhou et al., 2009)
Nanowire	Resistor	254	4	1800	~9.4	(Lin et al., 2009)
Nanowires arrays	Resistor	370	3	3367	~5.2	(Bera & Basak, 2010)
Nanowires film	Resistor	365	8	-	~4.7	(Liu et al., 2010a)
Nanorods	Resistor	360	10	80	-	(Manekkathodi et al., 2010)
Nanowires arrays	Resistor	369	2.5	24200	~5.4	(Dhara & Giri, 2011a)
Nanorod	Resistor	370	20	-	~3.0	(Chang et al., 2011)
Nanowires	*n-i-n* junction	365	-5	1345	-	(Kim et al., 2011a)
Nanorods (interdigitated)	Resistor	379	5	12.1mA/W	-	(Guo et al., 2011a)
Nanowires arrays	Resistor	360	5	7600	~2.2	(Bera & Basak, 2011)
Nanowire	FET	365	0.4	10^6	~1.8	(Park et al., 2011)
NWs network	Resistor	254	5	52	~3	(Kim et al., 2011b)

Table 1. The performance characteristics of ZnO NRs/nanowires based photodetectors reported in the literature.

5.1 Dark I–V characteristics

Recently we have shown that presences of native surface defects (oxygen vacancies) could be identified from the dark I–V curves (Dhara & Giri, 2011a). The charge-depletion layer induced by surface adsorption of oxygen molecules completely controls the charge transport in NRs, if the diameter of the NRs is comparable to the depletion layer thickness. Another important step in determination of the electrical properties of NRs is the metal–NRs interface through metal electrode. In addition to the above factors, in case of nanowires network structures, charge transport is also determined by the nanowire– nanowire contacts. The dark current-voltage (I-V) characteristic of the as-grown NRs shown in Fig. 13 shows a linear behavior up to a certain bias voltage. A linear fit to log-log plot of the I-V data could give detailed information about the mechanism of current conduction process. There is a crossover point of linear fitting around at 8V. Below this bias voltage, the power dependence of current on voltage is exactly 1, indicating an Ohmic conduction region, beyond which the current dependence on voltage is greater than one. This is most likely to be space charge limited current (SCLC) (Rose, 1955; Ohkubo et al., 2008), which arise from charge carriers trapped at the surface defect states that contribute to the current at higher bias voltage. In case of SCLC, power dependence greater than two could be observed depending on the energy distribution of trap centers. As the RTA treated NRs have very less native defects due to structural improvement and release of built in stress, a linear curve is expected. The NRs RTA treated at 800°C shows exactly a linear behavior, as expected.

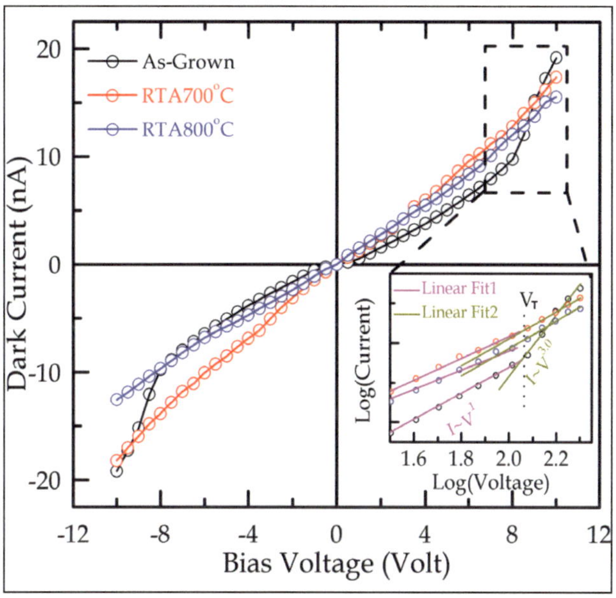

Fig. 13. The dark current-voltage characteristics of as-grown and RTA-treated ZnO NWs processed at 700°C and 800°C. Inset shows the magnified view of the selected region in log-log scale.

5.2 Spectral dependence of photodetection

Wavelength dependent PC studies of the ZnO NRs shows that it gives very high PC when it is exposed to the UV light (Fig. 14). The maximum PC is obtained at the excitation of 369 nm light, which is the band gap wavelength of ZnO NRs. The observed strong peak at 369 nm in the PC spectra is due to the band-edge absorption followed by generation of photocarriers (electron-hole pair). The small hump-like peaks in the visible region are due to the generation of carriers from the native defect states. In this case, the photosensitivity (photo-to-dark current ratio) is ~4500, which is quite low. By structural improvement or heterostructure formation, above hump could be eliminated and a visible-blind ZnO NRs based photodetectors with high sensitivity could be made.

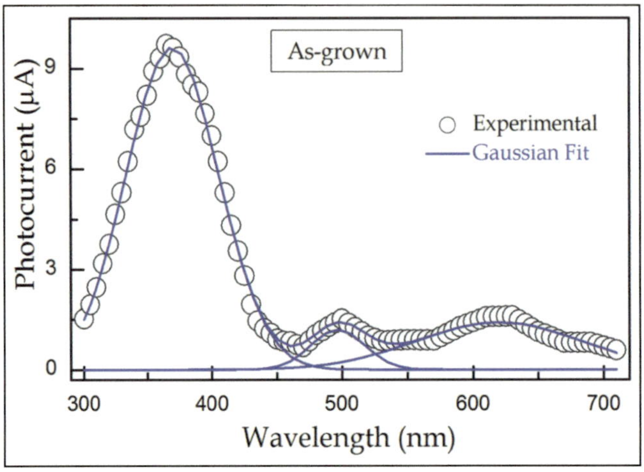

Fig. 14. The photocurrent spectra of the ZnO NRs measured at 2.5 V bias.

5.3 Photoresponse

The photoresponse behaviour of the ZnO NRs measured under the excitation of 365 nm UV light is shown in Fig. 15. It is seen that PC initially grows very fast and then slowly increased with time and finally saturated (mechanism is explain in next sub-section). The time-dependent PC growth and decay curves are fitted with the following equation (Dhara & Giri, 2011a),

$$I_{ph}(t) = I + A_1(1 - e^{-t/\tau_1}) - A_2 e^{-t/\tau_2} \tag{1}$$

$$I_{ph}(t) = I_{ph}(\infty) + A_3 e^{-t/\tau_3} + A_4 e^{-t/\tau_4} \tag{2}$$

where I, A_1, A_2, A_3 and A_4 are positive constants and $I_{ph}(\infty)$ refers to the photocurrent after infinitely long time of the decay experiment, which essentially is the dark current. The first exponential term in the growth and decay equations corresponds to the electron-hole generation and recombination processes and the last exponential term represents the oxygen adsorption process. Calculated time constants from fittings are τ_1 = 25.7 s and τ_2 = 347.9 s for PC growth, τ_3 =19.3 and τ_4 =316.0 s for PC decay, respectively. The photoresponse time for the as-grown NRs is very slow due to the presence of intrinsic defects/trap centres.

5.4 Photodetection mechanism of ZnO NRs

It is known that, the photoresponse of the ZnO NRs consists of two parts: a rapid process of photogeneration and recombination of electron-hole pairs, and a slow process of surface adsorption and photodesorption of oxygen molecules (Dhara & Giri, 2011a). The oxygen plays a crucial role in the photoresponse of ZnO. In dark condition, oxygen molecules from the air are easily stuck on the NRs surface by adsorption process and trapped electrons [$O_2(g)+e^- \rightarrow O_2^-$] available on the surface near the Zn lattice and decreased the conductivity (Kind et al., 2002), which is shown schematically in Fig. 16(a). This process leads to the formation of depletion layer near the surface resulting in the band bending of the conduction band (C.B) and the valence band (V.B). Formation of large number of ionized oxygen on the NWs surface enhances the band bending, resulting in a very low conductivity. During the UV illumination, electron-hole pairs are generated [$h\nu \rightarrow e^- + h^+$] by light absorption. Now these electrons/holes easily cross the depletion layers and contribute to the photoconduction process. At the same time, holes take part in the oxidization of ionized oxygen ($O_2^- + h^+ \rightarrow O_2(g)$, photodesorption process) and release one oxygen gas molecule by electron-hole recombination process (Fig. 16(b)). Then few of the released oxygen molecules are re-adsorbed on the surface and decrease the free electron carriers. The energy band diagram during UV illumination is shown in (c). After certain time electron-hole generation rate and oxygen re-adsorption rate becomes constant resulting in a steady photocurrent. It is known that adsorption process is slower than the photodesorption process. Therefore, during UV illumination, not all the holes are recombine with the electrons present in the ionized oxygen. As a result, excess holes are available for recombination with the exciton related free electrons. During photocurrent decay, the exciton related electron-hole recombination dominates, which corresponds to the faster decay component, so the photocurrent initially decreases very rapidly. With the surface re-adsorption of oxygen, the photocurrent comes to the initial value very slowly.

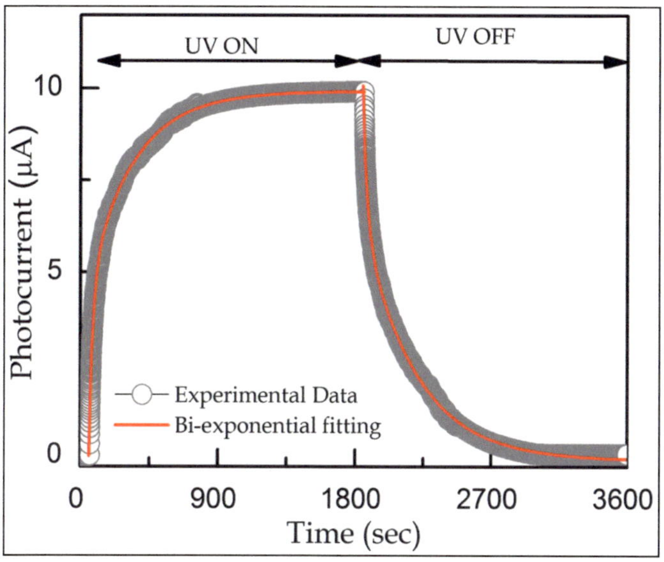

Fig. 15. The Photocurrent growth and decay behaviors (photoresponse) of as-grown ZnO NRs.

5.5 Effect of structural improvement

We have shown that a fivefold enhancement of photosensitivity in the UV region and faster photoresponse could be obtained from the ZnO NWs/NRs by structural improvement using RTA processing (Dhara & Giri, 2011a). The photocurrent growth and decay rates (photoresponse) from RTA-treated NWs are improved by a factor of approximately 2. After RTA at 800°C, the PC at 369 nm reaches a maximum value of 84.1 μA (Fig. 17) from that of 9.6 μA for the as-grown NWs, results in a sensitivity value of 24.2 × 10^3, leading to an enhancement factor of five. The PC growth and decay time constants are improved to 12.3 and 107.0 s for growth, 13.6 and 118.4 s for decay. The RTA processing substantially removes the surface defect-related trap centers and modified the surface of the ZnO NWs, resulting in enhanced PC and faster photoresponse. During RTA processing, the NRs recrystallize and structural quality is improved by releasing built-in stress and removing the defect states. Due to reduction of surface defects, the PC in the visible region drastically decreased. Therefore RTA processed photodetector is fully visible-blind and only sense the UV light. The high photosensitivity even in low light intensity is an indication of very low value of detection limit.

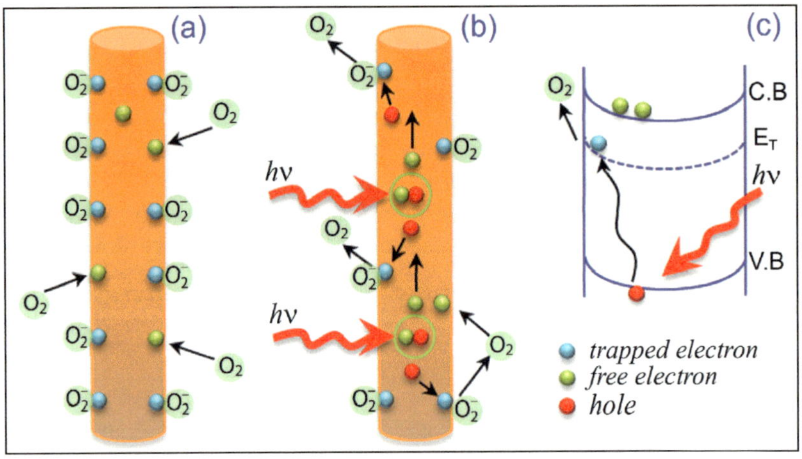

Fig. 16. A schematic of photoresponse mechanism of ZnO NRs: (a) at dark condition and (b) during UV illumination. (c) Schematic energy band diagram of photoresponse process during UV illumination.

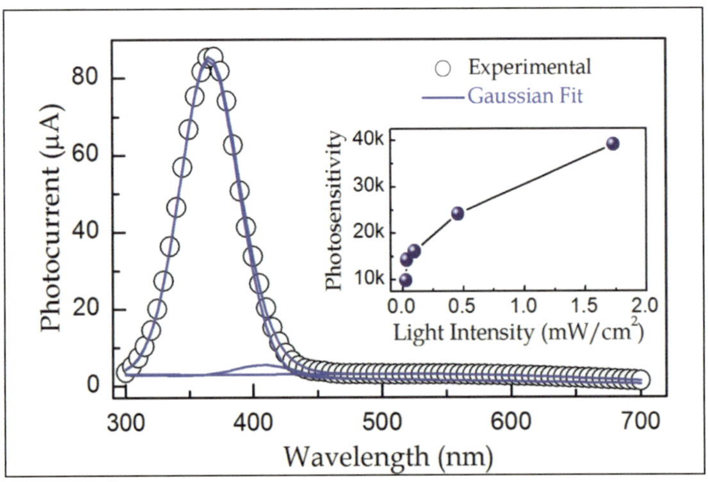

Fig. 17. The photocurrent spectra of the ZnO NRs after RTA treated at 800°C.

5.6 Effect of heterostructure by surface capping

Here we present the effect of surface capping on the ZnO NRs with anthracene on the enhancement of PC and photoresponse. Although the dark current is almost doubled after the capping, however a significant improvement in the photosensitivity is obtained. Compared to the as-grown case, the photoresponse time becomes much faster for the ZnO/anthracene system with response and reset times within second (Dhara & Giri, unpublished). The reasons for choosing anthracene are that it can act as UV sensitive material to enhance the photosensitivity and it can influence the oxygen adsorption process. After anthracene capping the maximum photocurrent increases to 50 μA from 4 μA, which is for as-grown NRs (Fig. 18). Photosensitivity value is also increased to 4183, leading to the six-fold enhancement. Very high photosensitivity and low dark current are the basic requirements for efficient photodetection. Due the surface capping, the surface of the NRs becomes modified and in this case thickness of the depletion layer is much lower than the case of as-grown NRs. When the sizes of the nanostructures are comparable to the space charge layer, surface depletion greatly affects the density and the mobility of the carriers in ZnO NWs rather than the contact potential (Li et al., 2007). Here anthracene layer is very thin and the carriers easily tunnel through the layer to the electrodes, resulting in increment in dark current.

The photoresponse spectrum (Fig. 19) measured at 360 nm shows very fast response with response and reset times of 1.5 and 1.6 s, respectively (the response and reset time can be defined as 1-1/e, (63%) of the maximum photocurrent increased and 1/e, (37%) of the maximum photocurrent decreased, respectively). In contrast, the as-grown NRs have response and reset time about 7.2 and 63.2 s, respectively. Therefore anthracene capped ZnO NRs heterostructure has five times faster response time and about forty times faster reset time, which is very good for real time application. Here, the anthracene has strong absorption in the UV region, when it is excited with UV light the photoexcited charge carriers transfer to the conduction band of ZnO NRs. This process results in high photocurrent and consequent higher photosensitivity. Modification/reduction of surface defects related traps by anthracene capping is also responsible for the obtained very high photocurrent.

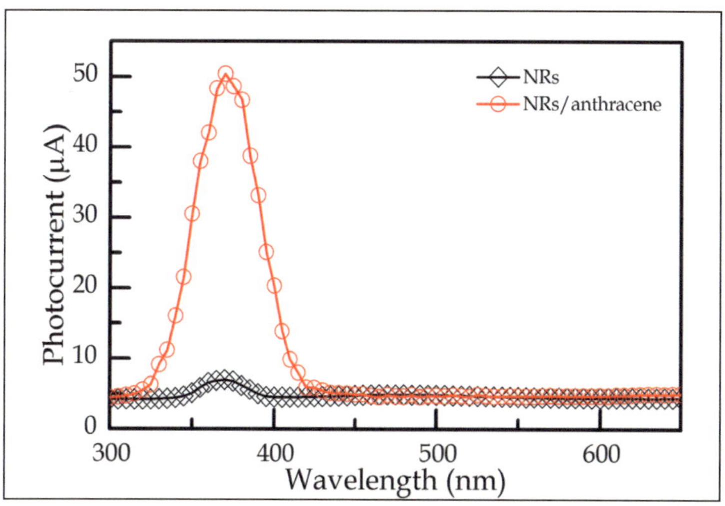

Fig. 18. Photocurrent spectra of the as-grown ZnO NRs and ZnO/anthracene based NRs heterostructure.

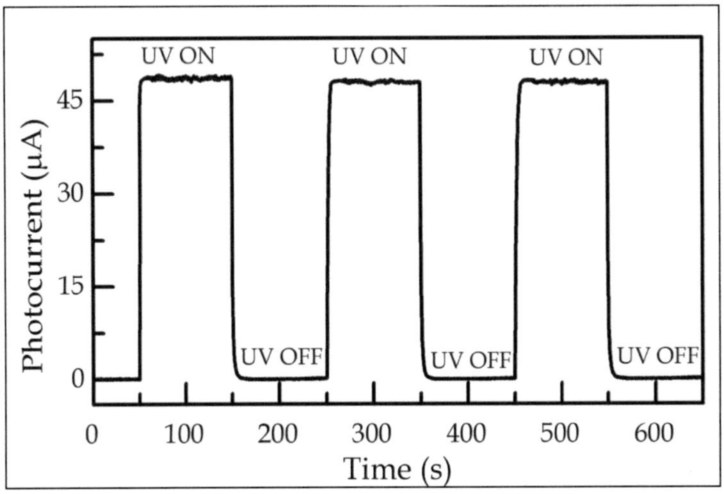

Fig. 19. Photoresponse behavior (at 360 nm) of the ZnO/anthracene based NRs heterostructure in a repetitive UV light "ON" and "OFF" conditions.

5.7 Metal NPs decorated ZnO NRs heterostructures

In case of Au/ZnO heterostructures, Au NPs and ZnO NRs interface actively interplay on the photodetection process. Au/ZnO heterostructure shows nearly linear dark I-V characteristic with reduced dark current. The decrement of current is more prominent in the lower bias voltage and at higher bias voltage it reached very close to the dark current of as-grown NWs. It is known that when a metal is brought in contact with semiconductor, it induces band bending due to the equilibrium of Fermi level (Liao et al., 2007; Dayeh et al., 2007). Depending on the difference in the work-functions of metal and semiconductor, two types of contacts have been formed at the metal semiconductor interface. Therefore, the decrement in dark current is due to the large upward band bending and formation of Schottky barrier at the interface between Au and ZnO. Because, Au has larger work-function, 5.47 eV (Lide, 2009) than ZnO, 4.65 eV (Aguilar et al., 2009). As a result, no electron will be transfer from Au to the conduction band of ZnO. However, at a lower bias voltage few numbers of electrons can pass from ZnO to the Al electrode, giving very low dark current. At higher bias voltage, which is greater than the Schottky barrier height, all the electrons can flow to external circuit giving almost equal dark current as for the case of as-grown NRs.

As expected, the as-grown NRs shows low PC in the UV region (Fig. 20.), whereas it is significantly enhanced after Au NPs decoration. Here seven times enhancement in the photosensitivity is obtained. The broad photocurrent peak from the Au/ZnO heterostructures in the visible region is due to the Au NPs absorption related carrier transfer to the conduction band of ZnO. As the dark current of the Au/ZnO heterostructures is lower than the as-grown ZnO NRs, a lower PC is expected. However, the obtained PC is quite high. In this case, with respect to the vacuum level, the energy level of oxygen vacancy defect states (5.43 eV) and Fermi level of Au (5.47 eV) is very close to each other (Lide, 2009). Therefore, the electrons from this defect states can transfer to the Fermi level of Au, which increases the electron density at the Fermi level of Au. The Au NPs are excited by incident light in the UV-violet region due to interband transition and in the green region due to the surface Plasmon resonance (SP band) (Garcia, 2011). After that the excited energetic electrons are stay in higher energy states, and these are so active that they can escape from the surface of the NPs and can transfer to the conduction band of ZnO. Under bias, these electrons along with electrons generated by band edge absorption of ZnO contributed to the current conduction process. Therefore, the obtained enhanced photocurrent in the UV as well as in the visible region is due to the increase of electron density in the conduction band of ZnO by ZnO band edge absorption and electron transport via Au NPs (Dhara & Giri, unpublished).

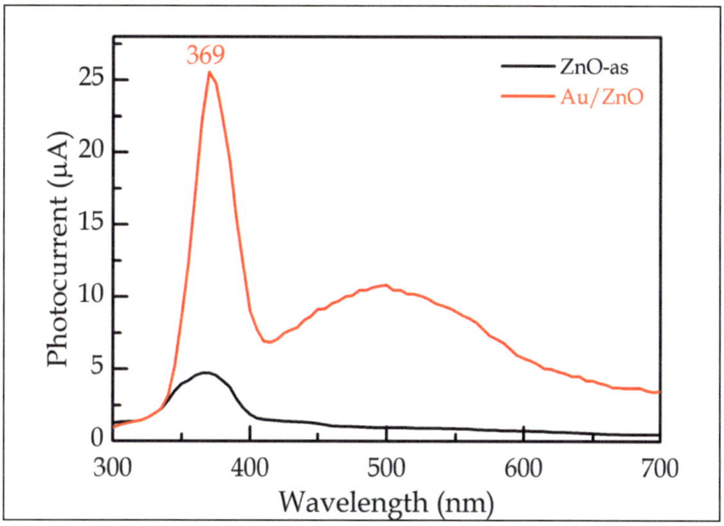

Fig. 20. Photocurrent spectra of the as-grown NRs and Au NPs decorated Au/ZnO NRs heterostructures.

A faster photoresponse is obtained from the Au/ZnO heterostructures and it is almost independent of the thickness of the Au layer. The response and reset times are as fast as 26.5 s and 70.0 s. In this case due to the decoration of Au NPs, surface of the ZnO NWs becomes modified and facilitated to increase the adsorption and desorption process, resulting in faster photoresponse.

On the contrary, Ti NPs decorated heterostructures show a linear I-V behavior with very high dark current. The dark current gradually increases with increase in Ti coverage. Due to the lower work-function of Ti (4.26 eV) (Lide, 2009) compared to ZnO, an Ohmic contact is formed with downward band bending. In this case, more numbers of electrons can easily transfer from Ti to the conduction band of ZnO at the interface. Then under the bias these electrons contributes to the current conduction process results in high dark current.

The PC in the UV region is drastically enhanced at an excitation wavelength of 369 nm. The Ti/ZnO heterostructure gives PC of 65.5 μA. Although the obtained PC from Ti/ZnO heterostructure is very high, but due to higher dark current the photosensitivity values are low. The photoresponse behaviours of the Ti decorated ZnO NRs show very fast response with response and reset time within few seconds. The fastest response time of 5.5 s and reset time of 7.7 s is obtained. These data indicate a significant improvement in the photoresponse process and much faster photoresponse could be obtained from Au or Ti decorated ZnO NRs heterostructures.

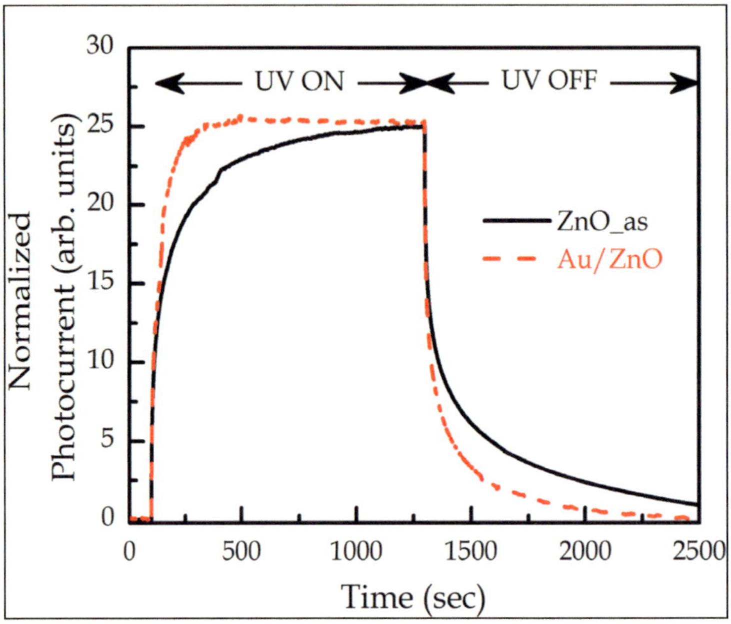

Fig. 21. Photoresponse behaviours of the as-grown NRs and Au NPs decorated Au/ZnO NRs heterostructures under the illumination of 365 nm UV light.

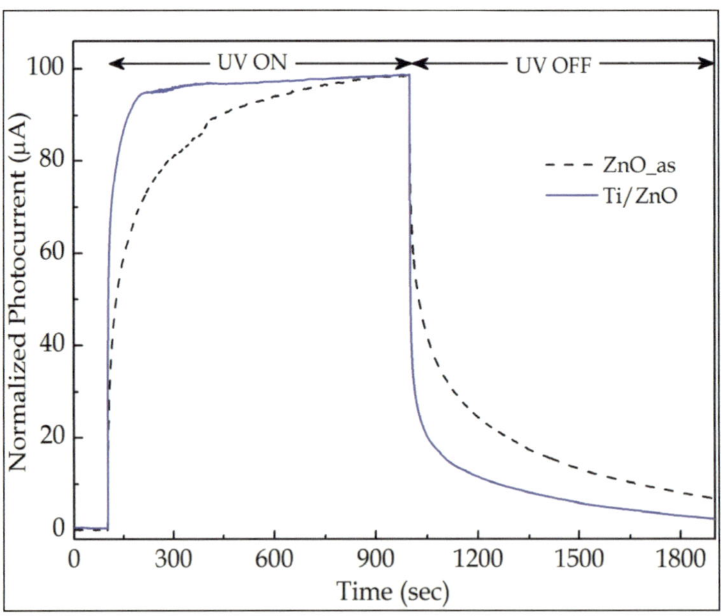

Fig. 22. Photoresponse behaviours of the as-grown NRs and Ti NPs decorated Ti/ZnO NRs heterostructure under the illumination of 365 nm UV light.

6. Summary

Here we reviewed our recent achievement on the controlled growth of vertically aligned ZnO NRs arrays and their heterostructures for the applications of efficient UV photodetection. We provided a summary of effects of several growth parameters on different growth methods for the well aligned ZnO NRs arrays. ZnO NRs arrays grown by three different methods; mechanosynthesis, vapour-liquid-solid and aqueous chemical methods are presented here. It is shown that the combined effects of ZnO seed layer and Au catalyst are favourable for the growth of well aligned ZnO NRs arrays. Three different types of ZnO NRs based heterostructures were fabricated; one with surface capping of anthracene and others with surface decoration of Au and Ti NPs with suitable sizes. Photodetection behaviours of the different systems are studied by dark I-V characteristics, wavelength dependent photocurrent and photoresponse. The results demonstrate that ZnO NRs heterostructures are indeed excellent candidates for UV photodetectors with very high sensitivity and faster response for real time sensing applications. Possible mechanisms of improved photodetection behaviours from different systems are also presented. These understanding will help to design and fabricate a ZnO NRs heterostructure based efficient UV photodetectors. An up-to-date summary of important results by several research groups worldwide on the ZnO NRs/NWs heterostructures based UV photodetectors is presented in Table 1. Our approaches show comparable significant improvement over several reports on ZnO NRs based photodetectors.

7. References

Aguilar, C. A.; Haight, R.; Mavrokefalos, A.; Korgel, B. A. & Chen, S. (2009). Probing electronic properties of molecular engineered zinc oxide nanowires with photoelectron spectroscopy. *ACS. Nano*, Vol.3, pp. 3057-3062

Ahn, S. E.; Lee, J. S.; Kim, H.; Kim, S.; Kang, B. K.; Kim, K. H. & Kim, G. T. (2004). Photoresponse of sol-gel-synthesized ZnO nanorods. *Appl. Phys. Lett.*, Vol.84, pp. 5022-5024

Alvi, N. H.; Riaz, M.; Tzamalis, G.; Nur, O. & Willander, M. (2010). Fabrication and characterization of high-brightness light emitting diodes based on n-ZnO nanorods grown by a low-temperature chemical method on p-4H-SiC and p-GaN. *Semicond. Sci. Technol.*, Vol.25, pp. 065004

Ao, W.; Li, J.; Yang, H.; Zeng, X. & Ma, X. (2006). Mechanochemical synthesis of zinc oxide nanocrystalline. *Powder Technol.*, Vol.168, pp. 148-151

Bera, A. & Basak, D. (2009a). Effect of surface capping with poly(vinyl alcohol) on the photocarrier relaxation of ZnO nanowires. *ACS Appl. Mater. Interface*, Vol.1, pp. 2066-2070

Bera, A. & Basak, D. (2009b). Role of defects in the anomalous photoconductivity in ZnO nanowires. *Appl. Phys. Lett.*, Vol.94, pp. 163119

Bera, A. & Basak, D. (2010). Photoluminescence and Photoconductivity of ZnS-Coated ZnO Nanowires. *ACS. Appl. Mater. Interfaces*, Vol.2, pp. 408-412

Bera, A. & Basak, D. (2011). Pd-nanoparticle-decorated ZnO nanowires: ultraviolet photosensitivity and photoluminescence properties. *Nanotechnol.* , Vol.22, pp. 265501

Bjo¨rk, M. T.; Ohlsson, B. J.; Sass, T.; Persson, A. I.; Thelander, C.; Magnusson, M. H.; Deppert, K.; Wallenberg, L. R. & Samuelson, L. (2002). One-dimensional heterostructures in semiconductor nanowhiskers. *Appl. Phys. Lett.*, Vol.80, pp. 1058-1060

Boyle, D. S.; Govender, K. & O'Brien, P. (2002). Novel low temperature solution deposition of perpendicularly orientated rods of ZnO: Substrate effects and evidence of the importance of counter-ions in the control of crystallite growth. *Chem. Commun.*, pp. 80–81

Breedon, M.; Rix, C. & Kalantar-zadeh, K. (2009). Seeded growth of ZnO nanorods from NaOH solutions. *Mater. Lett.*, Vol.63, pp. 249–251

Chakraborty, R.; Dhara, S. & Giri, P. K. (2011). Effect of rapid thermal annealing on microstructure and optical properties of ZnO nanorods. *Int. J. Nanosci.*, Vol.10, pp. 65-68

Chang, H.; Sun, Z.; Ho, K. Y.; Tao, X.; Yan, F.; Kwok, W. M. & Zheng, Z. (2011). A highly sensitive ultraviolet sensor based on a facile in situ solution-grown ZnO nanorod/graphene heterostructure. *Nanoscale*, Vol.3, pp. 258–264

Chen, C. C.; Yeh, C. C.; Chen, C. H.; MY, M. Y. Y. & Liu, H. L. (2001). Catalytic growth and characterization of gallium nitride nanowires. *J. Am. Chem. Soc.*, Vol.123, pp. 2791-2798

Chen, L.-Y.; Wu, S.-H. & Yin, Y.-T. (2009). Catalyst-free growth of vertical alignment ZnO nanowire arrays by a two-stage process. *J. Phys. Chem. C*, Vol.113, pp. 21572-21576

Cheng, C.; Wang, T.-L.; Feng, L.; Li, W.; Ho, K. M.; Loy, M. M. T.; Fung, K. K. & Wang, N. (2010a). Vertically aligned ZnO/amorphous-Si core-shell heterostructured nanowire arrays. *Nanotechnol.*, Vol.21, pp. 475703

Cheng, C. W.; Sie, E. J.; Liu, B.; Huan, C. H. A.; Sum, T. C.; Sun, H. D. & Fan, H. J. (2010b). Surface plasmon enhanced band edge luminescence of ZnO nanorods by capping Au nanoparticles. *Appl. Phys. Lett.*, Vol.96, pp. 071107

Cui, J. B.; Daghlian, C. P.; Gibson, U. J.; Pusche, R.; Geithner, P. & Ley, L. (2005). Low-temperature growth and field emission of ZnO nanowire arrays. *J. Appl. Phys.*, Vol.97, pp. 44315

Dayeh, S. A.; Soci, C.; Yu, P. K. L.; Yu, E. T. & Wang, D. (2007). Influence of surface states on the extraction of transport parameters from InAs nanowire field effect transistors. *Appl. Phys. Lett.*, Vol.90, pp. 162112

Dhara, S. & Giri, P. K. (2011a). Enhanced UV photosensitivity from rapid thermal annealed vertically aligned ZnO nanowires. *Nanoscale Res. Lett.*, Vol.6, pp. 504

Dhara, S. & Giri, P. K. (2011b). Quick single-step mechanosynthesis of ZnO nanorods and their optical characterization: milling time dependence. *Appl. NanoSci*, Vol.1, pp. 165-171, doi:10.1007/s13204-011-0026-z

Dhara, S. & Giri, P. K. (2011c). Shape evolution in one-dimensional ZnO nanostructure grown from ZnO nanopowder source: vapour-liquid-solid vs. vapour-solid growth mechanisms. *Int. J. Nanosci.*, Vol.10, pp. 75-79

Dhara, S. & Giri, P. K. (unpublished).

Ding, J.; Miao, W. F.; McCormick, P. G. & Street, R. (1995). Mechanochemical synthesis of ultrafine Fe powder. *Appl. Phys. Lett.* , Vol.67, pp. 3804-3806

Duan, X. & Lieber, C. M. (2000). General synthesis of compound semiconductor nanowires. *Adv. Mater.*, Vol.12, pp. 298-302

Fang, Y. J.; Sha, J.; Wang, Z. L.; Wan, Y. T.; Xia, W. W. & Wang, Y. W. (2011). Behind the change of the photoluminescence property of metal-coated ZnO nanowire arrays. *Appl. Phys. Lett.*, Vol.98, pp. 033103

Feng, L.; Liu, A.; Liu, M.; Ma, Y.; Wei, J. & Man, B. (2010). Synthesis, characterization and optical properties of flower-like ZnO nanorods by non-catalytic thermal evaporation. *J. Alloys Comp.*, Vol.492, pp. 427-432

Gao, P. X.; Ding, Y. & Wang, Z. L. (2003). Crystallographic-orientation aligned ZnO nanorods grown by tin catalyst. *Nano Lett.*, Vol.3, pp. 1315-1320

Garcia, M. A. (2011). Surface plasmons in metallic nanoparticles: fundamentals and applications. *J. Phys. D: Appl. Phys.* , Vol.44, pp. 283001

Gargas, D. J.; Eugenia, M.; Molares, T. & Yang, P. (2009). Imaging single ZnO vertical nanowire laser cavities using UV-laser scanning confocal microscopy. *J. Am. Chem. Soc.*, Vol.131, No.6, pp. 2125-2127. doi:10.1021/ja8092339

Giri, P. K.; Bhattacharya, S.; Singh, D. K.; Kesavamoorthy, R.; Panigrahi, B. K. & Nair, K. G. M. (2007). Correlation between microstructure and optical properties of ZnO nanoparticles synthesized by ball milling. *J. Appl. Phys.*, Vol.102, pp. 093515

Giri, P. K.; Dhara, S. & Chakraborty, R. (2010). Effect of ZnO seed layer on the catalytic growth of vertically aligned ZnO nanorod arrays. *Mater. Chem. Phys.*, Vol.122, pp. 18-22

Givargizov, E. I. (1975). Fundamental aspects of VLS growth. *J. Cryst. Growth*, Vol.31, pp. 20-30

Greene, L. E.; Yuhas, B. D.; Law, M.; Zitoun, D. & Yang, P. (2006). Solution-Grown Zinc Oxide Nanowires. *Inorg. Chem.*, Vol.45, pp. 7535-7543

Gudiksen, M. S.; Lauhon, U. J.; Wang, J.; Smith, D. C. & Lieber, C. M. (2002). Growth of nanowire superlattice structures for nanoscale photonics and electronics. *Nature*, Vol.415, pp. 617–620

Gudiksen, M. S. & Lieber, C. M. (2000). Diameter-selective synthesis of semiconductor nanowires. *J. Am. Chem. Soc.*, Vol.122, pp. 8801-8802

Guo, L.; Zhang, H.; Zhao, D.; Yao, B.; Li, B.; Zhang, Z. & Shen, D. (2011a). The growth and the ultraviolet photoresponse properties of the horizontal growth ZnO nanorods *Mat. Lett.*, Vol.65, pp. 1495–1498

Guo, Z.; Andreazza-Vignolle, C.; Andreazza, P.; Sauvage, T.; Zhao, D. X.; Liu, Y. C.; B.Yao; Shen, D. Z. & Fan, X. W. (2011b). Tuning the growth of ZnO nanowires. *Physica B*, Vol.406, pp. 2200–2205

He, J. H.; Hsu, J. H.; Wang, C. W.; Lin, H. N.; Chen, L. J. & Wang, Z. L. (2006). Pattern and feature designed growth of ZnO nanowire arrays for vertical devices. *J. Phys. Chem. B*, Vol.110, pp. 50-53

Hejazi, S. R. & Hosseini, H. R. M. (2007). A diffusion-controlled kinetic model for growth of Au-catalyzed ZnO nanorods: Theory and experiment. *J. Cryst. Growth*, Vol.309, pp. 70–75

Heo, Y. W.; Varadarajan, V.; Kaufman, M.; Kim, K.; Norton, D. P.; Ren, F. & Fleming, P. H. (2002). Site-specific growth of ZnO nanorods using catalysis-driven molecular-beam epitaxy. *Appl. Phys. Lett.*, Vol.81, pp. 3046–3048

Huang, M. H.; Mao, S.; Feick, H.; Yan, H.; Wu, Y.; Kind, H.; Weber, E.; Russo, R. & Yang, P. (2001a). Room-temperature ultraviolet nanowire nanolasers. *Science*, Vol.292, pp. 1897-1899

Huang, M. H.; Wu, Y.; Feick, H.; Tran, N.; Weber, E. & P. Yang 13 (2001b). Catalytic growth of Zinc Oxidenanowires by vapor transport. *Adv. Mater.*, Vol.13, pp. 113-116

Kim, D. C.; Jung, B. O.; Lee, J. H.; Cho, H. K.; Lee, J. Y. & Lee, J. H. (2011a). Dramatically enhanced ultraviolet photosensing mechanism in a n-ZnO nanowires/i-MgO/n-Si structure with highly dense nanowires and ultrathin MgO layers. *Nanotechnol.*, Vol.22, pp. 265506

Kim, D. C.; Kong, B. H. & Cho, H. K. (2009). Synthesis and growth mechanism of catalyst free ZnO nanorods with enhanced aspect ratio by high flow additional carrier gas at low temperature. *J. Phys. D: Appl. Phys.*, Vol.42, pp. 065406

Kim, K.-P.; Chang, D.; Lim, S. K.; Lee, S.-K.; Lyu, H.-K. & Hwang, D.-K. (2011b). Effect of TiO2 Nanoparticle Modification on Ultraviolet Photodetection Properties of Al-Doped ZnO Nanowire Network. *J. J. Appl. Phys.*, Vol.50, pp. 06GF07

Kind, H.; Yan, H. Q.; Messer, B.; Law, M. & Yang, P. D. (2002). Nanowire ultraviolet photodetectors and optical switche. *Adv. Mater.*, Vol.14, pp. 158–160

Kirkham, M.; Wang, X.; Wang, Z. L. & Snyder, R. L. (2007). Solid Au nanoparticles as a catalyst for growing aligned ZnO nanowires: A new understanding of the vapour-liquid-solid process. *Nanotechnol.*, Vol.18, pp. 365304

Law, M.; Green, L. E.; Jhonson, J. C.; Saykally, R. & Yang, P. (2005). Nanowire dye-sensitized solar cells. *Nature Mater*, Vol.4, pp. 455-495

Law, M.; Greene, L. E.; Radenovic, A.; Kuykendall, T.; Liphardt, J. & Yang, P. (2006). ZnO-Al_2O_3 and ZnO-TiO_2 core-shell nanowire dye-sensitized solar cells. *J. Phys. Chem. B* Vol.110, pp. 22652-22663

Li, C.; Fang, G.; Li, J.; Ai, L.; Dong, B. & Zhao, X. (2008). Effect of seed layer on structural properties of ZnO nanorod arrays grown by vapor-phase transport. *J. Phys. Chem. C* Vol.112, pp. 990-995

Li, C.; Fang, G.; Su, F.; Li, G.; Wu, X. & Zhao, X. (2006). Synthesis and photoluminescence properties of vertically aligned ZnO nanorod–nanowall junction arrays on a ZnO-coated silicon substrate *Nanotechnol.*, Vol.17, pp. 3740

Li, C. C.; Du, Z. F.; Li, L. M.; Yu, H. C.; Wan, Q. & Wang, T. H. (2007). Surface-depletion controlled gas sensing of ZnO nanorods grown at room temperature. *Appl. Phys. Lett.*, Vol.91, pp. 032101

Li, Q. H.; Gao, T.; Wang, Y. G. & Wang, T. H. (2005). Adsorption and desorption of oxygen probed from ZnO nanowire films by photocurrent measurements. *Appl. Phys. Lett.*, Vol.86, pp. 123117

Li, S.; Zhang, X.; Yan, B. & Yu, T. (2009). Growth mechanism and diameter control of well-aligned small-diameter ZnO nanowire arrays synthesized by a catalyst-free thermal evaporation method. *Nanotechnol.*, Vol.20, pp. 495604

Li, S. Y.; Lee, C. Y. & Tseng, T. Y. (2003). Copper-catalyzed ZnO nanowires on silicon (100) grown by vapor–liquid–solid process. *J. Cryst. Growth* Vol.247, pp. 357-362

Liao, Z.-M.; Liu, K.-J.; Zhang, J.-M.; Xu, J. & Yu, D.-P. (2007). Effect of surface states on electron transport in individual ZnO nanowires. *Phys. Lett. A* Vol.367, pp. 207-210

Lide, D. R. (2009) CRC Handbook of Chemistry and Physics. CRC, Boca Raton

Lin, C.-A.; Tsai, D.-S.; Chen, C.-Y. & He, J.-H. (2011). Significant enhancement of yellow–green light emission of ZnO nanorod arrays using Ag island films. *Nanoscale*, Vol.3, pp. 1195-1199

Lin, D.; Wu, H.; Zhang, W.; Li, H. & Pan, W. (2009). Enhanced UV photoresponse from heterostructured Ag-ZnO nanowires. *Appl. Phys. Lett.* , Vol.94, pp. 172103

Lin, H. Y.; Cheng, C. L.; Chou, Y. Y.; Huang, L. L.; Chen, Y. F. & Tsen, K. T. (2006). Enhancement of band gap emission stimulated by defect loss. *Opt. Express*, Vol.16, pp. 2372-2379

Liu, J.; Ahn, Y. H.; Park, J.-Y.; Koh, K. H. & Lee, S. (2009). Hybrid light-emitting diodes based on flexible sheets of mass-produced ZnO nanowires. *Nanotechnol.*, Vol.20, pp. 4452063

Liu, J.; Park, J.; Park, K. H.; Ahn, Y.; Park, J. Y.; Koh, K. H. & Lee, S. (2010a). Enhanced photoconduction of free-standing ZnO nanowire films by L-lysine treatment. *Nanotechnol.*, Vol.21, pp. 485504

Liu, K. W.; Chen, R.; Xing, G. Z.; Wu, T. & Sun, H. D. (2010b). Photoluminescence characteristics of high quality ZnO nanowires and its enhancement by polymer covering. *Appl. Phys. Lett.*, Vol.96, pp. 023111

Lyu, S. C.; Zhang, Y. & Lee, C. J. (2003). Low-temperature growth of ZnO Nanowire array by a simple physical vapor-deposition method. *Chem. Mater.*, Vol.15, pp. 3294-3299

Lyu, S. C.; Zhang, Y.; Ruh, H.; Lee, H. J.; Shim, H. W.; Suh, E. K. & Lee, C. S. (2002). Low temperature growth and photoluminescence of well-aligned zinc oxide nanowires. *Chem. Phys. Lett.*, Vol.363, pp. 134-138

Mancheva, M.; Iordanova, R. & Dimitriev, Y. (2011). Mechanochemical synthesis of nanocrystalline $ZnWO_4$ at room temperature. *J. Alloys Comp.* , Vol.509, pp. 15-20

Manekkathodi, A.; Lu, M. Y.; Wang, C. W. & Chen, L. J. (2010). Direct growth of aligned zinc oxide nanorods on paper substrates for low-cost flexible electronics. *Adv. Mater.*, Vol.22, pp. 4059-4063

McCluskey, M. D. & Jokela, S. J. (2009). Defects in ZnO. *J. Appl. Phys.*, Vol.106, pp. 071101

Ohkubo, I.; Tsubouchi, K.; Kumigashira, H.; Ohnishi, T.; Lippmaa, M.; Matsumoto, Y.; Koinuma, H. & Oshima, M. (2008). Trap-controlled space-chargelimited current mechanism in resistance switching at Al/$Pr_{0.7}Ca_{0.3}MnO_3$ interface. *Appl. Phys. Lett.*, Vol.92, pp. 22113

Pacholski, C.; Kornowski, A. & Weller, H. (2002). Self-assembly of ZnO: from nanodots to nanorods. *Angew. Chem. Int. Ed.*, Vol.41, pp. 1188-1191

Pan, Z. W.; Dai, Z. R. & Wang, Z. L. (2001). Nanobelts of semiconducting oxides. *Science*, Vol.291, pp. 1947-1949

Park, J. Y.; Yun, Y. S.; Hong, Y. S.; Oh, H.; Kim, J.-J. & Kima, S. S. (2005). Synthesis, electrical and photoresponse properties of vertically well-aligned and epitaxial ZnO nanorods on GaN-buffered sapphire substrates. *Appl. Phys. Lett.*, Vol.87, pp. 123108

Park, W.; Jo, G.; Hong, W. K.; Yoon, J.; Choe, M.; Lee, S.; Ji, Y.; Kim, G.; Kahng, Y. H.; Lee, K.; Wang, D. & Lee, T. (2011). Enhancement in the photodetection of ZnO nanowires by introducing surface-roughness-induced traps. *Nanotechnol.*, Vol.22, pp. 205204

Park, W. I.; Kim, D. H.; Jung, S. W. & Yi, G. C. (2002). Metalorganic vapor-phase epitaxial growth of vertically well-aligned ZnO nanorods. *Appl. Phys. Lett.*, Vol.80, pp. 4232–4234

Patra, S.; Satpati, B. & Pradhan, S. K. (2011). Quickest single-step mechanosynthesis of CdS quantum dots and their microstructure characterization. *J. Nanosci. Nanotechnol.*, Vol.11, No.6, pp. 4771-4780

Pearton, S. J.; Norton, D. P.; Ip, K.; Heo, Y. W. & Steiner, T. (2005). Recent progress in processing and properties of ZnO. *Prog. Mater. Sci.*, Vol.50, pp. 293–340

Porter, H. L.; Cai, A. L.; Muth, J. F. & Narayan, J. (2005). Enhanced photoconductivity of ZnO films Co-doped with nitrogen and tellurium. *Appl. Phys. Lett.*, Vol.86, pp. 211918

Pullar, R. C.; Farrah, S. & Alford, N. M. (2007). $MgWO_4$, $ZnWO_4$, $NiWO_4$ and $CoWO_4$ microwave dielectric ceramics. *J. Euro. Cera. Soc.* , Vol.27, pp. 1059-1063

Rose, A. (1955). Space charge limited currents in solids. *Phys. Rev.*, Vol.97, pp. 1538–1544

Soci, C.; Zhang, A.; Xiang, B.; Dayeh, S. A.; Aplin, D. P. R.; Park, J.; Bao, X. Y.; Lo, Y. H. & Wang, D. (2007). ZnO nanowire UV photodetectors with high internal gain. *Nano Lett.* , Vol.7, pp. 1003-1009

Song, J. & Lim, S. (2007). Effect of seed layer on the growth of ZnO nanorods. *J. Phys. Chem. C*, Vol.111, pp. 596-600

Tak, Y. & Yong, K. (2005). Controlled growth of well-aligned ZnO nanorod array using a novel solution method. *J. Phys. Chem. B* Vol.109, pp. 19263-19269

Tsuzuki, T. & McCormick, P. G. (2001). ZnO nanoparticles synthesised by mechanochemical processing. *Scripta Mater.*, Vol.44, pp. 1731-1734

Tsuzuki, T. & McCormick, P. G. (2004). Mechanochemical synthesis of nanoparticles. *J. Mat. Sci.*, Vol.39, pp. 5143-5146

Um, H.-D.; Moiz, S. A.; Park, K.-T.; Jung, J.-Y.; Jee, S.-W.; Ahn, C. H.; Kim, D. C.; Cho, H. K.; Kim, D.-W. & Lee, J.-H. (2011). Highly selective spectral response with enhanced responsivity of n-ZnO/p-Si radial heterojunction nanowire photodiodes *Appl. Phys. Lett.*, Vol.98, pp. 033102

Vayssieres, L. (2003). Growth of arrayed nanorods and nanowires of ZnO from aqueous solutions. *Adv. Mater.*, Vol.15, pp. 464-466

Verges, M. A.; Mifsud, A. & Serna, C. J. (1990). Formation of rodlike zinc-oxide microcrystals in homogeneous solutions. *J. Chem. Soc., Faraday Trans.* , Vol.86, pp. 959-963

Wagner, R. S. & Ellis, W. C. (1964). Vapor-liquid-solid mechanism of single crystal growth. *Appl. Phys. Lett.*, Vol.4, pp. 89-90

Wang, Q. P.; Zhang, D. H.; Xue, Z. Y. & Hao, X. T. (2002a). Violet luminescence emitted from ZnO films deposited on Si substrate by rf magnetron sputtering. *Appl. Surf. Sci.*, Vol.201, pp. 123–128

Wang, Y.; Zhang, L.; Liang, C.; Wang, G. & Peng, X. (2002b). Catalytic growth and photoluminescence properties of semiconductor single-crystal ZnS nanowires. *Chem. Phys. Lett. A*, Vol.357, pp. 314-318

Wei, Y.; Wu, W.; Guo, R.; Yuan, D.; Das, S. & Wang, Z. L. (2010). Wafer-scale high-throughput ordered growth of vertically aligned ZnO nanowire arrays. *Nano Lett.*, Vol.10, pp. 3414-3419

Wu, J. J.; Wen, H. I.; Tseng, C. H. & Liu, S. C. (2004). Well-aligned ZnO nanorods via hydrogen treatment of ZnO films. *Adv. Funct. Mater. Lett.*, Vol.14, pp. 806–810

Wu, W.-Y.; Chen, M.-T. & Ting, J.-M. (2009). Growth and characterizations of ZnO nanorod/film structures on copper coated Si substrates. *Thin Solid Films*, Vol.518, pp. 1549–1552

Wu, Y.; R.Fan & Yang, P. (2002a). Block-by block growth of single-crystalline Si/Si-Ge superlattice nanowires. *Nano Lett.*, Vol.2, pp. 83-86

Wu, Y.; Yan, H.; Huang, M.; Messer, B.; Song, J. H. & Yang, P. (2002b). Inorganic semiconductor nanowires: rational growth, assembly, and novel properties. *Chem. Eur. J.*, Vol.8, pp. 1260-1268

Wu, Y. & Yang, P. (2000). Germanium nanowire growth via simple vapor transport. *Chem. Mater.*, Vol.12, pp. 605-607

Wu, Y. & Yang, P. (2001). Direct observation of vapor-liquid-solid nanowire growth. *J. Am. Chem. Soc.*, Vol.123, pp. 165-166

Xu, S.; Adiga, N.; Ba, S.; Dasgupta, T.; Wu, C. F. J. & Wang, Z. L. (2009). Optimizing and improving the growth quality of ZnO nanowire arrays guided by statistical design of experiments. *ACS Nano*, Vol.3, pp. 1803–1812

Xu, S.; Qin, Y.; Xu, C.; Wei, Y.; Yang, R. & Wang, Z. L. (2010). Self-powered nanowire devices *Nat. Nanotechnol.*, Vol.5, pp. 366-373

Yang, Q.; Xin Guo; Wang, W.; Zhang, Y.; Xu, S.; Lien, D. H. & Wang, Z. L. (2010). Enhancing sensitivity of a single ZnO micro-/nanowire photodetector by piezo-phototronic effect. *ACS Nano*, Vol.4, No.10, pp. 6285-6291

Yao, B. D.; Chan, Y. F. & Wang, N. (2002). Formation of ZnO nanostructures by a simple way of thermal evaporation. *Appl. Phys. Lett.*, Vol.81, pp. 757-759

Yeong, K. S.; H., M. K. & Thong, J. T. L. (2007). The effects of gas exposure and UV illumination on field emission from individual ZnO nanowires. *Nanotechnol.*, Vol.18, pp. 185608

Yuan, H. & Zhang, Y. (2004). Preparation of well-aligned ZnO whiskers on glass substrate by atmospheric MOCVD. . *J. Cryst. Growth*, Vol.263, pp. 119–124

Zhang, Y. J.; Zhang, Q.; Wang, N. L.; Yan, Y. J.; Zhou, H. H. & Zhu, J. (2001). Synthesis of thin Si whiskers (nanowires) using $SiCl_4$. *J. Cryst. Growth & Design*, Vol.226, pp. 185-191

Zhao, D.; Andreazza, C.; Andreazza, P.; Ma, J.; Liu, Y. & Shen, D. (2005). Buffer layer effect on ZnO nanorods growth alignment. *Chem. Phys. Lett.*, Vol.408, pp. 335-338

Zhou, J.; Gu, Y. D.; Hu, Y. F.; Mai, W. J.; Yeh, P. H.; Bao, G.; Sood, A. K.; Polla, D. L. & Wang, Z. L. (2009). Gigantic enhancement in response and reset time of ZnO UV nanosensor by utilizing Schottky contact and surface functionalization. *Appl. Phys. Lett.*, Vol.94, pp. 191103

2

Structure, Morphology, and Optical Properties of the Compact, Vertically-Aligned ZnO Nanorod Thin Films by the Solution-Growth Technique

Chu-Chi Ting
Graduate Institute of Opto-Mechatronics Engineering, National Chung Cheng University, Chia-Yi, Taiwan, R.O.C.

1. Introduction

ZnO is a direct band gap semiconductor with hexagonal wurzite crystal structure (a = 0.325 nm, c = 0.520 nm), and has a wide band gap of 3.37 eV at 300 K (Kligshirn, 1975), large exciton binding energy of 60 meV (Özgür et al., 2005), and high refractive index ($n_{550\ nm}$ = 2.01). ZnO thin films have attracted many researchers to study because of its good optical and electrical characterizations for the applications to light-emitting diodes (Saito et al., 2002), field emitters (Zhu et al., 2003), and solar cells (Lee et al., 2000).

There are many methods for the fabrications of ZnO films such as metal-organic chemical vapor deposition (Yang et al., 2004), laser ablation (Henley et al., 2004), and sputtering (Jeong et al., 2003). However, most of technologies are correlated to the vacuum and high-temperature processes, which results in the high cost. In recent years, the solution-growth route has been used to fabricate the ZnO nanorod thin films (Vayssieres, 2001, 2003; Li et al., 2005; Tak & Yong, 2005; Lee et al., 2007). Vayssieres *et al.* developed the large three-dimensional (3D) and highly oriented porous microrod or nanorod array of n-type ZnO semiconductor by the equimolar (0.1 M) aqueous solution of zinc nitrate [$Zn(NO_3)_2\ 6H_2O$] and methenamine ($C_6H_{12}N_4$) at low temperature. The crystallographic faces of well-aligned single-crystalline hexagonal rods are perpendicularly grown along the [001] direction onto the substrate, resulting in the formation of very large uniform rod arrays (Vayssieres, 2001, 2003). Tak and Yong demonstrated that uniform ZnO nanorods were grown on the zinc-coated silicon substrate by the aqueous solution method containing zinc nitrate and ammonia water. Although the growth mechanism of ZnO nanorods in an organic amine solution has not completely been understood, there are several parameters influencing the growth characteristics (i.e., width, length, growth rate, and preferred orientation) of ZnO nanorods such as growth temperature, growth time, zinc ion concentration, pH of solution, and ZnO seed-layer morphology, which can be applied to control the tailored growth dimensions and orientation of ZnO nanorods (Li et al., 2005; Lee et al., 2007; Tak & Yong, 2005; Vayssieres, 2001, 2003).

It is noted that the surface morphology of ZnO nanorod thin films developed by Vayssieres *et al.* exhibited hexagonal-shaped nanorods and many unfilled inter-columnar voids between nanorods (Vayssieres et al., 2001). However, this kind of hexagonal surface morphology is obviously different from that of other oxide films (*e.g.*, TiO_2, SiO_2, SnO_2, and ZrO_2) fabricated by other solution-growth routes such as chemical bath deposition (CBD) and liquid phase deposition (LPD) (Kishimoto et al., 1998; Lin et al., 2006; Mugdur et al., 2007; Tsukuma et al., 1997). In general, the films synthesized by CBD or LPD exhibits the spheroidal grain morphology. We found that hexagonal-shaped ZnO nanorod thin films with less voids can be synthesized under specific processing parameters and their optical properties are similar to that of ZnO films prepared by sputtering methods. Although there are extensive reports on the structural and physical properties of ZnO nanorod thin films prepared by solution methods, few reports are available on the preparations and characteristic investigations of high packing-density ZnO nanorod thin films.

In this chapter, we fabricated the dense and well-aligned ZnO nanorod thin films by the simple solution method. Structural and optical properties of the resulting ZnO nanorod thin films were systematically examined in terms of the structural evolution of the films at different zinc ion concentrations, growth temperatures, growth time, growth routes, and ZnO seed-layer morphology. We believe that the dense and well-aligned ZnO nanorod thin films fabricated by solution-growth method can satisfy the basic requirement of optical-grade thin films, and has the merits of low temperature, large scale, and low cost.

2. Fabrication of the solution-growth ZnO nanorod thin films

2.1 Fabrication of ZnO seed layers

The ZnO-coated glass substrate acted as the seed layer for the growth of well-aligned ZnO nanorods in aqueous solution. The ZnO seed-layer thin films were fabricated by sol-gel spin-coating technology. 2-methoxyethanol (2-MOE, $HOC_2H_4OCH_3$, 99.5%, Merck) and monoethanolamine (MEA, $HOC_2H_4NH_2$, ≥ 99%, Merck) with molar ratio of Zn/2-MOE/MEA= 1/21/1 were first added to zinc acetate [$Zn(CH_3COO)_2$, 99.5%, Merck], followed by stirring for 10 h to achieve the sol-gel ZnO precursor solution. Then the ZnO precursor solution was spin-coated on silica glass substrates (Corning, Eagle 2000). The as-deposited sol-gel films were first dried at 100 °C/10 min, pyrolyzed at 400 °C/10 min, and further annealed at 400-800 °C/1 h to achieve the seed-layer ZnO thin films with an average grain sizes of 20-100 nm and a thickness of ~90 nm.

2.2 Fabrication of ZnO nanorod thin films

For the fabrication of solution-grown ZnO nanorod thin films, the ZnO seed-layer substrates were deposited in the Zn^{2+} aqueous solutions which were compose of the mixture of zinc nitrate [$Zn(NO_3)_2$ $6H_2O$, ≥ 99%, Merck], hexamethylenetetramine (HMT, $C_6H_{12}N_4$, ≥ 99 %, Merck), and H_2O with molar ratio of Zn/HMT/H_2O=0.1-1/1/1000 to make 0.005-0.05 M zinc ion solutions. The growth temperatures and time were precisely controlled at 55-95 °C and 1.5-6 h, respectively. The multiple-stepwise and one-step solution-growth routes were employed to the growth of the ZnO nanorod thin films. Figure 1 depicts the schematic flowchart of the multiple-stepwise and one-step solution-growth routes for the fabrication of ZnO nanorod thin films. For example, for the ZnO nanorod thin film grown at 75 °C/6 h by the multiple-stepwise route, the ZnO seed-layer substrate was first immersed in the growth solution, and then the growth solution was heated at 75 °C for 1.5 h. After ZnO nanorods growth, the ZnO nanorod thin film was removed from the solution and we immediately put it in another new growth solution, and then the growth solution was heated at 75 °C for another 1.5 h. The same process was repeated 2-4 times and the total growth time was accumulated from 3 to 6 h. On the other hand, the substrate was immersed in the growth solution at 75 °C for continuous 6 h for the one-step route.

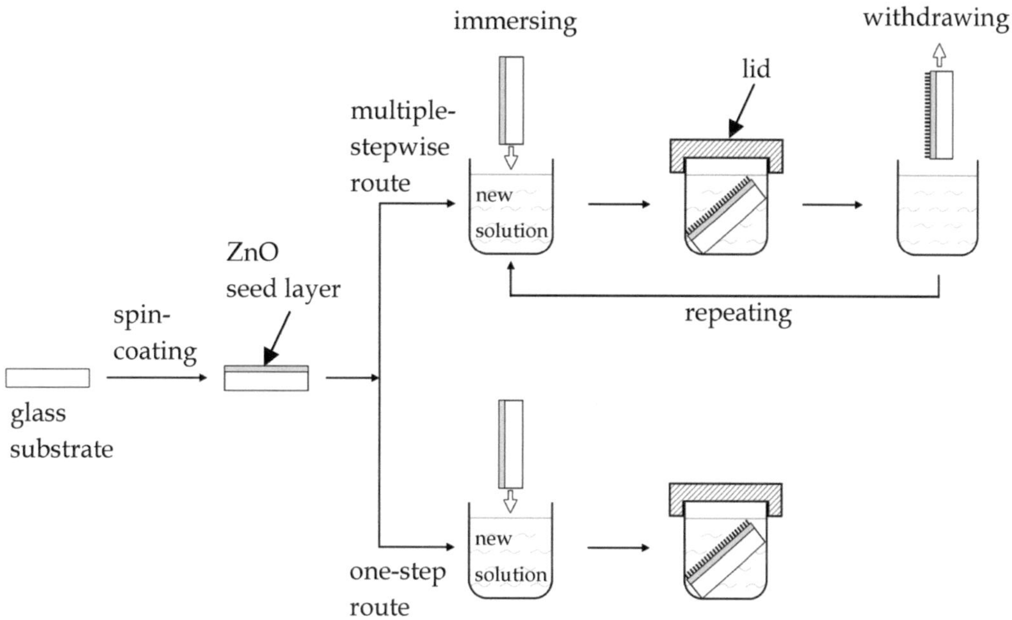

Fig. 1. Schematic flowchart of the multiple-stepwise and one-step solution-growth routes for the fabrication of ZnO nanorod thin films.

2.3 Measurement of physical properties

The crystal structure was detected by an X-ray diffractometer (Shimadzu, XRD 6000). Scanning electron microscope (Hitachi, S4800-I) was used for microstructural examination. The thickness of ZnO films was measured by the α-step profile meter (KLA-Tencor, Alpha-Step IQ). Transmission spectra in the UV and visible ranges were determined on a Shimadzu UV-2100 spectrophotometer. Samples were excited by using a 325 nm He-Cd laser with an output power of 4 mW at room temperature. the UV and visible fluorescence was detected by spectrophotometer (Horiba Jobin-yvon, iHR 550) equipped with a photomultiplier tube detector (Hamamatsu, 7732P-01) at room temperature.

3. Structure, morphology, and optical properties of the compact, vertically-aligned ZnO nanorod thin films

3.1 Film morphology

In our experiments, the zinc ion concentrations were adjusted from 0.005 to 0.05 M, the growth temperatures were controlled from 55 to 95 °C, the growth time was selected in the range of 1.5 to 6 h, the grain sizes of ZnO seed layer varied from 20 to 100 nm, and two kind of growth routes, i.e., multiple-stepwise and one-step route, were used. However, the most compact and densest ZnO nanorod thin film with the thickness of ~800 nm can only be fabricated under very specific conditions, i.e., 0.05 M, 75 °C, 6 h, multiple-stepwise route, and ZnO seed layer with an average grain size of ~20 nm. Figs. 2(a)-(j) illustrate the top-view and cross-sectional scanning electron microscopy (SEM) imagines of ZnO nanorod thin

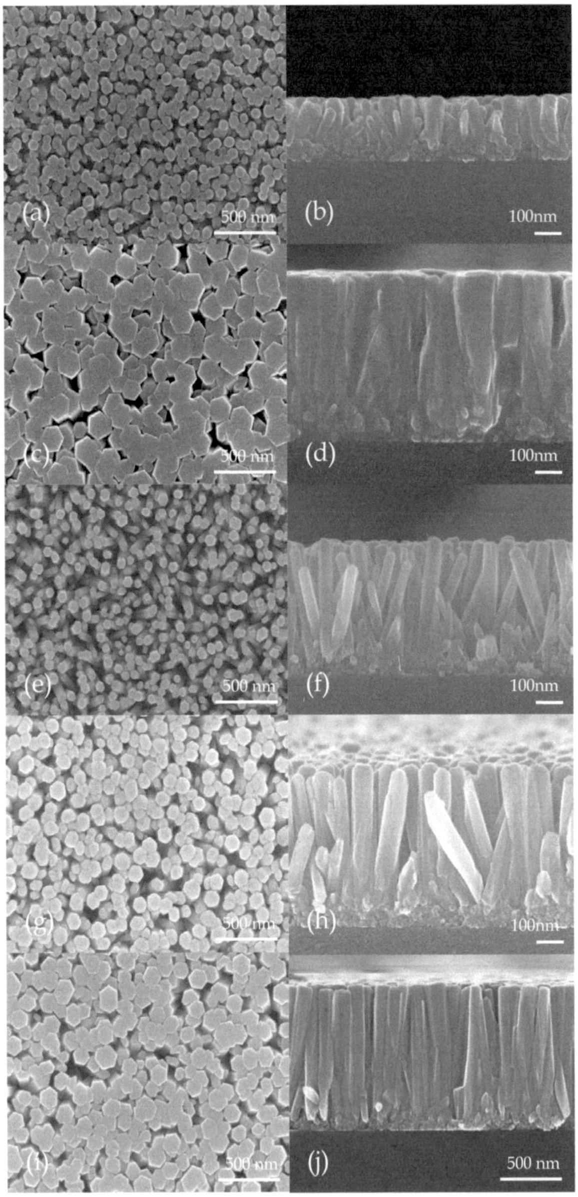

Fig. 2. Top-view and cross-sectional SEM imagines of ZnO nanorod thin films fabricated under the conditions of 0.05 M, seed-layer grain size of ~20 nm, and (a, b) 75 °C/1.5 h (multiple-stepwise route), (c, d) 75 °C/6 h (multiple-stepwise route), (e, f) 95 °C/1.5 h (multiple-stepwise route), (g, h) 75 °C/4.5 h (one-step route), and (i, j) 75 °C/6 h (one-step route).

films fabricated under the conditions of 0.05 M zinc ion concentration, ZnO seed layer with an average grain size of ~20 nm, different growth temperatures/time, and different solution-growth routes (one-step and multiple-stepwise routes). Obviously, the surface morphology of ZnO nanorod thin film fabricated by multiple-stepwise route at 75 °C/6 h exhibits larger aggregated hexagonal grains and more compact structure than others', as shown in Figs. 2(c) and 2(d). Cross-sectional SEM image also exhibits well-developed and larger fused columnar grains, which is very similar to the sputtered thin films (Mirica et al., 2004). However, for the ZnO nanorod thin film fabricated at 95 °C/1.5 h, the film is obviously composed of a large bundle of the ZnO nanorods and most of nanorods do not fuse together, as shown in Figs. 2(e) and 2(f), which resulted in the formation of lots of unfilled inter-columnar volume between nanorods. In addition, some ZnO nanorods do not vertically align very well and they are inclined to the substrate surface.

Figure 3 shows the average diameters and lengths versus growth time and temperatures of ZnO nanorods prepared under the conditions of 0.05 M, one-step route, multiple-stepwise route, and ZnO seed layer with an average grain size of ~20 nm. The diameter and length of ZnO nanorod thin films fabricated by multiple-stepwise route at 95 °C/6 h are ~240 and ~2300 nm, respectively, which is obviously larger than that of ZnO nanorod thin films fabricated by multiple-stepwise or one-step route at 75 °C/6 h. Therefore, the higher growth temperature can induce ZnO nanorods with larger diameter and length, consistent with others' investigations (Li et al., 2005; Lee et al., 2007; Tak & Yong, 2005; Vayssieres, 2001, 2003).

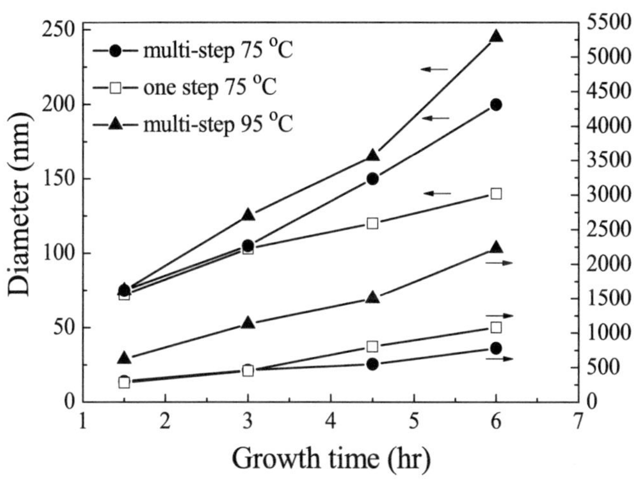

Fig. 3. Average diameters and lengths of ZnO nanorod thin films fabricated under the conditions of 0.05 M, seed-layer grain size of ~20 nm, different growth methods (one-step route and multiple-stepwise route), growth temperatures, and growth time.

For the ZnO nanorod thin films fabricated at 75 °C/1.5 h, short nanorods with the diameters of 60-80 nm and the height of ~200 nm are very crowded and combined each other at side faces, as shown in Figs. 2(a) and 2(b). Further increase in growth time to 6 h causes the highly c-axis-oriented hexagonal ZnO grains (as shown in Figure 4 in the next section) to coalesce and form larger aggregated hexagonal grains with the average diameter of ~200 nm and the height of ~800 nm, resulting in the reduction of unfilled inter-columnar volume and voids [see Figs. 2(c) and 2(d)].

Compared the SEM images of ZnO nanorod thin films fabricated by multiple-stepwise route at 75 °C/6 h [Figs. 2(c) and 2(d)] with that fabricated by one-step route at 75 °C/6 h [Figs. 2(i) and 2(j)], the former exhibited the larger aggregated hexagonal grains and fused columnar structure with the average diameter of ~200 nm and the height of ~800 nm; however, the latter exhibited the smaller aggregated hexagonal grains with the average diameter of ~140 nm and the height of ~1100 nm.

3.2 Crystal structure

Figure 4 shows the X-ray diffraction (XRD) patterns of ZnO nanorod thin films fabricated under growth temperatures, growth time, multiple-stepwise route, and the ZnO seed layer with an average grain size of ~20 nm. Obviously, all of the XRD patterns exhibits only one diffraction peak and the peak position at ~34.53-34.57°, i.e. (002) is the characteristic of wurzite ZnO (JCPDS No. 36-1451). Hence, these ZnO nanorod thin films possess highly preferred orientation with c-axis normal to the substrate.

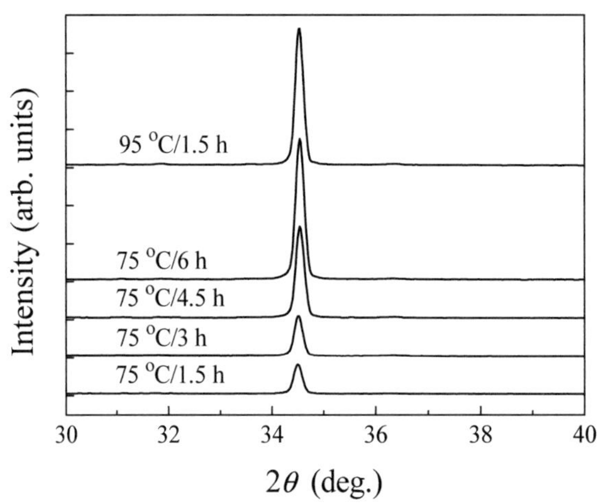

Fig. 4. XRD patterns of ZnO nanorod thin films fabricated under the conditions of 0.05 M, seed-layer grain size of ~20 nm, multiple-stepwise route, and different growth temperatures/time.

The diffraction intensity of ZnO nanorod thin film prepared at 75 °C / 6 h is similar to that of ZnO nanorod thin film prepared at 95 °C/1.5 h, which implies that they have similar crystallinity because of similar thickness (~ 800 nm) between these two samples. Although the 75 °C growth temperature is much lower than 95 °C, these coalesced and aggregated hexagonal nanorods fabricated at 75 °C still possess good crystallinity in comparison with the uncoalesced and well-shaped hexagonal nanorods fabricated at 90 °C and possessing the single crystalline nature (Li et al., 2005). However, the photoluminescence (PL) spectra show that ZnO nanorod thin film prepared at 75 °C/6 h had more oxygen defects as compared with that prepared at 95 °C/1.5 h, and this phenomenon will be discussed in the section of optical properties.

In addition, the (002) peak position of ZnO nanorod thin films prepared at 75 °C/1.5-6 h deviates from the randomly orientated ZnO powder value (34.42°) and shifts toward higher value, indicating the compressive stress existing in these extremely c-axis-oriented ZnO nanorod thin films (Sagar et al., 2007). The (002) peak position progressively varies from 34.50° to 34.54° by increasing growth time, which means that the compressive stress increases with the increase of thickness and aggregated hexagonal grain size. After calculation, the strains vary from -0.21 to -0.32% (Puchert et al., 1996).

3.3 Grown mechanisms of compact, vertically-aligned ZnO nanorod thin films

Some growth characteristics such as average diameters and lengths of ZnO nanorods could be determined by some significant parameters such as the morphology of a zinc metal seed layer, pH, growth temperature, and concentration of zinc salt in aqueous solution (Tak & Yong, 2005). Li *et al.* proposed the growth mechanism of ZnO nanorods fabricated by the aqueous solution method. The proposed mechanism includes three steps: (1) fine and independent ZnO nanorods grew and bundled together. (2) fine ZnO nanorods coalesced. (3) single large dimension hexagonal ZnO nanorod was formed (Li et al., 2005). Lee *et al.* systematically examined that the degree of alignment of dense ZnO nanorod arrays synthesized via a two-step seeding and solution-growth process was significantly influenced by the ZnO seed layer roughness. The highly c-axis aligned and dense ZnO nanorods can be obtained during the roughness of ZnO seed layer was ≦ 2 nm (Lee et al., 2007).

Vayssieres pointed that the diameter of ZnO nanorods could increase 10 times from 100-200 nm to 1000-2000 nm when the zinc ion concentration increased from 0.001 M to 0.01 M (Vayssieres, 2003). The higher zinc ion concentration can accelerate a smaller bundle of ZnO nanorods to coalesce together and form larger dimension ZnO nanorods for reducing the surface energy (Li et al., 2005). Hence, the zinc ion concentration can obviously influence the diameter of ZnO nanorods. For the one-step route, the growth solution is limited in a closed system. When the growth time increases, the zinc ions will be gradually depleted and the zinc ion concentration on the top of nanorods should be less than the initial solution, which reduces the lateral aggregation rate of hexagonal nanorods, induces the continuous growth of nanorods in vertical direction, and results in the nanorods with smaller diameter and larger length. However, multiple-stepwise route can supply and maintain the zinc ion concentration and accelerate the lateral coarsening growth of nanorods, which leads to the aggregation of hexagonal nanorods and the formation of close-packed columnar structure with larger diameter and shorter length. The growth mechanism of ZnO nanorod thin film prepared at 75 °C/1.5-6 h (multiple-stepwise route) is depicted in Figure 5. In addition, the formation of ZnO nanorods can be attributed to the following reaction equations (Li et al., 2005).

$$(CH_2)_6 N_4 + 6H_2O \rightarrow 6HCHO + 4NH_3 \quad (1)$$

$$NH_3 + H_2O \rightleftarrows NH_4^+ + OH^- \quad (2)$$

$$2OH^- + Zn^{2+} \rightarrow ZnO_{(s)} + H_2O \quad (3)$$

On the other hand, the (002) plane in ZnO structure has the highest atomic density and possesses the lowest surface free energy. Therefore, the growth of a preferred *c*-axis oriented ZnO nanorod thin films can be easily driven at such low growth temperature. Additionally,

Lee *et al.* pointed that the surface morphology of ZnO seed layer can also significantly influence the prefer-oriented growth of ZnO nanorods (Lee et al., 2007). The smaller surface roughness of ZnO seed layer can induce the growth of ZnO nanorod with highly c-axis preferred orientation. In our system, when the grain size of ZnO seed layer is larger than 20 nm, the (100) and (101) diffraction peaks can be detected (XRD patterns are not shown here), which indicates that some ZnO nanorods do not vertically align very well and are inclined to the substrate surface. The ZnO seed layer with larger grains has higher roughness and can induce the formation of inclined ZnO nanorods and more unfilled inter-columnar voids between ZnO nanorods, as described in some published literatures. (Lee et al., 2007; Zhao et al., 2006) This phenomenon results in the ZnO nanorod thin films with lower densification and transmittance. The influence of ZnO seed-layer morphology on the preferred orientation of resulting ZnO nanorod thin films will be the subject of a separate study in the future.

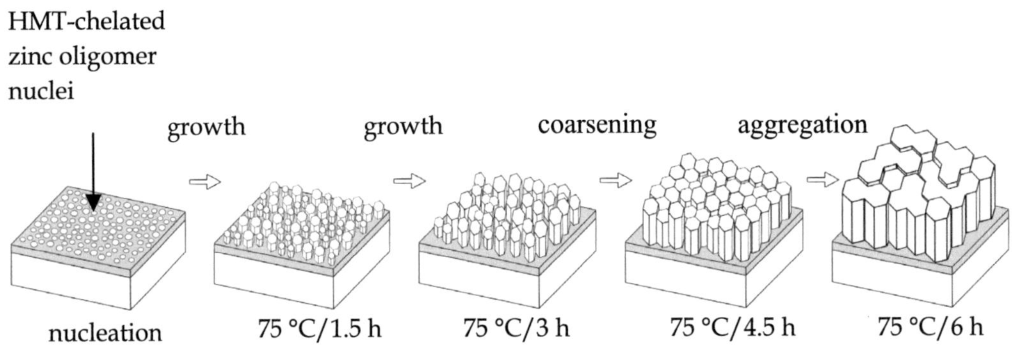

Fig. 5. Growth mechanism of ZnO nanorod thin film prepared at 75 °C/1.5-6 h (multiple-stepwise route).

3.4 Optical properties

3.4.1 Optical transmittance spectra

Figures 6(a)-6(c) show the optical transmittance spectra of ZnO nanorod thin films fabricated at 75 °C/1.5-6 h (multiple-stepwise route), 75 °C/1.5-6 h (one-step route), and 95 °C/1.5-6 h (multiple-stepwise route), respectively. The obvious interference fluctuation in the transmission spectra of ZnO nanorod thin films fabricated at 75 °C/1.5-6 h (multiple-stepwise route) are due to the interference phenomena of multiple reflected beams between the three interfaces: air-ZnO nanorods film, ZnO nanorods film-silica glass, and silica glass-air. The average visible transmittance calculated in the wavelength ranging 400-800 nm of the ZnO nanorod thin films fabricated at 75 °C for 1.5, 3, 4.5, and 6 h are 87.9, 87.5, 84.9, and 84.7%, respectively. Generally, there are three factors influencing the transmittance of ZnO nanorod thin films: (a) surface roughness, (b) defect centers, and (c) oxygen vacancies (Mohamed et al., 2006). In our system, the decrease of transmittance for the ZnO nanorod thin films fabricated at 75 °C for 1.5, 3, 4.5, and 6 h with the 100-800 nm in thickness could be related to two factors. One is the thicker ZnO nanorod thin films had larger hexagonal grain size and larger surface roughness. The other is the higher absorption effect for thicker films. The absorption coefficient can increase with the present of oxygen vacancies which is disclosed by the PL spectra (Figure 9) in the next section. Moreover, it is interesting to note that Figure 6(a) clearly indicates the red-shift in the fundamental absorption edge with the increase of film thickness. The sharp absorption edge at wavelengths of approximately 370 nm is very close to the intrinsic band gap of ZnO (3.37 eV) and the red-shift of absorption edge will be also discussed in the later part.

No obvious interference fluctuations in the transmission spectra were observed in the ZnO nanorod thin films fabricated at 75 °C/4.5 and 6 h (one-step route), and 95 °C/1.5-6 h (multiple-stepwise route), as shown in Figs. 6(b) and 6(c). Based on the SEM photographs [Figs. 2(e)-(j)], these films are composed of a bundle of the ZnO nanorods with smaller diameter, and these ZnO nanorods do not coalesce together very well, which results in the formation of lots of unfilled inter-columnar volume and coarse surface in these ZnO nanorod thin films. In addition, some ZnO nanorods do not vertically align very well and they are inclined to the substrate surface. Therefore, the low transmittance and no fluctuation could be attributed to the incident light experiencing multiple random scattering between unfilled inter-columnar voids, inclined ZnO nanorods, and perpendicular ZnO nanorods in the poor-quality ZnO nanorod films. This effect leads to the destruction of the interference of multiple reflections, no obvious interference fluctuations in the transmission spectra and lower transmittance.

3.4.2 Refractive index and packing density

The refractive index (n) of the ZnO nanorod thin films were derived from the transmittance spectra using Swanepoel's method (Swanepoel, 1983). For those ZnO nanorod thin films with no obvious interference fluctuations in the transmission spectrum, the refractive index of can not be derived by Swanepoel's method. Figure 7 shows that the refractive index of ZnO nanorod thin films fabricated at 75 °C are strongly dependent on the growth time.

The refractive indices (n at λ = 550 nm) of the ZnO nanorod thin films fabricated at 75 °C for 3, 4.5, and 6 h are 1.70, 1.71, and 1.74, respectively. The increase in n of the ZnO nanorod thin films with rising growth time is considered as a result of the increase in compactness and crystallinity, which is consistent with previous XRD and SEM investigations.

In order to evaluate the extent of porosity presenting in the ZnO nanorod thin films, the packing density (P) was evaluated using the following Bragg–Pippard formula which is more suitable for the film with columnar or cylindrical grains (Harris et al., 1979).

$$n_f^2 = \frac{(1-P)n_v^4 + (1+P)n_v^2 n_b^2}{(1+P)n_v^2 + (1-P)n_b^2} \tag{4}$$

where P is expressed as the packing density. The n_f, n_v and n_b are the refractive indices of the porous films, the voids (n_v=1or empty voids) and the bulk materials, respectively.

After calculation, Figure 8 shows the variation of packing densities with growth time for the ZnO nanorod thin films grown at 75 °C. The packing densities of the ZnO nanorod thin films fabricated at 75 °C for 3 and 6 h increase from 0.81 to 0.84. The packing density increases with the increase of thickness and refractive index, and reaches to a maximum value at a film thickness of ~800 nm, which could be attributed to the significant reduction in the porosity and increase in the crystallinity [supporting SEM photographs, Figs. 2(a)-(d), and XRD pattern, Figure 4].

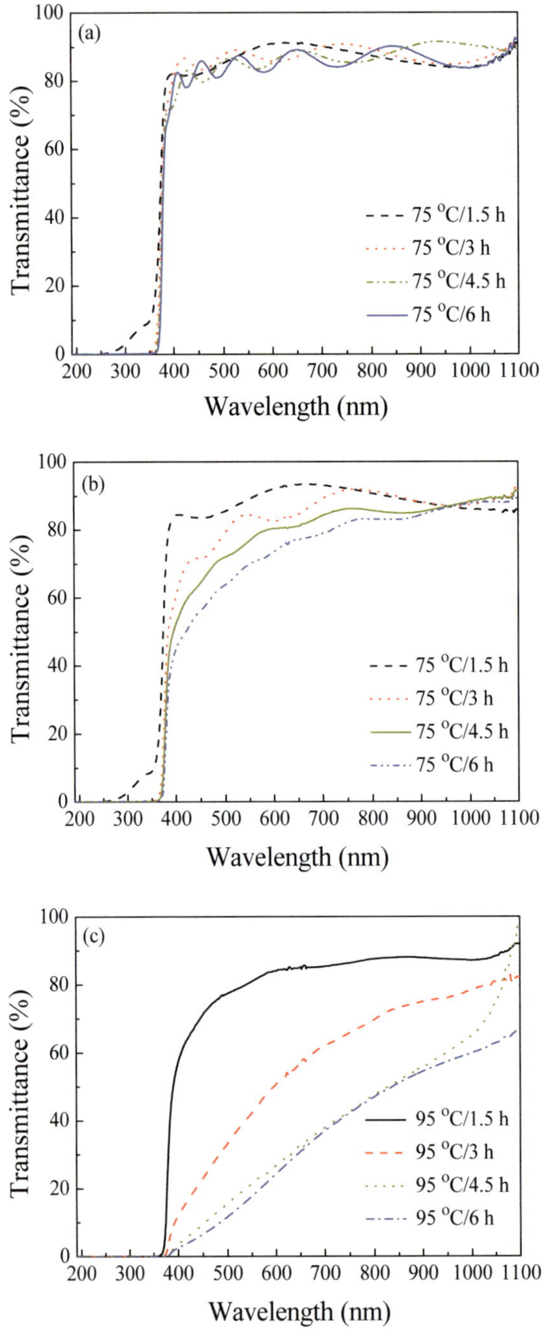

Fig. 6. Optical transmittance spectra of ZnO nanorod thin films fabricated under the conditions of 0.05 M, seed-layer grain size of ~20 nm, and (a) 75 °C/1.5-6 h (multiple-stepwise route), (b) 75 °C/1.5-6 h (one-step route), and (c) 95 °C/1.5-6 h (multiple-stepwise route).

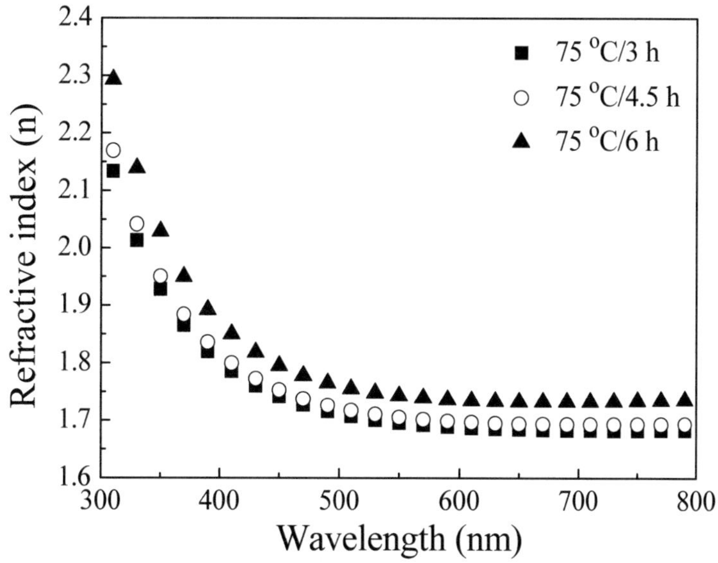

Fig. 7. Wavelength dependence of refractive index for ZnO nanorod thin films fabricated under the conditions of 0.05 M, seed-layer grain size of ~20 nm, multiple-stepwise route, and 75 °C/different growth time.

Because of the demands of compactness and high transmittance, most of the commercialized optical thin films are made by reactive sputtering technology under the high-vacuum and high-temperature condition. For comparisons, refractive indexes and packing densities of the sputtered ZnO films are quoted from some published reports. According to an investigation by Moustaghfir *et al.*, the refractive index (n at λ = 633 nm) and packing density of the radio frequency (r.f.) magnetron reactive sputtered ZnO film (a thickness of ~800 nm) fabricated under the sputtering conditions of a working pressure of 1 Pa, a r.f. power density of 0.89 Wcm^{-2}, Ar-O_2 ratio 95: 5, and the substrate temperature of room temperature (RT) were 1.89 and 0.93, respectively, and they could be enhanced to 1.91 and 0.94 by further annealing at 400 °C/1 h (Moustaghfir et al., 2003). Additionally, an earlier study of the r.f. magnetron reactive sputtered ZnO film (a thickness of 1000 nm) fabricated under the sputtering conditions of a working pressure of 1.33×10^{-2} m bar, a r.f. power of 500 W, Ar-O_2 ratio 40: 60, and the substrate temperature of room temperature (RT) by Mehan *et al.* revealed that the refractive indexes (n at λ = 550 nm) were 1.980 (n_{eb}: extra ordinary refractive index) and 1.963 (n_{ob}: ordinary refractive index), as well as packing densities were 0.986 and 0.978 (extra ordinary refractive index of bulk ZnO, n_{eb} = 2.006, and ordinary refractive index of bulk ZnO, n_{ob} = 1.990), respectively (Mehan et al., 2004). Although lots of parameters can influence the quality of sputtered ZnO films, such high refractive index and packing density may be the extreme values for the sputterred ZnO films. In our system, the optical transmittance (85 %), refractive index (1.74) and packing density (0.84) of optimum solution-growth ZnO nanorod thin film (a thickness of ~800 nm) is lower than that of the high-quality sputtered ZnO films. However, the solution-growth method is still a good technology for the fabrication of low-cost and low-temperature grown ZnO thin films.

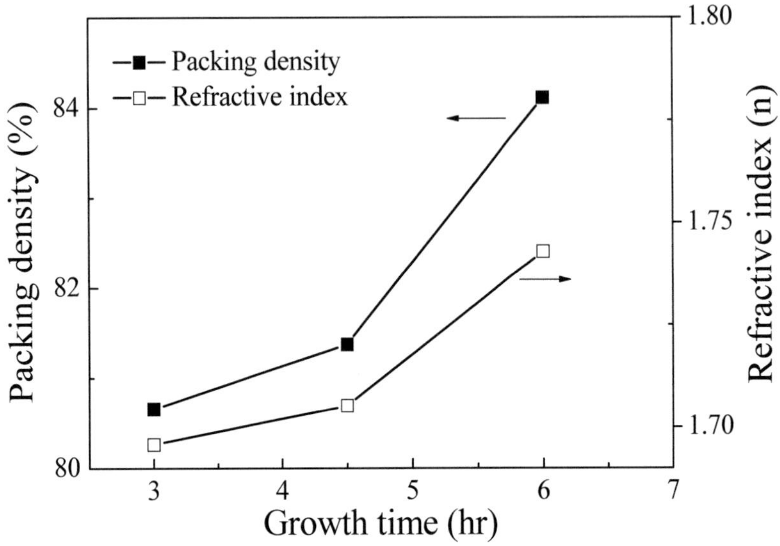

Fig. 8. Variation of packing densities and refractive indexes as a function of growth time for the ZnO nanorod thin films prepared under the conditions of 0.05 M, seed-layer grain size of ~20 nm, multiple-stepwise route, and 75 °C.

3.4.3 Photoluminescence spectra

Figure 9 shows the room temperature photoluminescence (PL) spectra of ZnO nanorod films fabricated at 75 °C/4.5 h, 75 °C/6 h, 95 °C/1.5 h, and 95 °C/3 h by multiple-stepwise route. The intense UV emission at 377-383 nm is due to the recombination of free excitons (Chen et al., 1998; Cho et al., 1999; Park et al., 2003). Obviously, the ZnO nanorod films prepared at 95 °C has more intense UV emission than that of ZnO nanorod films prepared at 75 °C. The intensity of UV emission is ascribed to film crystallinity, and the higher crystallinity possesses the higher intensity of UV emission (Wang & Gao, 2003). Compared the UV intensity of ZnO nanorod films prepared at 75 °C/6 h with that of the ZnO nanorod films prepared at 95 °C/1.5 h, these two films have similar thickness (~800 nm) and XRD diffraction intensities but the UV intensities are quite different. Therefore, the crystallinity of well-shaped hexagonal ZnO nanorod films prepared at 95 °C/1.5 h should be higher than that of ZnO nanorod films prepared at 75 °C/6 h even though the XRD diffraction intensities could not be used to make a judgment of crystallinity for these two films.

On the other hand, all of the PL spectra of ZnO nanorod films exhibit the obvious green-yellow emission at ~572 and ~600 nm, which are associated with the oxygen vacancies and oxygen interstitials, respectively (Ohashi et al., 2002; Studenikin et al., 1998; Wu et al., 2001). However, the ZnO nanorod films prepared at 75 °C had more intense green-yellow emission than that of ZnO nanorod films prepared at 95 °C, which indicates that the lower growth temperature could induce the formation of more oxygen vacancies and interstitials during the ZnO nanorods coarsen and aggregate together. In addition, the yellow emission of ZnO nanorod films prepared at 75 °C gradually dominated by increasing growth time, which indicates that the green emission and yellow emission compete with each other, and more oxygen interstitials are produced with increasing growth time and film thickness. The above-mentioned PL phenomena imply that our most compact and highly c-axis-oriented ZnO nanorod films still possess lots of oxygen vacancies and interstitials.

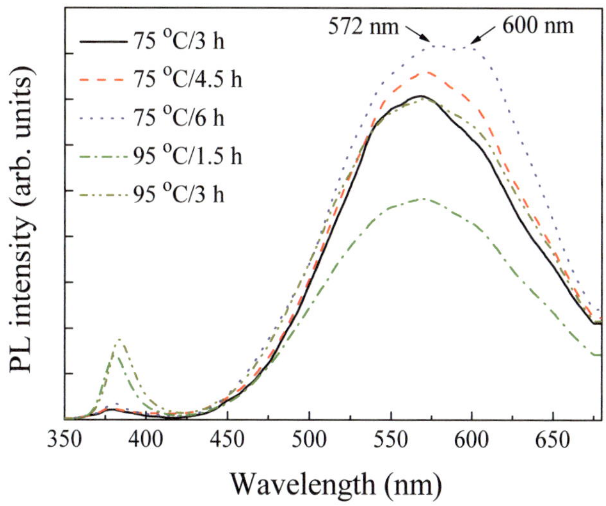

Fig. 9. Room temperature PL spectra of ZnO nanorod thin films fabricated under the conditions of 0.05 M, seed-layer grain size of ~20 nm, multiple-stepwise route, and different growth temperatures/time.

3.4.4 Optical band gap

The optical band gap (E_g) of the ZnO nanorod thin film which is a direct-transition-type semiconductor can be related to absorption coefficient (a) by

$$\alpha h\nu = \text{const} \cdot (h\nu - E_g)^{1/2} \tag{5}$$

Here we assume the absorption coefficient a=(1/d)ln(1/T), where T is the transmittance and d is the film thickness (Serpone et al., 1995; Tan et al., 2005). Figure 10 plots the relationship of $(ahv)^2$ versus photon energy (E) of the ZnO nanorod thin films fabricated under 75 °C/3-6 h and the extrapolated optical band gaps of the films are determined. When the growth time increases from 3 to 6 h, the values o f E_g decrease from 3.35 to 3.31 eV which gradually diverges from the intrinsic band gap of ZnO (3.37 eV). It is known that the energy band gap of a ZnO thin film could be affected by the residual strain (Mohamed et al., 2006; Puchert et al., 1996; Srikant & Clarke, 1997), defects (Burstein, 1954; Dong et al., 2007; Moss, 1954; Sakai et al., 2006), and grain size confinement (Prathap et al., 2008; Wang et al., 2003). For ZnO nanorod thin films fabricated at 75 °C for 3 to 6 h, the average grain sizes enlarge from ~105 to ~200 nm and the film thicknesses increase from ~460 to ~800 nm, which results in the variation of strain from -0.26 to -0.32%. In addition, the PL intensity of yellow emission gradually increases and the more oxygen interstitials are produced. Prathap *et al.* found the energy band gaps increased with the increase of film thickness and grain size in ZnS films fabricated by thermal evaporation (Prathap et al., 2008). Wang *et al.* also observed that the peak position of free excitonic emission redshifted from 3.3 to 3.2 eV with an increase of grain size from 21 to 64 nm, which could be attributed to the quantum confinement effect

(Wang et al., 2003). Although many factors influence the variation of energy band gap, in our system the energy band gaps increasing with the increase of film thickness might be related to the dependence of enhanced strain, enlarged grain size and more oxygen interstitials.

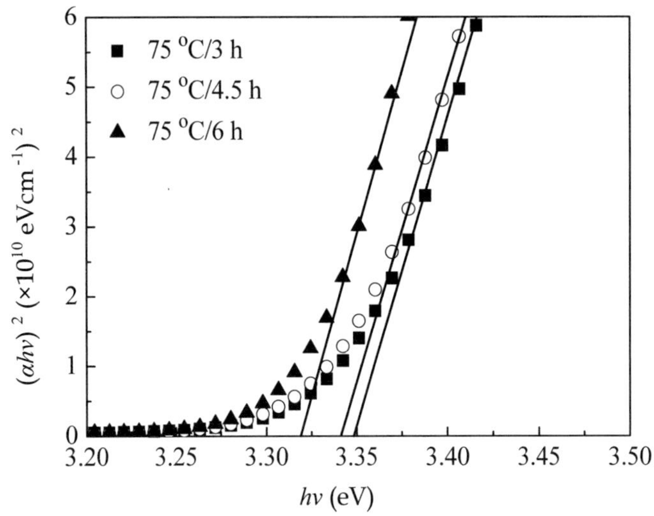

Fig. 10. $(\alpha h\nu)^2$ as a function of photon energy for the ZnO nanorod thin films prepared under the conditions of 0.05 M, seed-layer grain size of ~20 nm, multiple-stepwise route, 75 °C, and different growth time.

4. Conclusion

Highly c-axis-oriented ZnO nanorods thin films were obtained on silica glass substrates by a simple solution-growth technique. The fabrication of highly dense ZnO nanorod thin films are highly dependent on the different zinc ion concentrations, growth temperatures, growth time, growth routes, and ZnO seed-layer morphologies. The higher zinc ion concentrations, growth temperature, and growth time can induce ZnO nanorods with larger diameter and length. The most compact and vertically-aligned ZnO nanorod thin film with the thickness of ~800 nm and average hexagonal grain size of ~200 nm exhibits the extremely C-axis orientation, average visible transmittance 85%, refractive index 1.74, packing density 0.84, and energy band gap 3.31 eV, and it was fabricated under the optimum parameters: 0.05 M, 75 °C, 6 h, multiple-stepwise, and ZnO seed layer with an average grain size of ~20 nm. The photoluminescence spectrum indicates that the densest ZnO nanorod thin film possesses lots of oxygen vacancies and interstitials.

As we demonstrate here, the solution-growth technique is a non-vacuum, low-temperature, low-cost, large-scale, easily controlled process for the fabrication of high-quality, optical-grade ZnO thin films with highly compact ZnO nanorod arrays. In particular, this process can operate at low temperature without organic binders/surfactants or further heat treatment, and thus can be applied to flexible electronics.

5. Acknowledgement

The author would like to thank the National Science Council of the Republic of China for financially supporting this research under Contract No. NSC 96-2221-E-194-042-MY2.

6. References

Burstein, E. (1954). Anomalous Optical Absorption Limit in InSb. *Phys. Rev.,* Vol. 93, No. 3, (Feb 1954), pp. (632-633), ISSN 0031-899X.

Chen, Y., Bagnall, D.M., Koh, H.J., Park, K.T., Hiraga, K., Zhu, Z., & Yao, T. (1998). Plasma Assisted Molecular Beam Epitaxy of ZnO on C-Plane Sapphire: Growth and Characterization. *J. Appl. Phys.,* Vol. 84, No. 7, (Oct 1998), pp. (3912-3918), ISSN 0021-8979.

Cho, S., Ma, J., Kim, Y., Sun, Y., Wong, G.K.L., Ketterson, J.B. (1999). Photoluminescence a and Ultraviolet Lasing of Polycrystalline ZnO Thin Films Prepared by the Oxidation of the Metallic Zn. *Appl. Phys. Lett.,* Vol. 75, No. 18, (Nov 1999), pp. (2761-2763), ISSN 0003-6951.

Dong, B.Z., Fang, G.J., Wang, J.F., Guan, W.J., & Zhao, X.Z. (2007). Effect of Thickness on Structural, Electrical, and Optical Properties of ZnO: Al Films Deposited by Pulsed Laser Deposition. *J. Appl. Phys.,* Vol. 101, No. 3, (Feb 2007), pp. (033713-1-033713-7), ISSN 0021-8979.

Harris, M., Macleod, H.A., Ogura, S., Pelletier, E., & Vidal, B. (1979). In-Situ Ellipsometric Monitor with Layer-by-Layer Analysis for Precise Thickness Control of EUV Multilayer Optics. *Thin Solid Films,* Vol. 57, No. 1, (Feb 1979), pp. (173-178), ISSN 0040-6090.

Henley, S.J., Ashfold, M.N.R., & Cherns, D. (2004). The Growth of Transparent Conducting ZnO Films by Pulsed Laser Ablation. *Surf. Coat. Technol.,* Vol. 177-178, (Jan 2004), pp. (271-276), ISSN 0257-8972.

Jeong, S.H., Kim, B.S., & Lee, B.T. (2003). Photoluminescence Dependence of ZnO Films Grown on Si(100) by Radio-Frequency Magnetron Sputtering on the Growth Ambient. *Appl. Phys. Lett.,* Vol. 82, No. 16, (Apr 2003), pp. (2625-2627), ISSN 0003-6951.

Kishimoto, H., Takahama, K., Hashimoto, N., Aoib, Y. & Deki, S. (1998). Photocatalytic Activity of Titanium Oxide Prepared by Liquid Phase Deposition (LPD). *J. Mater. Chem.,* Vol. 8, No. 9, (Feb 1998), pp. (2019-2024), ISSN 0959-9428.

Kligshirn, C. (1975). The Luminescence of ZnO under High One-and Two-Quantum Excitation. *Phys. Status Solidi B,* Vol. 71, No. 2, (Oct 1975), pp. (547-556), ISSN 0370-1972.

Lee, J.C., Kang, K.H., Kim, S.K., Yoon, K.H., Song, J.S., & Park, I.J. (2000). RF Sputter Deposition of the High-Guality Intrinsic and N-Type ZnO Window Layers for Cu(In,Ga)Se2-Based Solar Cell Applications. *Sol. Energy Mater. Sol. Cells,* Vol. 64, No. 2, (Sep 2000), pp. (185-195), ISSN 0927-0248.

Lee, Y.J., Sounart, T.L., Scrymgeour, D.A., Voigt, J.A., & Hsu, J.W.P. (2007). Control of ZnO Nanorod Array Alignment Synthesized Via Seeded Solution Growth. *J. Cryst. Growth.,* Vol. 304, No. 1, (Jun 2007), pp. (80-85), ISSN 0022-0248.

Li, Q., Kumar, V., Li, Y., Zhang, H., Marks, T.J., & Chang, R.P.H. (2005). Fabrication of ZnO Nanorods and Nanotubes in Aqueous Solutions. *Chem. Mater.*, Vol. 17, No. 5, (Mar 2005), pp. (1001-1006), ISSN 0897-4756.

Lin, J.M., Hsu, M.C., & Fung, K. Z. (2006). Deposition of ZrO2 Film by Liquid Phase Deposition. *J. Power Sources*, Vol. 159, No. 1, (Sep 2006), pp. (49-54), ISSN 0378-7753.

Mehan, N., Gupta, V., Sreenivas, K., & Mansinght, A. (2004). Effect of Annealing on Refractive Indices of Radio-Frequency Magnetron Sputtered Waveguiding Zinc Oxide Films on Glass. *J. Appl. Phys.*, Vol. 96, No. 6, (Sep 2004), pp. (3134-3139), ISSN 0021-8979.

Mirica, E., Kowach, G., Evans, P., & Dut, H. (2004). Morphological Evolution of ZnO Thin Films Deposited by Reactive Sputtering. *Cryst. Growth. Des.*, Vol. 4, No. 1, (Sep 2004) pp. (147-156), ISSN 1528-7483

Mohamed, S.H., El-Rahman, A.M.A., & Salem, A.M. (2006). Effect of rf Plasma Nitriding Time on Electrical and Optical Properties of ZnO Thin Films. *J. Phys. Chem. Solids.*, Vol. 67, No. 11, (Nov 2006) pp. (2351-2357), ISSN 0022-3697.

Moss, T.S. (1954). The Interpretation of the Properties of Indium Antimonide. *Proc. Phys. Soc. B*, Vol. 67, No. 10, (Oct 1954), pp. (775-782), ISSN 0370-1328

Moustaghfir, A., Tomasella, E., Amor, S.B., Jacquet, M., Cellier, J., & Sauvaget, T. (2003). Structural and Optical Studies of ZnO Thin Films Deposited by r.f. Magnetron Sputtering: Influence of Annealing. *Surf. Coat. Technol.*, Vol. 174-175, (Oct 2003), PP. (193-194), ISSN 0257-8972.

Mugdur, P.H., Chang, Y.-J., Han, S-.Y., Su, Y-W., Morrone, A.A., Ryu, S.O., Lee, T.-J., & Chang, C.-H. (2007). A Comparison of Chemical Bath Deposition of CdS from a Batch Reactor and a Continuous-Flow Microreactor. *J. The Electrochem. Soc.*, Vol. 154, No. 9, (Jul 2007), pp. (D482-D488), ISSN 0013-4651.

Özgür, Ü., Alivov, Y.I., Liu, C., Teke, A., Reshchikov, M.A., Doğan, S., Avrutin,V., Cho, S.-J., & Morkoç, H. J. (2005). A Comprehensive Review of ZnO Materials and Devices. *J. Appl. Phys.*, Vol. 98, No. 4, (Aug 2005), PP. (041301-1-041301-103), ISSN 0021-8979.

Puchert, M.K., Timbrell, P.Y., & Lamb, R.N. (1996). Postdeposition Annealing of Radio Frequency Magnetron Sputtered ZnO Films. *J. Vac. Sci. Technol. A*, Vol. 14, No. 4, (Aug 1996), PP. (2220-2230), ISSN 0734-2101.

Park, W.I., Jun, Y.H., Jung, S.W., & Yia, G.C. (2003). Excitonic Emissions Observed in ZnO Single Crystal Nanorods. *Appl. Phys. Lett.*, Vol. 82, No. 6, (Feb 2003), pp. (964-966), ISSN 0003-6951.

Prathap, P., Revathi1, N., Subbaiah, Y.P.V., & Reddy, K.T.R. (2008). Thickness Effect on the Microstructure, Morphology and Optoelectronic Properties of ZnS Films. *J. Phys.: Condens. Matter*, Vol. 20, No. 3, (Dec 2008), pp. (035205-1-035205-10), ISSN 0953-8984

Puchert, M.K., Timbrell, P.Y., & Lamb, R.N. (1996). Postdeposition Annealing of Radio Frequency Magnetron Sputtered ZnO Films. *J. Vac. Sci. Technol. A*, Vol. 14, No. 4, (Aug 1996), pp. (2220-2230), ISSN 0734-2101.

Sagar, P., Shishodia, P.K., Mehra, R.M., Okada, H., Wakahara, A., & Yoshidat, A. (2007). Photoluminescence and Absorption in Sol-Gel-Derived ZnO Films. *J. Lumin.*, Vol. 126, No. 2, (Oct 2007), pp. (800-806). ISSN 0022-2313.

Saito, N., Haneda, H., Sekiguchi, T., Ohashi, N., Sakaguchi, I., & Koumoto, K. (2002). Low-Temperature Fabrication of Light-Emitting Zinc Oxide Micropatterns Using Self-

Assembled Monolayers. *Adv. Mater.*, Vol. 14, No. 6, (Mar 2002), pp. (418-421), ISSN 935-9648.

Sakai, K., Kakeno, T., Ikari, T., Shirakata, S., Sakemi, T., Awai, K., & Yamamoto, T. (2006). Defect Centers and Optical Absorption Edge of Degenerated Semiconductor ZnO Thin Film Grown by a Reactive Plasma Deposition by Means of Piezoelectric Photothermal Spectroscopy. *J. Appl. Phys.*, Vol. 99, No. 4, (Feb 2006), pp. (043508-1-043508-7), ISSN 0021-8979.

Ohgaki, N., Sekiguchi, T., Aoyama, K., Ohgaki, T., Terada, Y., Sakaguchi, I., Tsurumi, T., & Haneda, H. J. (2001). Band-Edge Emission of Undoped and Doped ZnO Single Crystals at Room Temperature. *J. Appl. Phys.*, Vol. 91, No. 6, (Mar 2002), pp.(3658-3663), ISSN 0021-8979.

Serpone, N., Lawless, D., & Khairutdinov, Rr. (1995). Size Effects on the Photophysical Properties of Colloidal Anatase Ti02 Particles: Size Quantization or Direct Transitions in This Indirect Semiconductor?. *J. Phys. Chem.*, Vol. 99, No. 45, (Nov 1995), pp. (16646-16654). ISSN 0022-3654.

Srikant, V., Clarke, D.R. (1997). Optical Absorption Edge of ZnO Thin Films: The Effect of Substrate. *J. Appl. Phys.*, Vol. 81, No. 6357, (May 1997), pp. (6357-6364), ISSN 0021-8979.

Studenikin, S.A., Golego, N., & Cocivera, M. (1998). Fabrication of Green and Orange Photoluminescent, Undoped ZnO Films Using Spray Pyrolysis. *J. Appl. Phys.*, Vol. 84, No. 4, (Aug 1998), pp. (2287-2294), ISSN 0021-8979.

Swanepoel,R. (1983). Determination of the Thickness and Optical Constants of Amorphous Silicon. *J. Phys. E: Sci. Instrum.,* Vol. 16, No. 12, (Dec 1983), pp.(1214-1222), ISSN 0022-3735.

Tak, Y., Yong, K.J. (2005). Controlled Growth of Well-Aligned ZnO Nanorod Array Using a Novel Solution Method. *J. Phys. Chem. B*, Vol. 109, No. 41, (Oct 2005), pp. (19263-19269), ISSN 1089-5647.

Tan, S.T., Chen, B.J., Sun, X.W., Fan, W.J., Kwok, H.S., Zhang, X.H., & Chua, S.J. (2005). Blueshift of Optical Band Gap in ZnO Thin Films Grown by Metal-Organic Chemical-Vapor Deposition. *J. Appl. Phys.*, Vol. 98, No. 013505, (Jul 2005), pp. (013505-1-013505-5), ISSN 0021-8979.

Tsukuma, K., Akiyama, T., & Imai, H. (1997). Liquid Phase Deposition Film of Tin Oxide. *J. Non-Cryst Solids,* Vol. 210, No. 1, (Feb 1997), pp. (48-54), ISSN 0022-3093.

Vayssieres, L., Keis, K., Lindquist, S. E., & Hagfeldt, A. J. (2001). Purpose-Built Anisotropic Metal Oxide Material: 3D Highly Oriented Microrod Array of ZnO. *J. Phys. Chem. B*, Vol. 105 , No. 17, (May 2001), pp. (3350-3352), ISSN 1089-5647.

Vayssieres, L. (2003). Growth of Arrayed Nanorods and Nanowires of ZnO from Aqueous Solution**. *Adv. Mater.*, Vol. 15, No. 5, (Mar 2003), pp. (464-466), ISSN 1521-4095.

Wang, J. M., Gao, L. J. (2003). Wet Chemical Synthesis of Ultralong and Straight Single-Crystalline ZnO Nanowires and Their Excellent UV Emission Properties. *J. Mater. Chem.*, Vol. 13, No. 10, (Aug 2003), pp. (2551-2254), ISSN 0959-9428.

Wang, Y.G., Lau, S.P., Lee, H.W., Yu, S.F., Tay, B.K., Zhang, X.H., & Hng, H.H. (2003). Photoluminescence Study of ZnO Films Prepared by Thermal Oxidation of Zn Metallic Films in Air. *J. Appl. Phys.*, Vol. 94, No. 1, (Jul 2003), pp. (354-358), ISSN 0021-8979.

Wu, X.L., Siu, G.G., Fu, C.L., & Ong, H.C. (2001). Photoluminescence and Cathodoluminescence Studies of Stoichiometric and Oxygen-Deficient ZnO Films. *Appl. Phys. Lett.*, Vol. 78, No. 16, (Apr 2001), pp. (2285-2287), ISSN 0003-6951.

Yang, J.L., An, S.J., Park, W.I., Yi, G.C., & Choi, W. (2004). Photocatalysis Using ZnO Thin Films and Nanoneedles Grown by Metal-Organic Chemical Vapor Deposition**. *Adv. Mater.*, Vol. 16, No. 18, (Sep 2004), pp. (1661-1664), ISSN 1521-4095.

Zhao, J., Jin, Z.G., Li, T., & Liu, X.X. (2006). Nucleation and Growth of ZnO Nanorods on the ZnO-Coated Seed Surface by Solution Chemical Method. *J. Eur. Ceram. Soc.*, Vol. 26, No. 13, (Sep 2005), pp. (2769-2775), ISSN 0955-2219.

Zhu, Y.W., Zhang, H.Z., Sun, X.C., Feng, S.Q., Xu, J., Zhao, Q., Xiang, B., Wang, R.M., & Yua, D.P. (2003). Efficient Field Emission from ZnO Nanoneedle Arrays. *Appl. Phys. Lett.*, Vol. 83, No. 1, (Jul 2003), pp. (144-146), ISSN 0003-6951.

3

ZnO Nanorods: Synthesis by Catalyst-Free CVD and Thermal Growth from Salt Composites and Application to Nanodevices

Oleg V. Kononenko[1], Arkady N. Redkin[1], Andrey N. Baranov[3], Gennady N. Panin[1,2], Artem A. Kovalenko[4] and Anatoly A. Firsov[1]

[1]*Institute of Microelectronics Technology and High Purity Materials, RAS Chernogolovka, Moscow region,*

[2]*Quantum-Functional Semiconductor Research Center, Department of Physics, Dongguk University, Seoul,*

[3]*Chemistry Department, Moscow State University Moscow,*

[4]*Department of Materials Science, Moscow State University, Moscow,*

[1,3,4]*Russia*

[2]*South Korea*

1. Introduction

Zinc oxide (ZnO) is a unique functional semiconductor material with a wide band gap (3.37 eV), high binding energy of the exciton (60 meV) at room temperature, and an effective ultraviolet luminescence (Özgür et al., 2005). Materials based on ZnO can be used as chemical (Fan & Lu, 2005) and biological (Yang et al., 2009, Yeh et al., 2009) sensors, solar cells (Law et al., 2005, Wei et al., 2010), light emitting diodes (Park & Yi, 2004), laser (Huang et al., 2001, Govender et al., 2002), and composite materials (Hu et al., 2003). Zinc oxide is a biocompatible material with antiseptic properties. Quasi one-dimensional (1D) structures on the basis of zinc oxide are a promising material for nanoelectronics (Park et al., 2005).

In recent years, there has been an increasing interest in quasi one-dimensional nanocrystalline zinc oxide (nanorods, nanowires, and nanowhiskers), motivated by its perfect crystal structure and unusual properties due to size effects. In addition, single-crystal samples make it possible not only to raise the exciton density and create low-threshold gain media but also to reduce scattering losses. In this context, aligned arrays of single-crystal nanorods 20–200 nm in diameter and several microns in length are of special interest. Currently the optical properties of ZnO as a semiconductor material with good luminescent properties are of greatest interest. This is due to the possibility of applying materials on the basis of zinc oxide in the creation of new effective optoelectronic devices.

An effective approach for creation of nanodevices for light emitting and field emission is fabrication of vertically aligned ZnO 1D structures. Aligned ZnO nanocrystal arrays on substrates are commonly grown using thin metal film as a catalyst for 1D growth and different methods, such as vapor-liquid-solid (VLS) process (Zhao et al., 2003), metal-organic chemical vapor deposition (MOCVD) (Park & Yi, 2004) or sol-gel process (Krumeich et al., 1999). The drawback to this method is that the nanocrystals may be contaminated with the catalyst (Collins et al., 1957 and Oh et al 2008). Moreover, catalyst particles as a rule remain on the tips of the grown zinc oxide nanocrystals (Kim et al., 2005 and J. Park et al., 2003). Catalyst-Free Chemical Vapor Deposition (CVD) is a promising method for ZnO nanostructure synthesis of high structural and optical perfection. Up to now, few papers have been devoted on the subject of catalyst-free growth of ZnO nanowire and nanorod arrays at temperature below 600°C (W.I. Park et al., 2003; Umar et al., 2005; Liu et al., 2005; Wang et al., 2005).

It is well known that the dopants can control the electronic and luminescence properties of the material. In recent years, attention has also focused on spin-dependent phenomena in dilute magnetic zinc oxide in which stoichiometric fraction of the zinc atoms are replaced by transition metal atoms. The growth from the salt mixture is promising method for doping of ZnO nanorods by transition metals and acceptor dopants.

2. Catalyst-free CVD synthesis of ZnO nanorods

The growth of ZnO nanorods was carried out by the elemental vapor-phase synthesis at a reduced pressure in flow-type reactor, developed earlier (Red'kin et al., 2007). A high-purity metallic granulated zinc (99.99%) was placed in an alumina boat which was then inserted at the end of quartz ampoule sealed at one end. At the open end, the ampoule had a wide slit, below which substrates were mounted with their front sides up. The ampoule was introduced into a horizontal two-zone flow-type quartz reactor so that the zinc source was located in one of the zones (evaporation zone), and the substrates, in the other (growth zone).

The reactor was first evacuated with forevacuum pump for 1 h and then high-purity argon (99.999%) was introduced into the reactor without terminating of evacuation. As a result, the steady-state pressure in the reactor reached 10^3 Pa. Then the temperature in the growth zone (t_2) was raised to the working one, 500 - 550 °C. Next, the temperature in the evaporation zone (t_1) was raised to reach 610 - 620 °C, and then high-purity oxygen (99.999%) was introduced into the reactor. The total flow of argon–oxygen mixture was 66.7 sccm. The oxygen concentration in the mixture was 10 vol. %.

During the process zinc vapor from the evaporation zone reached the growth zone and reacted with oxygen. Zinc vapor flow was 10 - 12 g/h. Under the above conditions, zinc/oxygen molar ratio in the vapor phase was about 5/1. The synthesis time was 30 - 40 min.

The growth of arrays of ZnO well aligned nanorods is very sensitive to synthesis parameters. In earlier works we made an extensive study in order to determine the regularities and optimum conditions for elemental vapor synthesis of highly aligned ZnO nanorod arrays on (100) and (111) silicon and glass substrates (Red'kin et al., 2007, 2009). Note also, that neither 1D growth catalyst nor zinc oxide buffer layers were deposited preliminary onto the substrates. The vertical growth of nanorods was observed on single-crystalline substrates and on amorphous glass substrates, which indicates that a substrate crystallography exerts no appreciable effect on nanorod growth direction under the conditions of these experiments.

Electron microscopy examination of the samples showed that ZnO nanorods normally directed towards a substrate surface grew on all the substrates used. The degree of array misalignment primarily depends on their position in the growth zone. In the first part of the growth zone located closer to the zinc vapor source, arrays of nanorods are uniform in size and mainly directed normal to the substrate surface irrespective of a substrate type.

As a rule, nanorods with the most uniform vertical alignment grow at the beginning and in the middle of the growth zone. A more detail analysis of the arrays revealed a relationship between the density of nanorod positioning on a substrate and the degree of their ordering. In more ordered arrays on Si (100) and Si (111) the density of nanorods is by 20 - 25% higher than in less ordered arrays on substrates of the same type. Arrays of approximately equal density on various substrates have similar distribution of nanorod growth directions. As have shown rocking curves the deviation from the normal direction of growth in these samples do not differ very much. So, the density of ZnO nanocrystals per square unit on a substrate is one of the factors which determine the uniformity of nanorod direction distribution in an array. Nanorod diameter also plays a certain role. It was noted that arrays of thicker nanorods are more uniformly vertically aligned. In other words, the more closely packed the growing nanorods, the more uniform is their vertical growth.

Note that initial stages of ZnO nanocrystals nucleation are of important role. Earlier we supposed that the growth of ZnO nanorods proceeds by VLS process under the conditions of our experiment. In the initial step nanodrops of metal zinc condensed on the substrate surface because of the temperature difference in the evaporation and growth zones. Further they act as a 1D growth catalyst (Red'kin et al., 2009). This "self-catalysis", with metal zinc droplets acting as catalyst of 1D growth of ZnO nanocrystals, was also described elsewhere (Wei et al., 2005; Zha et al., 2008). According to this mechanism, the initial density of zinc nanodrops on a substrate determines the nanorod density in a grown array. It explains why the uniformity of vertical growth is, as a rule, higher in the part of the growth zone which is closer to the zinc evaporation zone. The relationship of ZnO nanorods density and thickness and uniformity of vertical nanorod growth was observed earlier (Reiser et al., 2007, Zhao et al., 2005, Lee et al., 2007), where a similar conclusion was made concerning the effect of nucleation density on predominant direction of ZnO nanocrystal growth and uniformity of arrays grown by different methods.

Another difference of the self-catalysis from Au catalyst growth is that the diameter of catalyst droplets can both to decrease, and to increase, depending on relationship of zinc and oxygen partial pressure. It allows us to control nanorod shape by varying growth parameters such as zinc source and substrate temperature, reactor pressure and oxygen concentration. Figures 1-5 shows various types of nanorod morphology, which can be obtained by changing growth parameters during nanorod synthesis. Diameter of nanorods can decrease (fig.1, 2) or increase (fig.3) or it can alternately occur (fig.4). If change Zn partial pressure occurs suddenly rather big Zn drops can break up to the small ones. It leads to that on the tip of nanorods several nanorods with smaller diameters start to grow (fig.5).

To estimate the crystal perfection of nanorods in the grown samples, we studied their cathode-luminescence (CL) spectra. CL spectra were measured in two modes: direct and panchromatic, which allowed a comparison of the emission areas and the electron microscopy image of the nanorods on the substrate. As is seen from fig.6, the measured emission was mainly from the top faces of ZnO nanorods. The CL spectra showed that the grown nanorods were of high quality, irrespective of the substrate type. The spectra of all samples (fig.7) contain only one strong UV band of the edge emission peaking at 382 nm due to free exciton recombination and no green and blue lines which are usually attributed to point defects in ZnO, such oxygen and zinc vacancies (Gruzintsev & Yakimov, 2005).

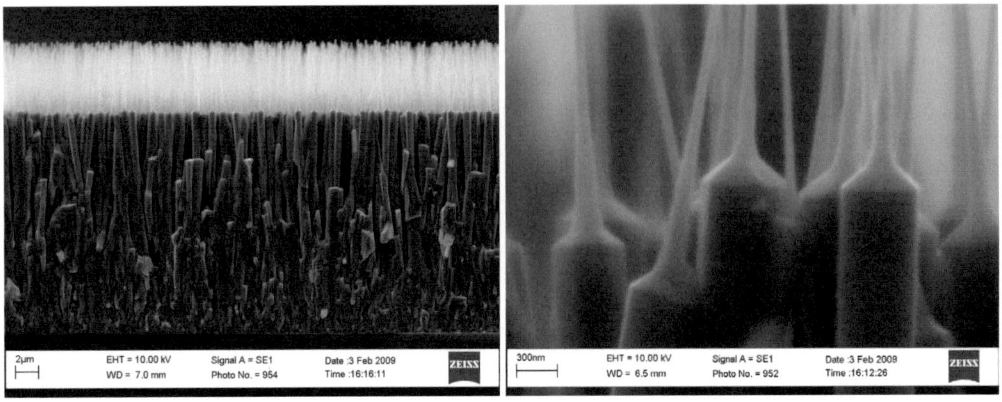

Fig. 1. (Left) scanning electron microscope (SEM) image of the cross-section of the ZnO nanorods arrays grown on Si (100) substrate. In the end of synthesis partial pressure of zinc was decreased. (Right) magnified SEM image of the ZnO nanorods shown in the left photo.

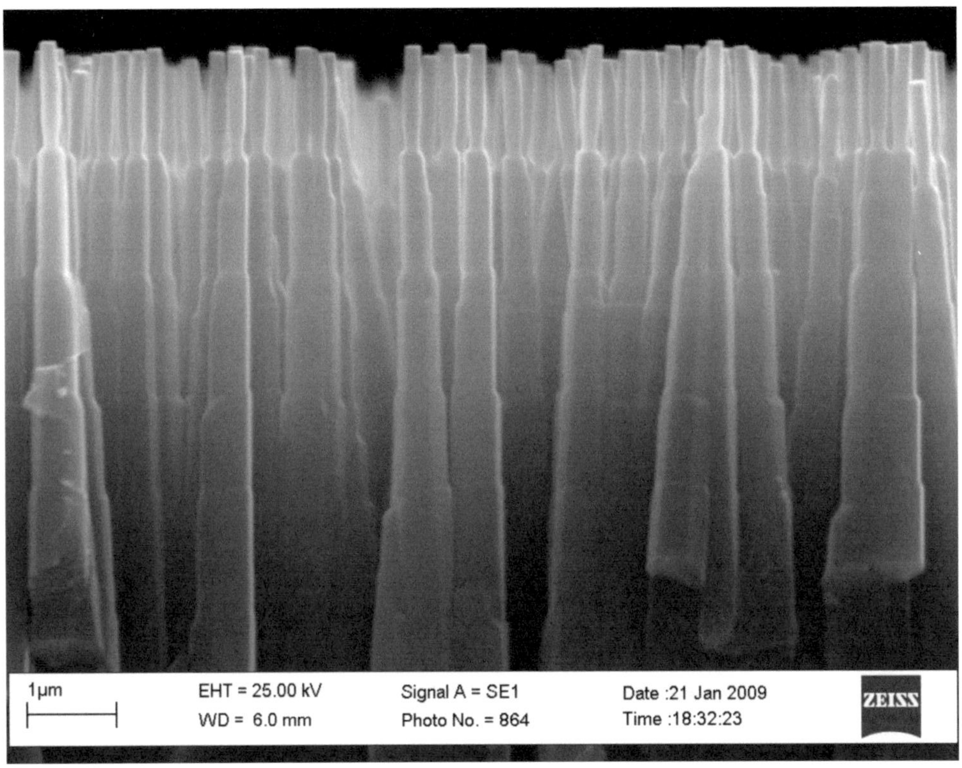

Fig. 2. SEM image of the cross-section of the ZnO nanorods arrays grown on Si (100) substrate. During synthesis zinc partial pressure was lowered in equal intervals.

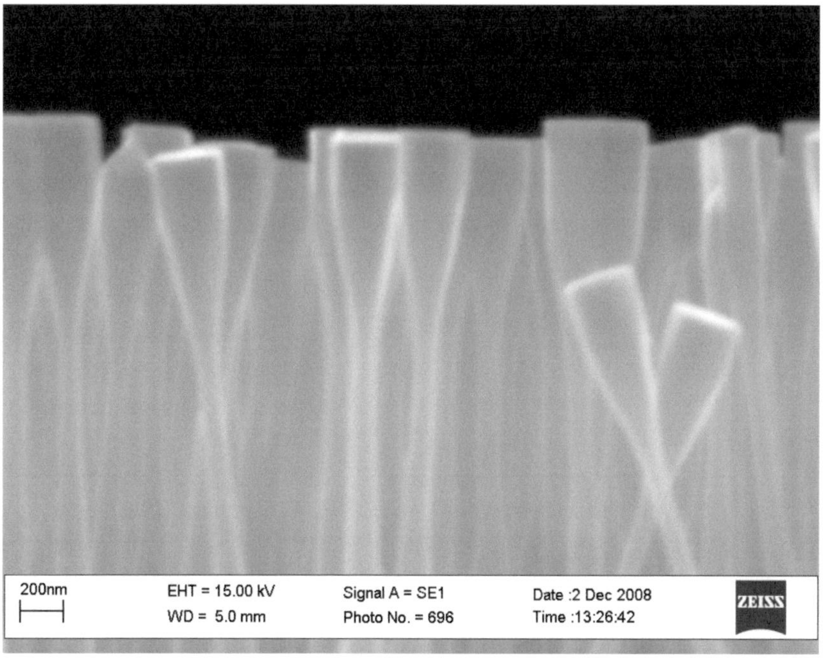

Fig. 3. SEM image of the cross-section of the ZnO nanorods arrays grown on Si (100) substrate. A partial pressure of oxigen was decreased near the end of synthesis.

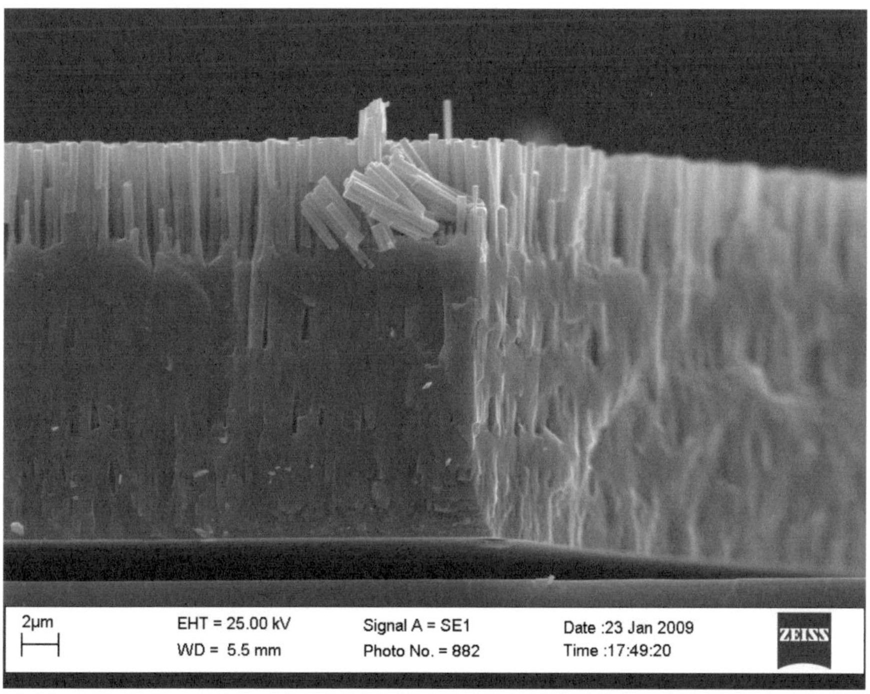

Fig. 4. SEM image of the cross-section of the ZnO nanorods arrays grown on Si (100) substrate. During synthesis zinc partial pressure was lowered and raised alternately.

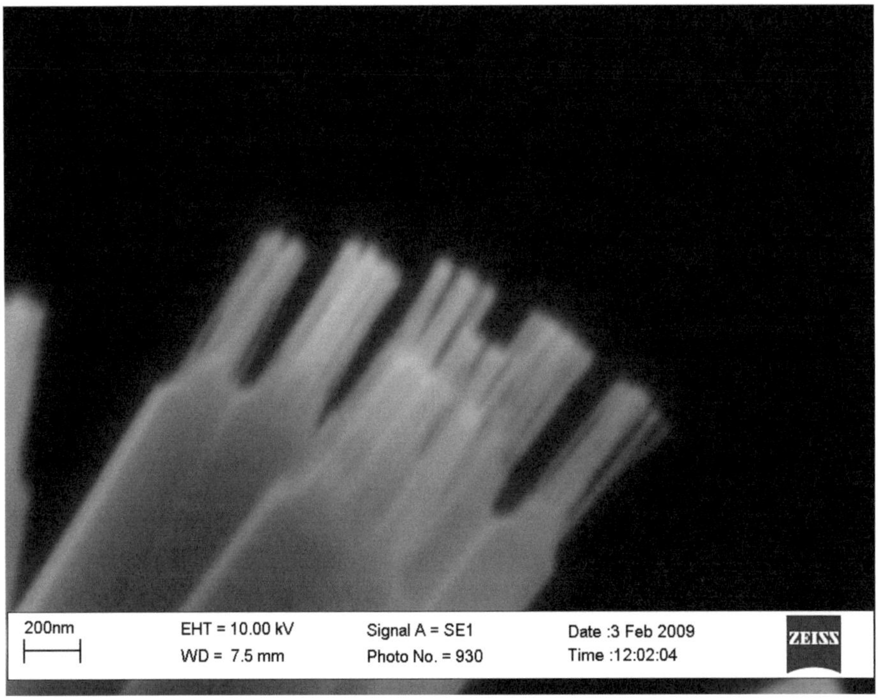

Fig. 5. SEM image of the cross-section of the ZnO nanorods arrays grown on Si (100) substrate. In the end of synthesis partial pressure of zinc was sharply lowered.

3. ZnO nanorod p-n homojunction

Stable and reproducible p-type ZnO for fabricating ZnO homojunction optoelectronic devices still remains challenging (Look et al., 2004).

ZnO single crystals and epitaxial films contain, as a rule, quite a few of intrinsic defects (usually higher than 10^{16} cm^{-3}). Majority of intrinsic defects are donors. When ZnO is doped with an acceptor, the compensation of donors takes place. The inversion of the n-type conductivity into the p-type one is possible only in samples with a low initial concentration of intrinsic defects under conditions when the compensation of the acceptors by the donor defects is minimized.

Sb was deposited by thermal evaporation onto the tops of aligned ZnO nanorod arrays grown by the catalyst-free elemental vapor-phase synthesis on n^+-Si(100) substrats. Then samples were annealed at 420°C in air atmosphere during different time.

CL spectrometry was used to measure the optical properties of Sb-doped ZnO nanorods (fig.8-10). For short-time annealed samples, a peak at 370.5 nm is dominant, which may be associated with donor-bound exciton emissions (fig.8). A peak at 380.3 nm may be associated with Sb-dopant. For medium-time annealed samples, both peaks are about of the same intensity (fig.9). For long-time annealed samples, a peak at 377.7 nm becomes dominant (fig.10). The peak positions are shifted to lower energies due to effects arising from the increased Sb-dopant concentration.

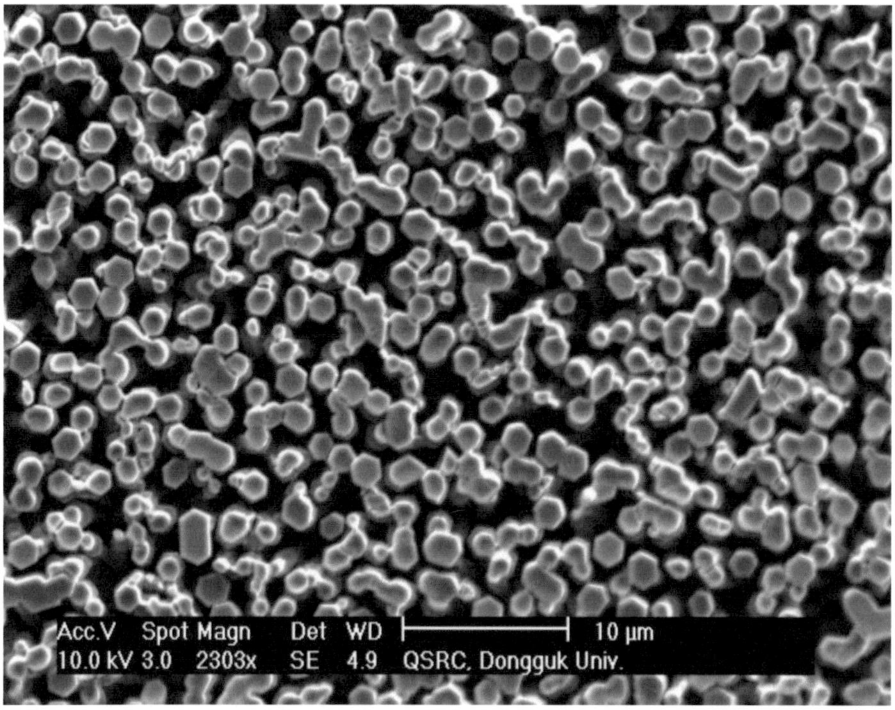

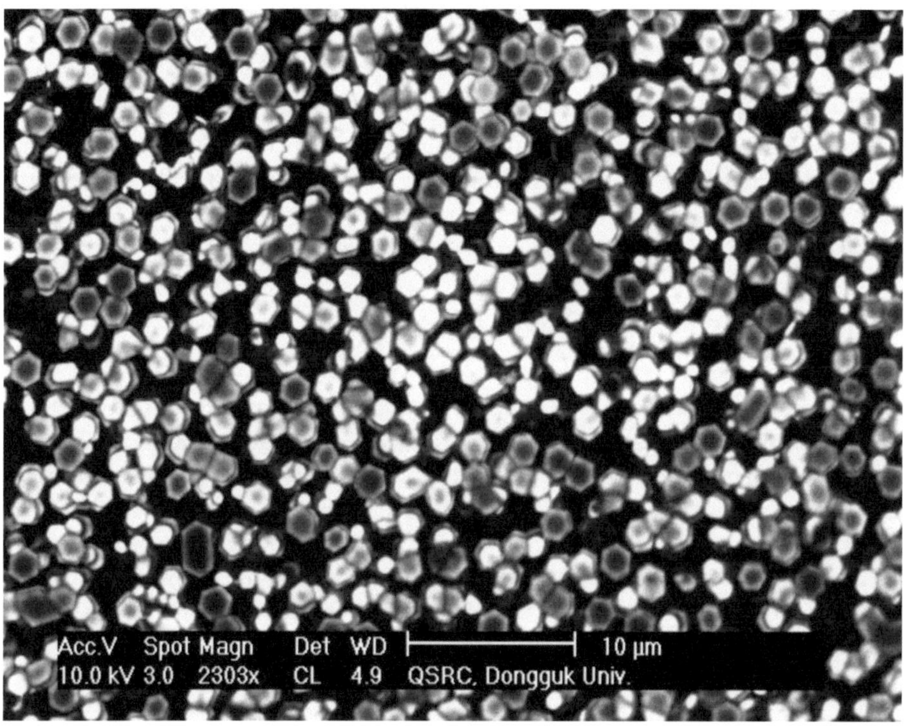

Fig. 6. SEM (top) and CL (bottom) images of a ZnO nanorod array grown by the elemental vapor-phase synthesis.

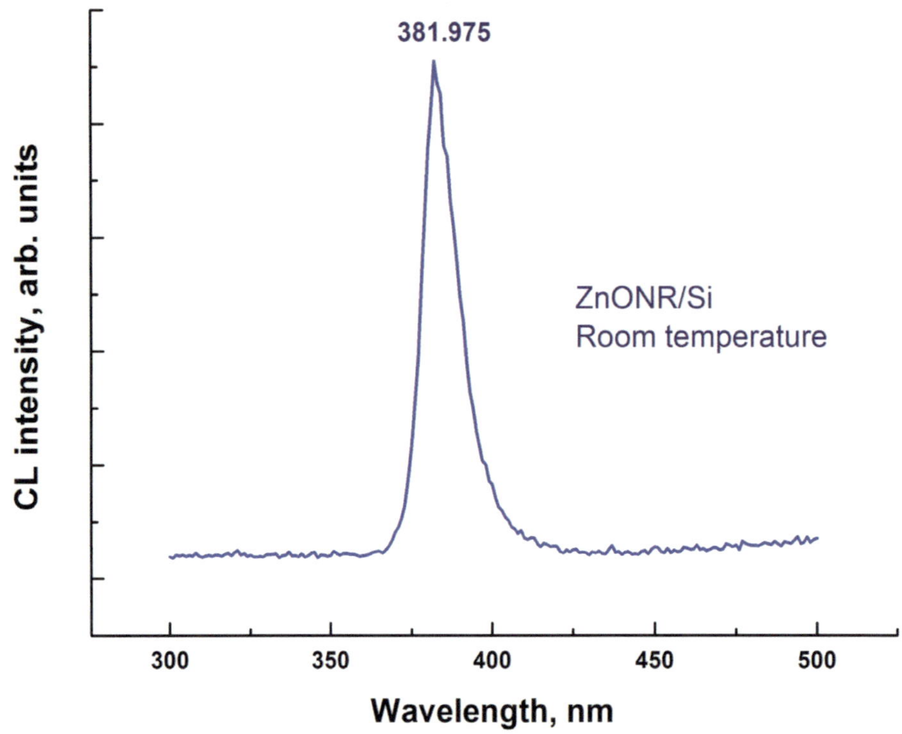

Fig. 7. CL spectrum of a ZnO nanorod array grown by the elemental vapor-phase synthesis.

Electrical contacts were prepared to Sb-doped ZnO nanorods and to n^+-Si (100) substrate. *I-V* characteristics were measured for all three samples. However, *I-V* curve only of the long-time annealed sample shows good rectification (fig.11). Blue light could be seen clearly from the diode by the naked eye (fig.12). The emission has been detected at ~6 V and increases with increasing the bias voltage (Kononenko et al., 2009).

4. ZnO nanorod p-n heterojunction

As the inversion of n-type conductivity to p-type conductivity in ZnO is a challenge so far, the other p-type wide band gap semiconductors and organic materials can be used for p-n junction fabrication on the basis of ZnO materials. Such semiconductors those are compatible with zinc oxide are $SrCu_2O_2$, $CuAlO_2$, GaN and NiO (Ohta & Hosono, 2004). The fabrication of hybrid heterojunction diodes have been reported in several works (Park & Yi, 2004, Zhang et al., 2009). These materials however require complex deposition techniques, such as the Molecular Beam Epitaxy, Chemical Vapor Deposition, Pulsed Laser Deposition. NiO and organic materials are an exception to this range. NiO can be synthesized by the thermal vacuum evaporation at the oxygen pressure of the order of 10^{-4}–10^{-5} Torr. In addition, a high enough (on the order of 10^{19} - 10^{20} cm^{-3}) hole concentration in the NiO films can be obtained. Organic materials can be deposited on ZnO materials by the spin coating. The fabrication of an inorganic/organic heterostructure diodes have been reported in references (Konenkamp et al., 2004, Konenkamp et al., 2005, Sun et al., 2008).

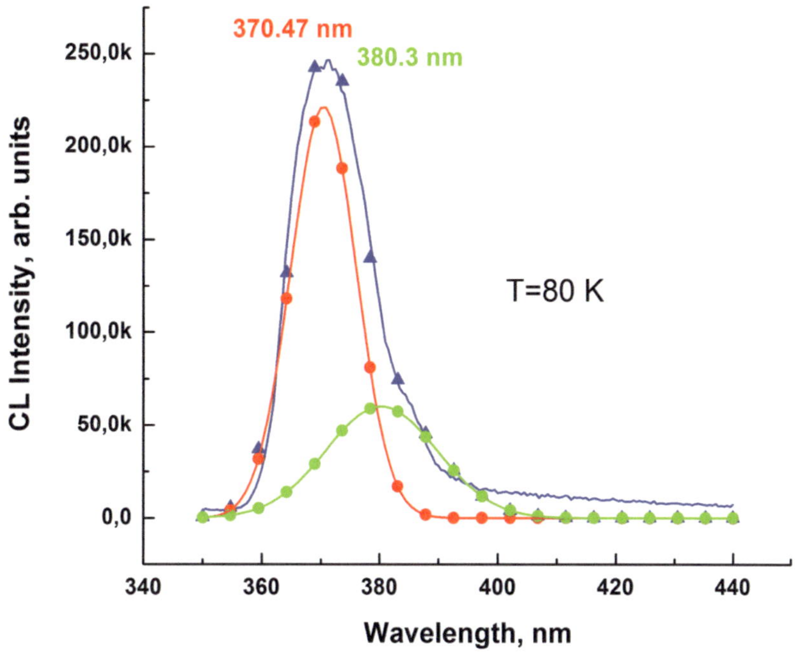

Fig. 8. CL spectrum of short-time annealed Sb-doped ZnO nanorod arrays grown on n+-Si (100) substrate.

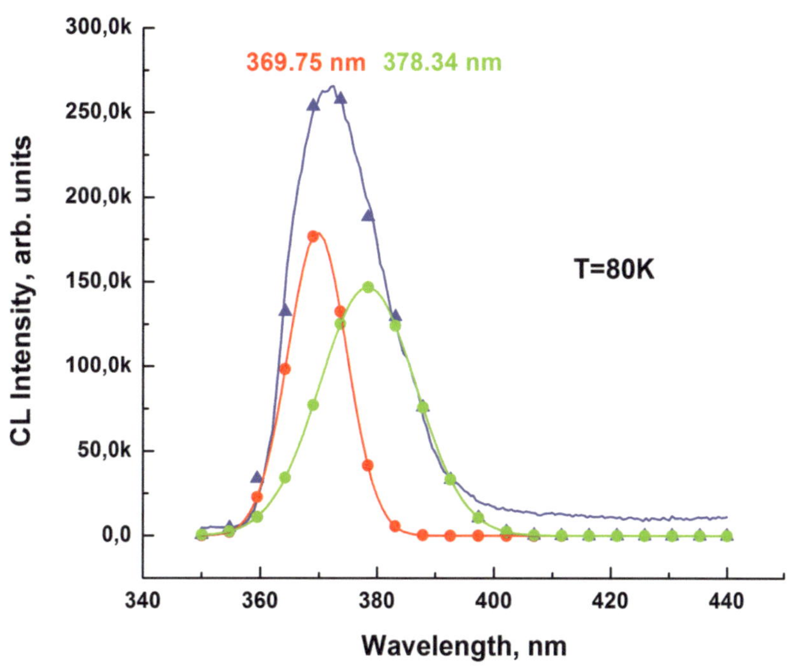

Fig. 9. CL spectrum of medium-time annealed Sb-doped ZnO nanorod arrays grown on n+-Si (100) substrate.

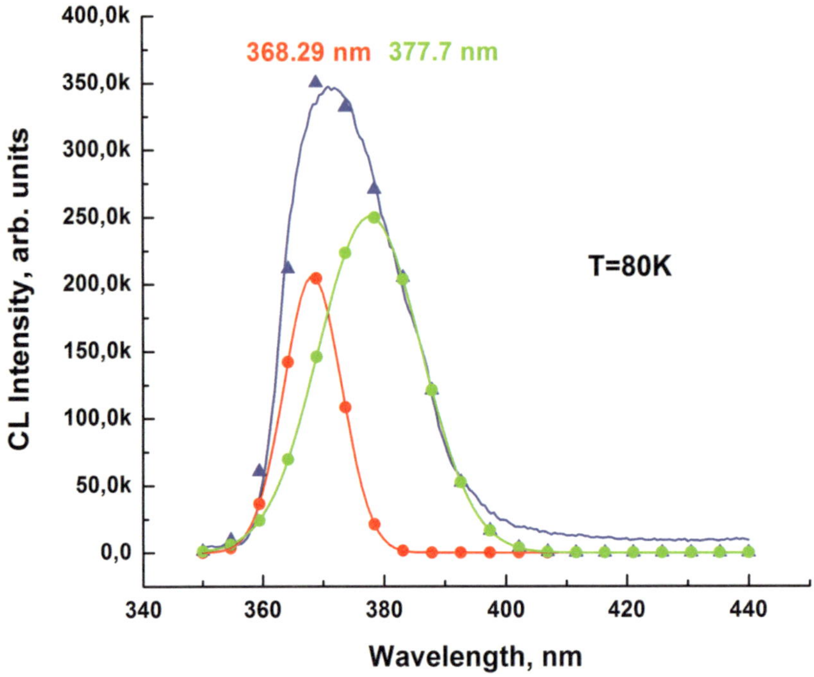

Fig. 10. CL spectrum of long-time annealed Sb-doped ZnO nanorod arrays grown on n^+-Si (100) substrate.

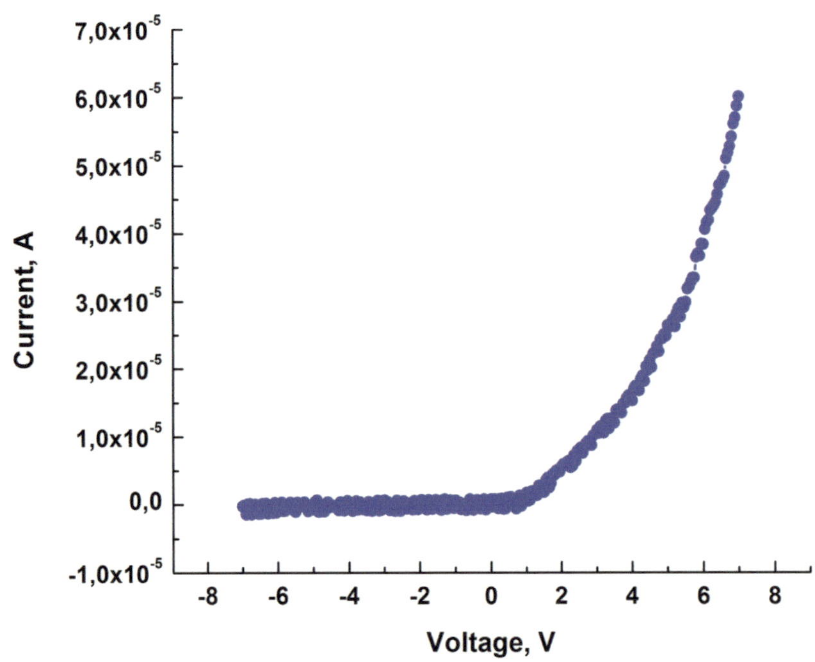

Fig. 11. The current-voltage characteristics of the Sb-doped ZnO nanorods diode.

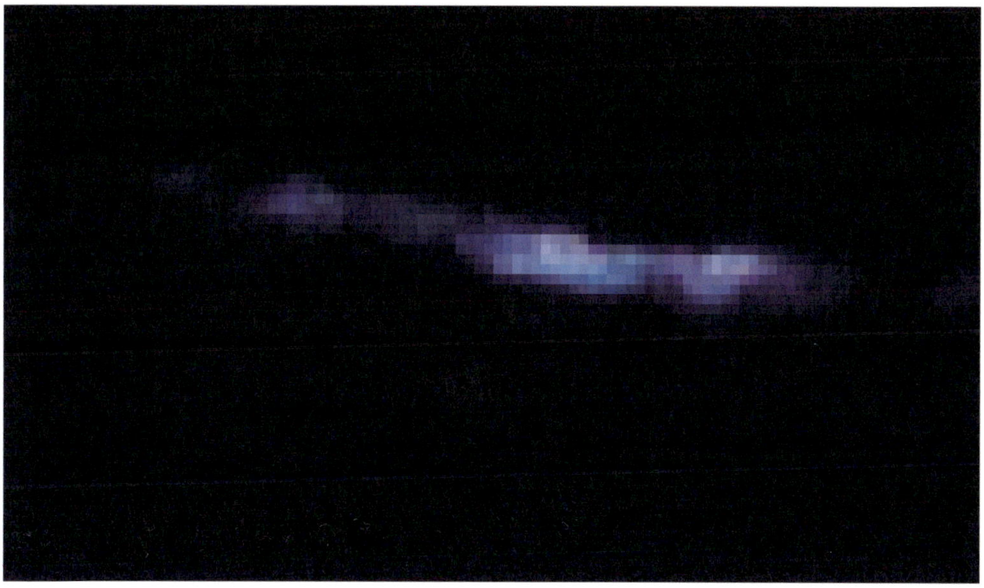

Fig. 12. The optical micrograph of blue emission from the Sb-doped ZnO nanorods diode.

Two types of vertical structures were fabricated on the basis of ZnO and NiO. NiO film was deposited by e-beam evaporation of high purity Ni in an oxygen atmosphere at a partial pressure of 10^{-5} Torr on Si (100) substrate. Vertically aligned ZnO nanorods were grown by the catalyst-free elemental vapor-phase synthesis on the NiO film. Indium electrodes were prepared on the top of nanorods and the NiO film. The other structure was fabricated as follows. Vertically aligned ZnO nanorods were grown on Si (100) substrate. Then NiO was deposited by e-beam evaporation of high purity Ni in an oxygen atmosphere at a partial pressure of 10^{-5} Torr. Ni electrode was deposited in vacuum onto NiO and indium electrode was deposited on ZnO.

An inorganic-organic hybrid heterostructure was fabricated as follows. Vertically aligned ZnO nanorods were grown on indium tin oxide (ITO) glass substrate. Poly(methyl methacrylate) (PMMA) was spin-coated on the grown nanorods to provide a smooth surface for subsequent thin film deposition. A drop of poly(3,4-ethylene-dioxythiophene) (PEDOT)/ polystyrenesulfonate(PSS) was spin coated on the top of ZnO nanorods at a spin speed of 2000 rpm. Finally, a copper wire was attached on the surface of the PEDOT/PSS film using silver paste as an electrode. The other copper wire was soldered by indium to ITO film.

4.1 NiO-ZnO nanorods p-n heterojunctions

We produced heterogeneous p-n junctions on the basis of the ZnO nanorods with n-type conductivity and NiO films with p-type conductivity having sufficiently good rectifying characteristics. Figure 13 shows the *I-V* characteristics of a *p*-type NiO/*n*-type ZnO nanorods heterojunction diode measured in dark and UV (420 nm wavelength) illumination at room temperature. Rectifying *I-V* characteristics were obtained with a forward threshold voltage of ~1.5 V. The diode showed a strong, reversible response to above band gap ultraviolet light, with the UV-induced current being approximately a factor of 2.5 larger than the dark current at a given voltage.

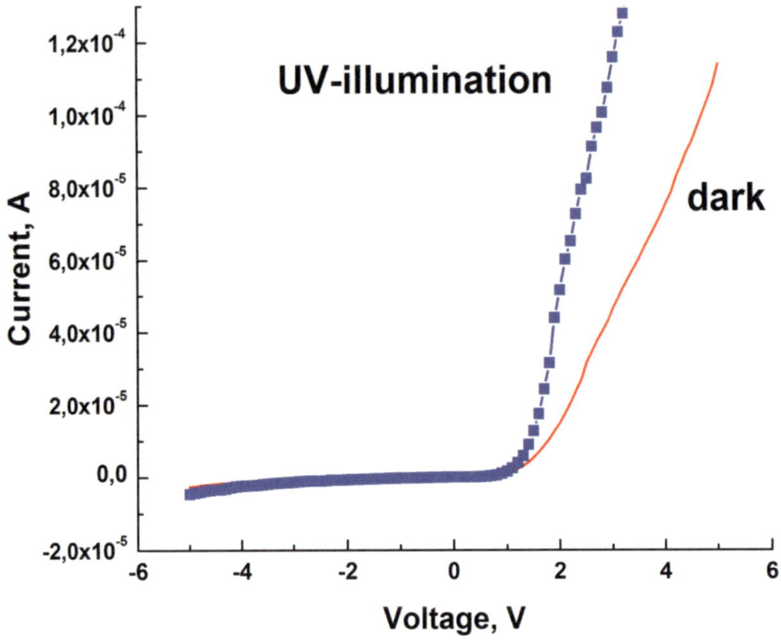

Fig. 13. Typical I-V characteristics of a pn-heterojunction diode under dark and UV-illumination conditions at room temperature. The threshold voltage is ~1.5 V.

Figure 14 (left) shows the *I-V* characteristic of a *n*-type ZnO nanorods/*p*-type NiO heterojunction diode measured in dark. Rectifying *I-V* characteristics were obtained with a forward threshold voltage of ~1.5 V. A structure of the heterojunction was investigated with the use of a focused ion beam system Strata 201 (FEI Company). A cross-section of the ZnO nanorods coated by NiO was prepared by FIB milling. Figure 14 (right) shows FIB image of the cross-section. It is seen in the picture that nickel oxide grown on ZnO nanorods have a fiber structure with fiber diameters of ~50 nm and fiber length of ~100 - 500 nm.

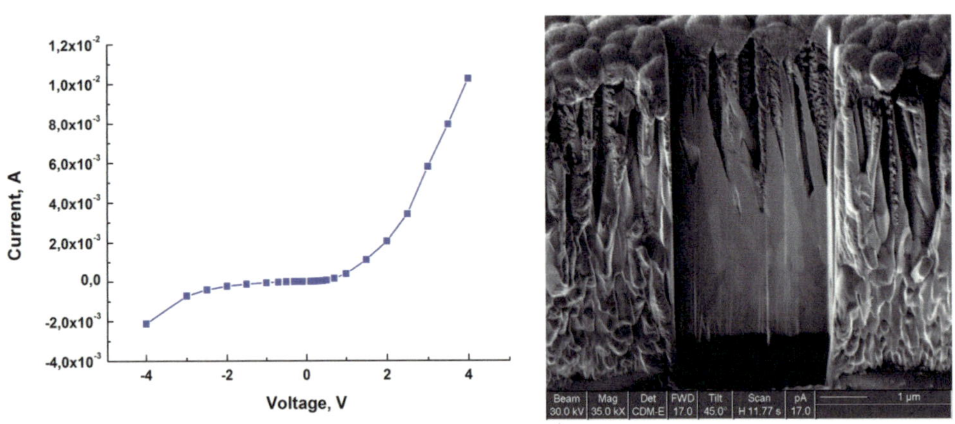

Fig. 14. (Left) Typical I-V characteristics of a pn-heterojunction diode under dark conditions at room temperature. The threshold voltage is ~1.5 V. (Right) FIB image of the cross-section of the ZnO nanorods coated by NiO, prepared by FIB milling.

4.2 Inorganic-organic hybrid diodes

Figure 15 shows the *I-V* characteristics of an inorganic-organic pn-heterojunction diode measured in dark and UV illumination at room temperature. Rectifying *I-V* characteristics were obtained with a forward threshold voltage of ~0.4 V. The inorganic-organic pn-heterojunction diode showed a strong, reversible response to above band gap ultraviolet light, with the UV-induced current being approximately a factor of 10 larger than the dark current at a given voltage in a negative branch and a factor of 1.7 in a positive branch. Electroluminescence was observed in such diodes. The emission has been detected at ~15 V.

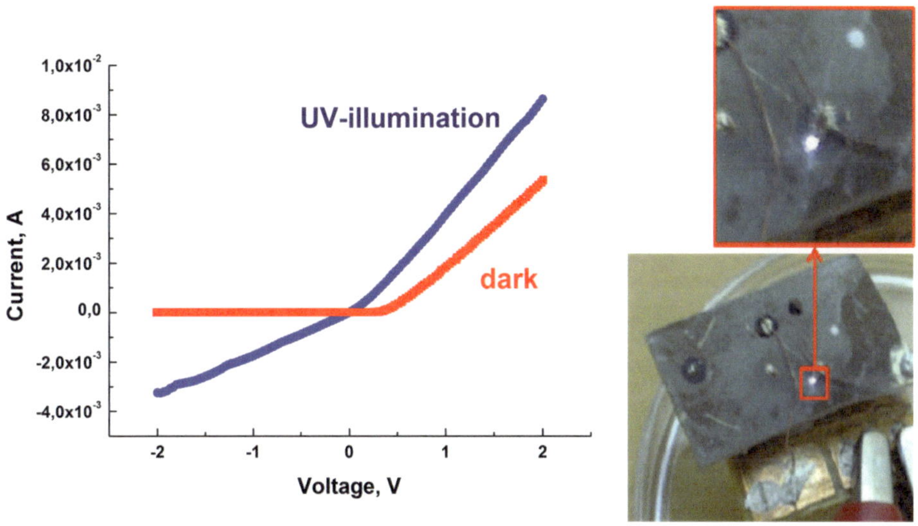

Fig. 15. *I-V* characteristics of the inorganic-organic pn-heterojunction diode under dark and UV-illumination conditions at room temperature (left). The optical micrograph of blue-white emission from the inorganic-organic diode (right).

5. Thermal growth of ZnO nanorods from salt composites

It is well known that the dopants can control the electronic and luminescence properties of the material. In recent years, attention has also focused on spin-dependent phenomena in dilute magnetic zinc oxide in which stoichiometric fraction of the zinc atoms are replaced by transition metal atoms. The growth from the salt mixture is promising method for doping of ZnO nanorods by transition metals and acceptor dopants.

ZnO nanorods were prepared by thermal growth from the solution processed precursor using a freeze drying and milling technique (Baranov et al., 2004). The details of nanorods growth were described earlier (Baranov et al., 2005). Synchrotron X-ray diffraction and high temperature SEM examinations were used to study crystallochemical and morphological evolution of a ZnO-NaCl system during thermal processing.

Aqueous solution of 0.5M $Zn(NO_3)_2$ and $Mn(CH_3COO)_2$ in a Zn/Mn mole ratio of 50/1 (sample Mn2) was vigorously mixed with excessive amount of 2M $(NH_4)_2CO_3$ solution. Aqueous solution for the second sample (MS) was prepared from $Zn(NO_3)_2$ $6H_20$, $Mn(CH_3COO)_2$ $4H_2O$ and $SnCl_2$ $2H_2O$ in a Zn/Mn/Sn mole ratio of 20/1/1. A few drops of concentrated nitric acid were added to avoid hydrolysis of $SnCl_2$ at dissolution. Then the solution was precipitated by excessive amount of 2M aqueous $(NH_4)_2CO_3$ solution. Precipitates were centrifuged, thoroughly washed by distilled water and freeze dried at P = 5 Pa using Alpha 2-4 (Christ, Germany) laboratory freeze drier. The main component of as-obtained powder samples was zinc carbonate hydroxide $Zn_2(OH)_2CO_3 \bullet xH_2O$ (ZCH) composed of 50 nm diameter amorphous nanofibres. Mixture of the Zn-containing precursor with NaCl/Li_2CO_3 salt composition in a 0.5/9/1 weight ratio was milled for 12 h in zirconia jar with zirconia balls using Pulverisette 5 (Fritsch) planetary mill at 700 rpm. Growth of ZnO nanorods was performed from the as-prepared powder in a muffle furnace at 700°C for two hours. Finally product was placed onto dense paper filter and salts were washed many times by water. Supernatant solution was checked on the absence of Cl^- anions by reaction with aqueous solution of $AgNO_3$.

For the synthesis of iron doped nanorods we prepared aqueous solutions of $Zn(NO_3)_2$ and $Fe(NO_3)_3$ salts in a Zn/Fe mole ratio of 100/3 (sample ZnO:FeLi). All other procedures were performed as previously described.

For the synthesis of antimony doped nanorods we dissolved ZnO in a concentrated nitric acid. $SbCl_3$ was added to the hot tartar acid solution and dissolved upon boiling in accordane with reaction:

$$SbCl_3 + 3H_2C_4H_4O_6 \rightarrow Sb(HC_4H_4O_6)_3 + 3HCl \, .$$

Then zinc and antimony contained solutions were mixed in order to obtain 0.01, 0.1, 0.3, 0.5, 0.7, 1 molar % of Sb. All other procedures were performed as previously described.

Elemental analysis was performed by mass-spectrometer Optima 3000XL ICP (Perkin Elmer). For analysis diluted solutions of analyzed samples were prepared (~1 mg of nanorods per 1000 ml of solution in 0.01 M hydrochloric acid). Atomization and ionization were reached by inductive coupled plazma. Calibration was made using standard solutions with known concentration of Sb, Fe, Mn, Li.

SEM (Philips SEM or Supra 50VP (LEO) measurements and cathodoluminescence (CL) spectra of the samples were taken by XL 30S FEG high-resolution scanning electron microscope (HRSEM) with a MonoCL system for CL spectroscopy, and energy dispersive X-ray analysis were used to examine the samples. The morphologies and size distributions of ZnO nanorods were examined by using the JEM-4010 high-resolution transmission electron microscope (HRTEM) at the accelerated voltage of 400kV. TEM with EDAX was employed for characterization of microstructure and distribution of Mn and Sn ions in the ZnO matrix. Magnetic properties of the samples were examined using a Quantum Design SQUID magnetometer in the temperature range from 5 K to 300 K. X-ray powder diffraction (XRD) data of the synthesized nanorods have been collected by a Rigaku D/MAX 2500 (CuK_{α}.radiation).

Figures 16 (top) and (bottom) show HRSEM images of Mn and MnSn-doped ZnO nanorods, respectively M2 and MS samples, synthesized from the NaCl-Li_2CO_3-containing salt mixture. The detailed SEM analysis revealed that as-prepared ZnO nanorods are straight in morphology and smooth on the surface. Both samples show the quite uniform nanorods with diameters ranging from 15 to 100 nm and from 0.5 to 5 μm in length. The size and the shape of the nanorods correlate well with the annealing temperature in the range 600-700°C. ZnO nanorods of the smallest size were formed at 600°C. Our experiments have shown that the resulting morphologies strongly depend on the growth temperature.

Fig. 16. High resolution scanning electron microscopy images of ZnO nanorods doped with Mn (top) and Mn and Sn (bottom).

Figure 17 shows XRD patterns of Mn (Fig. 17a) and MnSn-doped (Fig. 17b) ZnO nanorods. All the main peaks in Figures 17a and 17b match well with zincite (S.G. P63/mmc (186); JCPDS card 36-1451) although some minor unidentified impurities are obviously present in the samples.

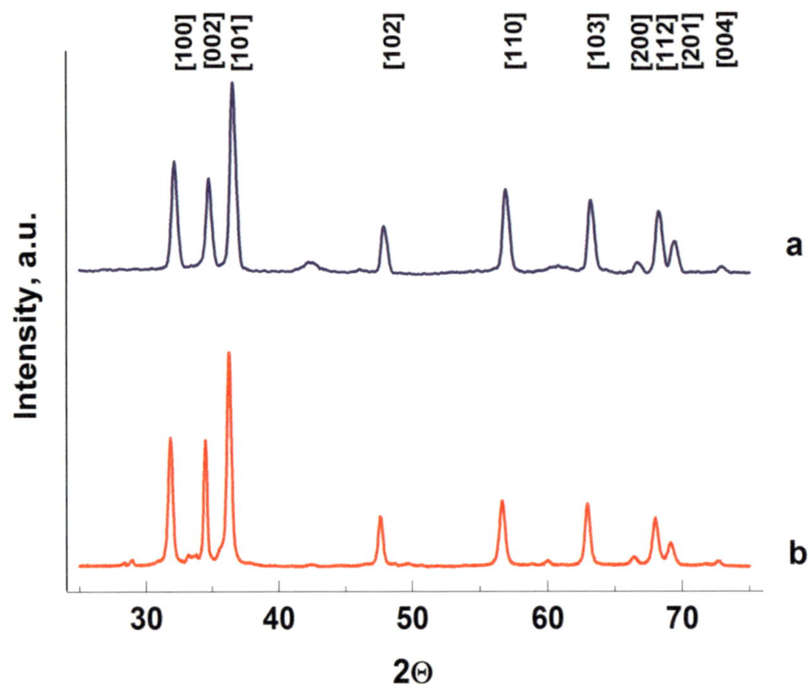

Fig. 17. XRD patterns of ZnO nanorods doped with (a) Mn and (b) Mn and Sn.

The lattice parameters of the samples a = 3.2472(2) Å and c=5.1981(8) (M2 sample), and a = 3.234(2) Å and c= 5.179(4) nm Å (MS sample) were calculated from XRD data. These parameters differ slightly from those of bulk ZnO (a = 3.2498(9) Å and c = 5.20661(15) Å). The lattice parameters for samples M2 and MS revealing that the doping of Mn and Mn/Sn, does not change the wurtzite structure of ZnO, but lead to moderate decreasing of lattice parameters indicating doping effect.

To assess stoichiometry of ZnO nanorods and distribution of Mn and Sn ions in samples, energy dispersive X-ray spectrometry of individual ZnO nanorods was performed using a high-resolution TEM. Manganese and tin contents did not exceed 1 and 0.3 at %, respectively, for the sample marked as MS. For the sample M2 manganese was not more than 0.5%. High-resolution TEM (Fig. 18) confirmed that the nanorods grow along the c-axis direction. Manganese and tin incorporate into the ZnO lattice instead of being precipitated, demonstrated neither second phase inside the nanorods, nor attachments at the nanorod surface. Both the selected area electron diffraction (SAED) pattern and HRTEM images revealed an ordered structure of ZnO hexagonal structure. The absence of any superstructure reflection in the SAED pattern images indicates the absence of any additional long- or short-range ordering in the samples. The typical single crystal defects like twins are obvious from the pictures and denoted by arrows.

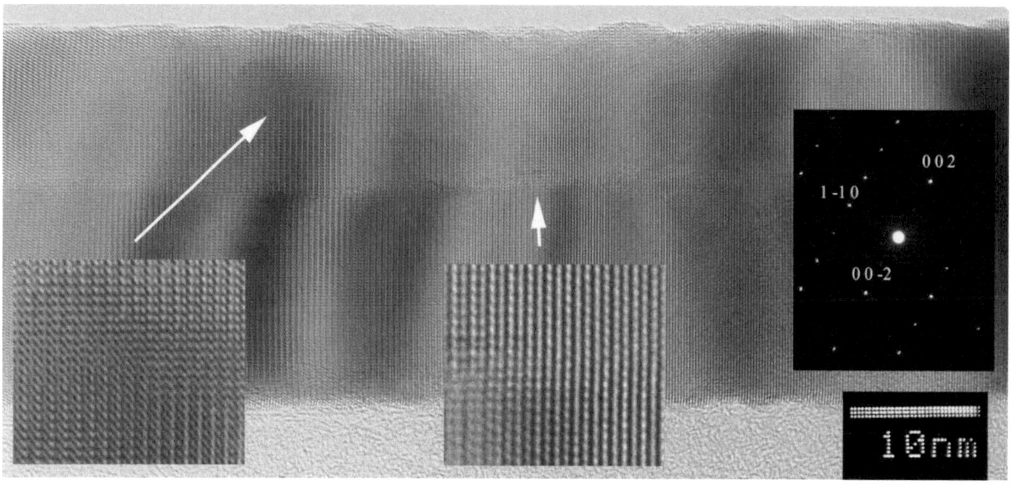

Fig. 18. High resolution transmission electron microscopy images of ZnO nanorods doped with Mn and Sn. Insets: selected area of electron diffraction viewed along [110] direction and the ZnO atom-scale resolution images.

6. Planar device structures from doped ZnO nanorods synthesized by the thermal growth from salt composites

6.1 Resistance switching in ZnO:Fe,Li nanorods

Individual ZnO nanorods doped by Fe, Li were configured as two terminal devices with the Al electrode-ZnO-Al electrode structure on an oxidized silicon substrate (fig.19). E-beam lithography was used to pattern electrodes contacting individual nanowires. Electrical transport properties of the nanorods were studied by applying to the electrodes quasi-dc voltage $0 \rightarrow 3 \rightarrow -3 \rightarrow 0$ with a constant sweep velocity.

The current-voltage characteristics of the ZnO:2%Fe,Li nanorods exhibits a rectifying behavior indicating the Schottky-like barrier formation and displays stable hysteresis (fig.20). The hysteresis at negative voltages is more pronounced than at positive voltages. The nanorod starts in the low resistive state when sweeping the voltage from zero to positive voltages. In the subsequent voltage sweep from positive to negative the nanorod shows an increased resistance. At negative voltage the nanorod resistance switches back from a high to a low resistance. The virgin nanorod shows a higher resistance than obtained in the subsequent cycles with a carrier injection. In contrast to abrupt resistance changes, a smooth resistance change is observed. It is likely, that electric-field domains are built and attenuated resulting in the observable switching effect. Moreover one has to take a nonuniform distribution of trapped charges and the surface band bending into account, which can be altered by applying voltage in forward or reverse directions. Two state resistive switching at RT can be realized using the ZnO nanorods. Emploing a positive voltage of +3V switches the nanorod device into a high impedance state. After applying negative voltage of -3 V the low impedance state is recovered. Between these write and erase voltages the state can be readout with 1.5V (fig.21) (Panin et al., 2007a, 2007b).

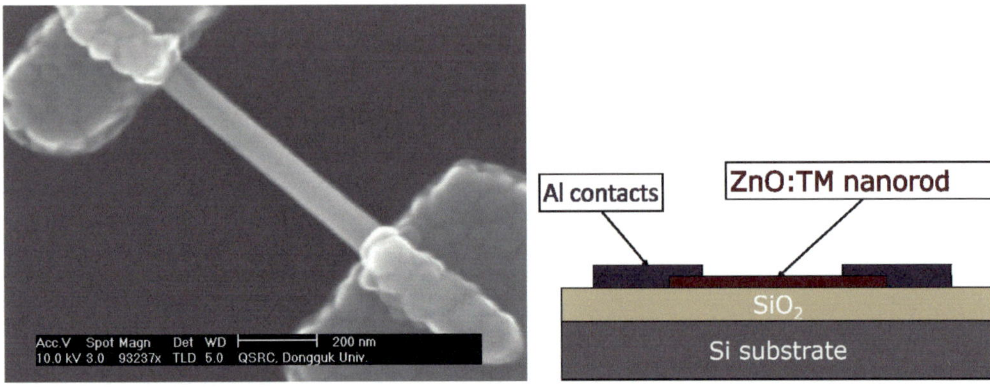

Fig. 19. A SEM image (left) and a scheme (right) of the nanorod with deposited Al contacts patterned by e-beam lithography.

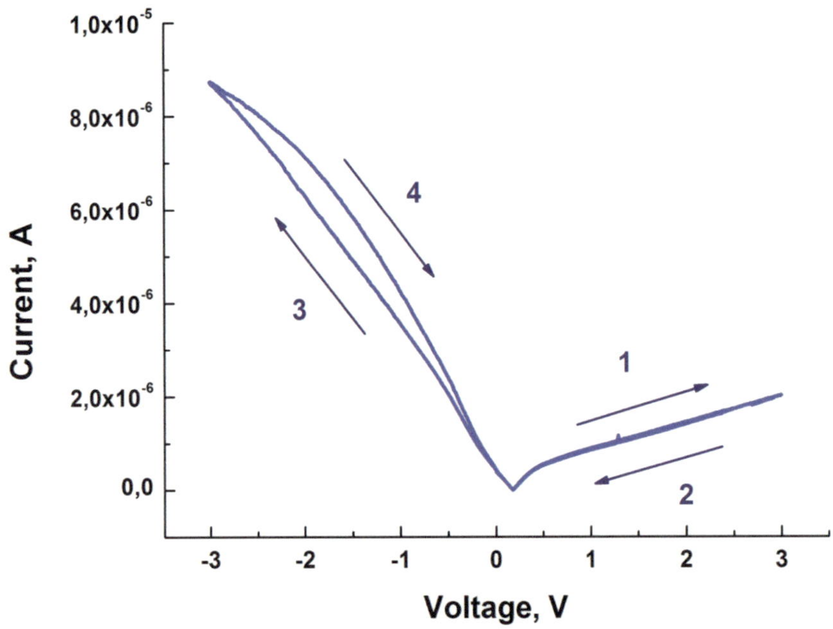

Fig. 20. *I-V* characteristics of a ZnO:2%Fe,Li nanowire for a voltage sweep from 0→3→-3→0 V.

6.2 Field effect transistors from Cr- and Sb-doped ZnO nanorods

Devices for transport measurements were fabricated from Cr- and Sb-doped ZnO nanorods using standard photo- and electron beam lithography and a lift-off technique (Kononenko et al., 2009). Nanorods were transferred onto the surface of thermally oxidized silicon chips with Au pads and lanes. E-beam lithography and a lift-off process were used to pattern e-beam evaporation deposited aluminum electrodes contacting a single nanorods. Two-terminal structures were used for measurements of *I-V* characteristics. n^+ Si(100) substrate is used as a gate. Diameters of both Cr-doped and Sb-doped nanorods were about 50 nm.

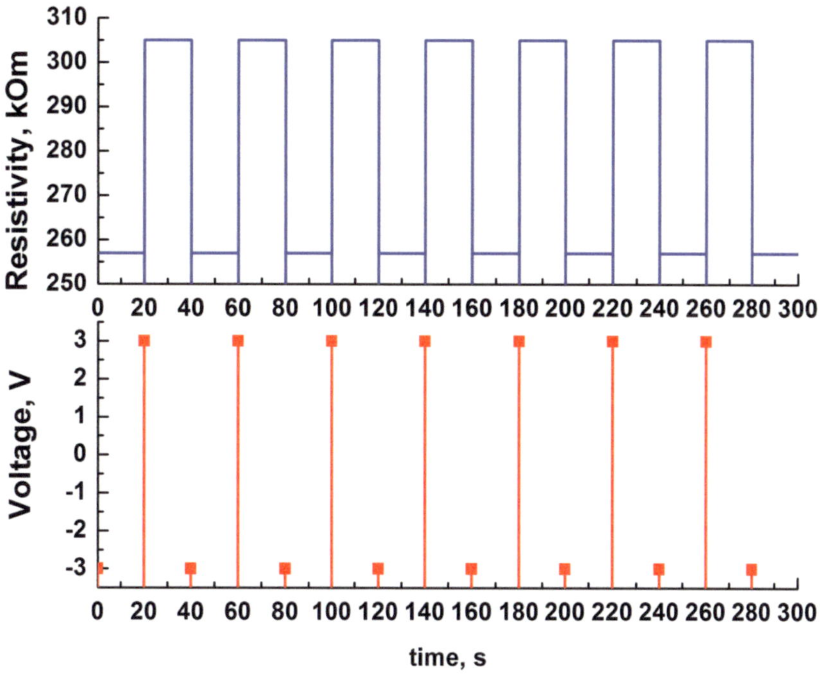

Fig. 21. Two state resistive switching of the ZnO nanowire device at RT. Applied voltage versus time (lower curve); readout resistance versus time (upper curve).

The electrical characteristics of Cr- and Sb-doped ZnO nanorods are shown in figures 22 and 23. The current-voltage curves were measured under different back gate voltages. Conductance of the nanorod increases with increasing back gate voltage which can be explained using energy band bending caused by back gating (Tans et al., 1998). The positive back gate potential increases electron concentration and bends the conduction band towards the Fermi level. Result of that is increasing of conductance. The negative back gate depletes the electron concentration. The conduction band bends from the Fermi level, yielding lower conductance. From the dependence of source-drain current on gate voltages we found that source-drain current increased with changing of voltages from negative to positive in both Cr-doped and Sb-doped nanorods. That is both transistors are n-channel.

We also observed from data of source-drain current versus gate voltages that threshold voltage is about +6V for the Cr-doped nanorods and that is about -7V for the Sb-doped nanorods. It indicates that first one is n-channel enhancement-mode FET and second one is n-channel depletion-mode FET. An on/off current ratio in both FETs as large as 10^5.

It is known that the electronic transport of nanowires can be strongly influenced by the surface effects due to such surface states and/or defects (Dayeh et al., 2007, Jones et al., 2007, Hanrath & Korgel 2005a, Hanrath & Korgel 2005b). Therefore, control of the density of surface states is a key factor in various device applications. Hong et al. have previously reported that ZnO nanowire FETs with n-channel depletion-mode and enhancement-mode transistors can be realized due to the difference of surface states and/or defects induced by the surface morphology of ZnO nanowire side walls (Hong et al., 2007). The realization of such nanowire transistors having different operational modes can lead to wide applications for the logic circuits (Park et al., 2005).

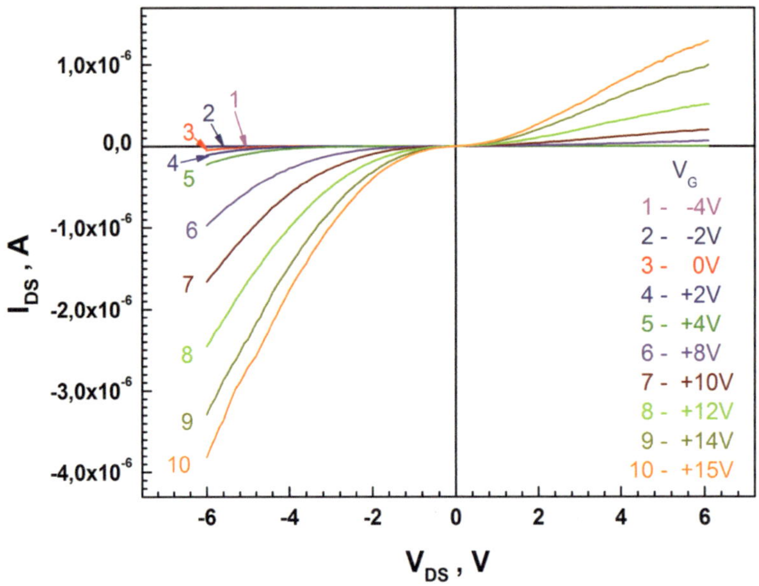

Fig. 22. Output characteristics (I_{DS}-V_{DS}) for Cr-doped nanorod n-channel enhancement-mode FET.

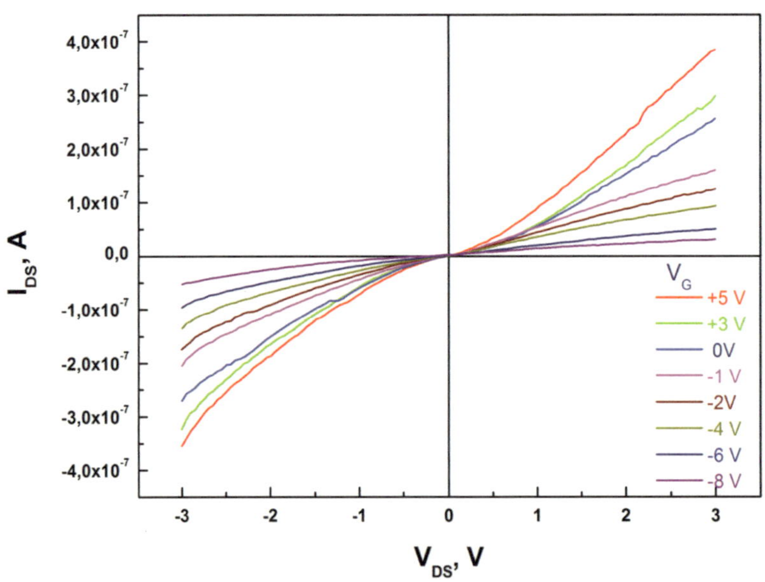

Fig. 23. Output characteristics (I_{DS}-V_{DS}) for Sb-doped nanorod n-channel depletion-mode FET.

The surface depletion can have a significant influence on the electronic transport behavior of ZnO nanorods since the depletion width can be comparable to the diameter size of nanorod.

The trapping of carrier electrons in trap states can cause electron depletion in the channel. The depletion width in small diameter nanorods can be comparable to the diameter size of nanorod. In our case Cr-doped nanorod is completely depleted under the no gate bias condition due to the larger depletion region than the nanorod diameter and Sb-doped nanorod is partially depleted under the no gate bias condition due to the smaller depletion region than the nanorod diameter. Since the diameters of both nanorods are the same, the surface states discrepancy can be connected with dopant concentration or dopant nature. Additional investigations is necessary to clarify this issue.

7. Conclusion

In conclusion, arrays of high crystalline and optical quality ZnO nanorods vertically oriented to a substrate surface can be grown on single-crystalline and amorphous substrates by catalyst-free elemental chemical vapor deposition at a reduced pressure and large excess of zinc vapor. A low level of intrinsic defects in such nanorods facilitates inversion of n-type conductivity to p-type one at doping by acceptor impurity. Homogeneous p-n junctions were formed in vertically aligned ZnO nanorods by diffusion of antimony deposited onto their tops during annealing. Inorganic and inorganic-organic heterogeneous p-n junctions were fabricated on the basis of vertically aligned ZnO nanorods. Blue-white emission was observed from the homogeneous p-n junction ZnO nanorod and inorganic-organic hybrid diodes.

The growth from the salt mixture is promising method for doping of ZnO nanorods by transition metals and acceptor dopants in order to control the electronic and luminescence properties of the nanorods. ZnO nanorods doped by transition metals and acceptor dopants can be used for nanoelectronic applications. Two-level resistive switching of ZnO:Li,Fe nanorods at room temperature were demonstrated. Si back gate Field Effect Transistors were fabricated from Cr- and Sb-doped ZnO nanorods. Transport properties of the Field Effect Transistors were investigated. We found that both type (Cr-doped and Sb-doped) of transistors are n-channel. Field Effect Transistor fabricated from ZnO:Cr nanorod demonstrated enhancement mode behavior and Field Effect Transistor fabricated from ZnO:Sb nanorod demonstrated depletion mode behavior. It can be important for design of ZnO nanorod logic circuits.

8. Acknowledgment

This work was supported by the Russian Ministry of Science and Education grant 02.740.11.5215 and the program of basic research of Presidium of the Russian Academy of Sciences “Foundation of basic research of nanotechnologies and nanomaterials”. The authors would like to thank Dr. A. Irzhak (Common Use Scientific Center of Materials Science, State Technological University "Moscow Institute of Steel and Alloys" (MISIS)) for investigation of specimens in the FIB system, Mr. V.T. Volkov for e-beam deposition of electrode metals.

9. References

Baranov, A.N., Chang, C.H., Shlyakhtin, O.A., Panin, G.N., Kang, T. . & Oh, Y.-J. (2004). In Situ Study of the ZnO-NaCl System During the Growth of ZnO Nanorods. *Nanotechnology,* Vol. 15, No 11, (Nov 2004), PP. (1613-1619), ISSN 0957-4484.

Baranov, A.N., Panin, G.N., Kang, T.W. & Oh, Y.-J. (2005). Growth of ZnO Nanorods from a Salt Mixture. *Nanotechnology* Vol. 16, No. 9, (Sep 2005), PP. (1918-1923), ISSN 0957-4484.

Collins, C.B., Carlson, R.O. & Gallagher, C. (1957). Properties of Gold-Doped Silicon. *Phys. Rev.* Vol. 105, No. 4, (Feb 1957), PP. (1168–1173), ISSN 0031-899X

Dayeh, S.A., Soci, C., Yu, P.K.L., Yu, E. T., Wang, D. (2007). Influence of Surface States on the Extraction of Transport Parameters from InAs Nanowire Field Effect Transistors. *Appl. Phys. Lett.*, Vol.90, No 16, (Apr 2007), PP. (162112-1-162112-3), ISSN 0003-6951.

Fan Z. & Lu, J.G. (2005). Gate-refreshable Nanowire Chemical Sensors. *Appl. Phys. Lett.*, Vol. 86, No. 12, (Mar 2005), PP. (123510-1-123510-3), ISSN 0003-6951.

Govender, K., Boyle, D.S., O'Brien, P., Binks, D., West, D. & Coleman, D. (2002). Room-temperature Lasing Observed from ZnO Nanocolumns Grown by Aqueous Solution Deposition. *Adv. Mater.*, Vol. 14, No. 17, (Sep 2002), PP. (1221–1224), ISSN: 0935-9648.

Gruzintsev, A.N. & Yakimov, E.E. (2005). Annealing Effect on the Luminescent Properties and Native Defects of ZnO. *Inorg. Mater.*, Vol. 41, No. 7, (Jul 2005), PP. (725–729), ISSN: 0020-1685.

Hanrath, T. & Korgel, B.A. (2005). Influence of Surface States on Electron Transport through Intrinsic Ge Nanowires. *J. Phys. Chem. B*, Vol. 109, No. 12, (Mar 2005), PP. (5518-5524), ISSN 1089-5647.

Hu, J.Q., Bando, Y. & Liu, Z.W. (2003). Synthesis of Gallium-Filled Gallium Oxide–Zinc Oxide Composite Coaxial Nanotubes. *Adv. Mater.*, Vol. 15, No. 12, (Jun 2003), PP. (1000-1003), ISSN: 0935-9648.

Hong, W.-K., Hwang, D.-K., Park, I.-K., Jo, G., Song, S., Park, S.-J., Lee, T., Kim, B.-J., Stach, E.A. (2007). Realization of Highly Reproducible ZnO Nanowire Field Effect Transistors With *n*-channel Depletion and Enhancement Modes. *Appl. Phys. Lett.* Vol. 90, No. 24, (Jun 2007), PP. (243103-1-243103-3), ISSN 0003-6951.

Huang, M.H., Mao, S., Feick, H., Yan, H.Q., Wu, Y.Y., Kind, H., Weber, E., Russo, R. & Yang, P.D. (2001). Room-temperature Ultraviolet Nanowire Nanolasers. *Science*, Vol. 292, No. 5523, (Jun 2001), PP. (1897–1899), ISSN 0036-8075.

Jones, F., Léonard, F., Talin, A.A. & Bell, N.S. (2007). Electrical Conduction and Photoluminescence Properties of Solution-grown ZnO Nanowires. *J. Appl. Phys.*, Vol. 102, No. 1, (Jul 2007), PP. (014305-1-014305-7), ISSN 0021-8979.

Kim, S.-W., Fujita, Sh. & Fujita Sh. (2005). ZnO Nanowires with High Aspect Ratios Grown by Metalorganic Chemical Vapor Deposition Using Gold Nanoparticles. *Appl. Phys. Lett.*, Vol. 86, No. 15, (Apr 2005), PP. (153119-1-153119-3), ISSN 0003-6951.

Konenkamp, R., Word, R. C. & Schlegel, C. (2004). Vertical Nanowire Light-emitting Diode. *Appl. Phys. Lett.*, Vol. 85, No. 24, (Dec 2004), PP. (6004-6006), ISSN 0003-6951.

Konenkamp, R., Word, R.C. & Godinez, M. (2005). Ultraviolet Electroluminescence from ZnO/polymer Heterojunction Light-emitting Diodes. *Nano Lett.*, Vol. 5, No. 10, (Oct 2005), PP. (2005–2008), ISSN 1530-6984.

Kononenko, O.V., Redkin, A.N., Panin, G.N., Baranov, A.N., Firsov, A.A., Levashov, V.I., Matveev, V.N. & Vdovin, E.E. (2009). Study of Optical, Electrical and Magnetic Properties of Composite Nanomaterials on the Basis of Broadband Oxide Semiconductors. *Nanotechnologies in Russia*, Vol. 4, Nos. 11–12, (Apr 2009), PP. (822–827), ISSN 1995-0780.

Krumeich, F., Muhr, H. J., Niederberger, M., Bieri, F., Schnyder, B. & Nesper, R. Morphology and Topochemical Reactions of Novel Vanadium Oxide Nanotubes. (1999). *J. Am. Chem. Soc.*, Vol. 121, No. 36, (Sep 1999), PP. (8324-8331), ISSN 0002-7863.

Law, M., Greene, L.E., Johnson, J.C., Saykally, R. & Yang, P.D. (2005). Nanowire Dye-sensitized Solar Cells. *Nat. Mater.*, Vol. 4, No. 6, (Jun 2005), PP. (455–459), ISSN 1476-1122.

Lee, Y.-J., Sounart, T.L., Scrymgeour, D.A., Voigt, J.A. & Hsu, J.W.P. (2007). Control of ZnO Nanorod Array Alignment Synthesized Via Seeded Solution Growth. *Journal of Crystal Growth,* Vol. 304, No. 1, (Jun 2007), PP. (80–85), ISSN 0022-0248.

Liu, F., Cao, P.J., Zhang, H.R., Shen, C.M., Wang, Z., Li, J.Q. & Gao, H.J. (2005). Well-aligned Zinc Oxide Nanorods and Nanowires Prepared Without Catalyst. *Journal of Crystal Growth,* Vol. 274, No. 1-2, (Jan 2005), PP. (126–131), ISSN 0022-0248.

Look, D.C., Claflin, B., Alivov, Y.I. & Park, S.J. (2004). The Future of ZnO Light Emitters. *Phys. Status Solidi A,* Vol. 201, No.10, (Aug 2004), PP. (2203–2212), ISSN 1862-6300.

Oh, S.H., van Benthem, K., Molina, S.I., Borisevich, A.Y., Luo, W., Werner, P., Zakharov, N.D., Kumar, D., Pantelides, S.T. & Pennycook, S.J. (2008). Point Defect Configurations of Supersaturated Au Atoms Inside Si Nanowires. *Nano Lett.,* Vol. 8, No. 4, (Apr 2008), PP. (1016–1019), ISSN 1530-6984.

Ohta H. & Hosono H. (2004). Transparent Oxide Optoelectronics, *Mater. Today,* Vol. 7, No. 6, (Jul 2004), PP. (42-51), ISSN 1369-7021.

Özgür, Ü., Alivov, Ya.I., Liu, C., Teke, A., Reshchikov, M.A., Doğan, S., Avrutin, V., Cho, S.-J. & Morkoç H.J. (2005). A Comprehensive Review of ZnO Materials and Devices. *J. Appl. Phys.*, Vol. 98, No. 4, (Aug 2005), PP. (041301-1-041301-103), ISSN 0021-8979.

Park, J., Choi, H.-H., Siebein, K. & Singh, R.K. (2003). Two-Step Evaporation Process for Formation of Aligned Zinc Oxide Nanowires. *J. Cryst. Growth,* Vol. 258, No.3–4, (Nov 2003), PP. (342–348), ISSN 0022-0248.

Park, W.I., Yi, G.-C., Kim, M.Y. & Pennycook, S.J. (2003). Quantum Confinement Observed in ZnO/ZnMgO Nanorod Heterostructures. *Adv. Mater.,* Vol. 15, (Mar 2003), No. 6, PP. (526-529), ISSN 0935-9648.

Park, W.I. & Yi, G. C. (2004). Electroluminescence in *n*-ZnO Nanorod Arrays Vertically Grown on *p*-GaN. *Adv. Mater.,* Vol. 16, No. 1, (Jan 2004), PP. (87-90), ISSN 0935-9648.

Park, W.I., Kim, J.S., Yi, G.-C. & Lee, Y.-J. (2005). ZnO Nanorod Logic Circuits. *Adv. Mater.,* Vol. 17, No. 11, (Jun 2005), PP. (1393-1397), ISSN 0935-9648.

Panin, G., Baranov, A.N., Kang, T.W., Kononenko, O.V., Dubonos, S.V., Min, S. K., and Kim H. J. (2007a). Electrical and Magnetic Properties of Doped ZnO Nanowires. *Mater. Res. Soc. Symp. Proc.* Vol. 957, PP. (0957-K04-06-1-0957-K04-06-6), ISBN 9781558999145, San Francisco, California, USA, April 9-13, 2007.

Panin, G., Baranov, A.N., Kononenko, O.V., Dubonos, S.V., and Kang, T.W. (2007b). Resistance Switching Induced by an Electric Field in ZnO:Li, Fe Nanowires. *AIP Conf. Proc.,* Vol. 893, PP. (743-744), ISBN: 978-0-7354-0397-0, Vienna, Austria, July 24-28, 2006.

Red'kin A.N., Makovei Z.I., Gruzintsev A.N., Dubonos S.V. & Yakimov E.E. (2007). Vapor Phase Synthesis of Aligned ZnO Nanorod Arrays from Elements. *Inorganic Materials,* Vol. 43, No. 3, (Mar 2007), PP. (253–257), ISSN 0020-1685.

Red'kin, A., Makovei, Z., Gruzintsev, A., Yakimov, E., Kononenko, O.V. & Firsov A.A. (2009). Elemental Vapor-Phase Synthesis of Nanostructured Zinc Oxide. *Inorg. Mat.*, Vol. 45, No. 11, (Nov 2009), PP. (1330–1336), ISSN 0020-1685.

Reiser, A., Ladenburger, A., Prinz, G.M., Schirra, M., Feneberg, M., Langlois, A., Enchelmaier, R., Li, Y., Sauer, R. & Thonke, K. (2007). Controlled Catalytic Growth and Characterization of Zinc Oxide Nanopillars on *a*-plane Sapphire. *J. Appl. Phys.*, Vol. 101, No. 5, (Mar 2007), PP. (054319-1-054319-9), ISSN 0021-8979.

Sun, X.W., Huang, J.Z., Wang, J.X. & Xu, Z. (2008). A ZnO Nanorod Inorganic/Organic Heterostructure Light-Emitting Diode Emitting at 342 nm. *Nano Lett.*, Vol. 8, No. 4, (Apr 2008), PP. (1219-1223), ISSN 1530-6984.

Tans, S.J., Verschueren, A.R. M. & Dekker C. (1998). Room-temperature transistor based on a single carbon nanotube. *Nature,* Vol. 393, No. 6680, (May 1998), PP. (49-52), ISSN 0028-0836.

Umar, A., Kim, S.H., Lee, Y.-S., Nahm, K.S. & Hahn, Y.B. (2005). Catalyst-free Large-quantity Synthesis of ZnO Nanorods by a Vapor–solid Growth Mechanism: Structural and Optical Properties. *Journal of Crystal Growth,* Vol. 282, No. 1-2, (Aug 2005), PP. (131–136), ISSN 0022-0248.

Wang, R. C., Liu, C. P., Huang, J. L. & Chen, S.-J. (2005). ZnO Hexagonal Arrays of Nanowires Grown on Nanorods. *Appl. Phys. Lett.,* Vol. 86, No. 25, (Jun 2005), PP. (251104-1-251104-3), ISSN 0003-6951.

Wei, M., Zhi, D. & MacManus-Driscoll, J.L. (2005). Self-catalysed Growth of Zinc Oxide Nanowires. *Nanotechnology,* Vol. 16, No. 8, (Aug 2005), PP. (1364–1368), ISSN 0957-4484.

Wei, Y.G., Xu, C., Xu, S., Li, C., Wu, W.Z. & Wang, Z.L. (2010). Planar Waveguide-nanowire Integrated Three-dimensional Dyesensitized Solar Cells. *Nano Lett.*, Vol. 10, No. 6, (Jun 2010), PP. (2092–2096), ISSN 1530-6984.

Yang, K., She, G.-W., Wang, H., Ou, X.-M., Zhang, X.-H., Lee, C.-S. & Lee, S.-T. (2009) ZnO Nanotube Arrays as Biosensors for Glucose. *J. Phys. Chem. C*, Vol. 113, No. 47, (Nov 2009), PP. (20169 - 20172), ISSN 1932-7447.

Yeh, P.-H., Li, Z. & Wang, Zh.L. (2009). Schottky-Gated Probe-Free ZnO Nanowire Biosensor. *Adv. Mater.*, Vol. 21, No. 48, (Dec 2009), PP. (4975-4978), ISSN 0935-9648.

Zhao, Q.X., Willander, M., Morjan, R.R., Hu, Q.H., Campbell, E.E.B. (2003). *Appl. Phys. Lett.* Vol. 83, No. 1, (Jul 2003), PP. (165-167), ISSN 0003-6951.

Zha, M., Calestani, D., Zappettini, A., Mosca, R., Mazzera, M., Lazzarini, L. & Zanotti, L. (2008). Large-area Self-catalysed and Selective Growth of ZnO Nanowires. *Nanotechnology,* Vol. 19, No. 32, (Aug 2008), PP. (325603-1-325603-5), ISSN 0957-4484.

Zhang, X.-M., Lu, M.-Y., Zhang, Y., Chen, L.-J. & Wang, Zh.L. (2009). Fabrication of a High-Brightness Blue-Light-Emitting Diode Using a ZnO-Nanowire Array Grown on p-GaN Thin Film. *Adv. Mater.,* Vol. 21, No. 27, (Jul 2009), PP. (2767-2770), ISSN 0935-9648.

Zhao, D., Andreazza, C., Andreazza, P., Ma, J., Liu, Y. & Shen, D. (2005). Buffer layer effect on ZnO nanorods growth alignment. *Chemical Physics Letters,* Vol. 408, No. 4-6, (Jun 2005), PP. (335-338), ISSN 0009-2614.

4

Collective Plasmonic States Emerged in Metallic Nanorod Array and Their Application

Masanobu Iwanaga
National Institute for Materials Science and
Japan Science and Technology Agency (JST), PRESTO
Japan

1. Introduction

Plasmons are well known as collective excitations of free electrons in solids. Simple unit structures are nanoparticles such as spheres, triangles, and rods. Optical properties of metallic nanoparticles were reported at the beginning of the 20th century (Maxwell-Garnett, 1904, 1906). Now it is well known that the resonances in metallic nanoparticles are described by Mie theory (Born & Wolf, 1999). It is interesting to note that the studies on nanoparticles were concentrated at the beginning of the century, at which quantum mechanics did not exist. Shapes and dimensions of metallic nanoparticles such as triangles and rods were clearly classified by their dark-field images after a century from the initial studies on nanoparticles (Kuwata et al., 2003; Murray & Barnes, 2007). Nanoparticles were revived around 2000 in the era of nanotechnology.

It may be first inferred that dimers and aggregations of metallic nanostructures have bonding and anti-bonding states stemming from Mie resonances in the nanoparticles. The conjecture was confirmed in many experimental studies (For example, Prodan et al., 2003; Liu et al., 2007; Liu et al., 2009). Dimer structures composed of a pair of nanospheres or nanocylinders are one of the most examined structures. At the initial stage of the dimer study, very high-enhancement of electric field at the gap was frequently reported based on a computational method of finite-difference time domain (FDTD), which is directly coded from classical electromagnetics or Maxwell equations. However, recent computations including nonlocal response of metal, which is quantum mechanical effect, disagree the very high-enhancement (García de Abajo, 2008; McMahon et al., 2010). Especially, as the gap is less than 5 nm, the discrepancies in cross section and extinction becomes prominent. While the physics in dimensions of nm and less obeys quantum mechanics in principal, many experimental and theoretical results show that classical electromagnetics holds quite well even in tens of nm scale. Thus, it is not yet conclusive where the boundary of classical electromagnetics and quantum mechanics exists in nm-scale plasmonics. It will be elucidated when further development of nanofabrication techniques will be able to produce nm-precision metallic structures with reliable reproducibility. Taking the present status of nanotechnology into account, we focus on structures, such as gap, of the dimension more than 5 nm, where classical electromagnetics holds well.

Contemporary nanofabrication technology can produce a wide variety of plasmonic structures, which are usually made of metals. In addition to unit structures such as nanoparticles, periodic structures are also produced, where surface plasmon polaritons (SPPs) are key resonances. Strictly, the SPPs in periodic structures are different from the original SPPs induced at ideally flat metal-dielectric interface. Periodic structures enable to reduce the original SPP into the first Brillouin zone; it is therefore reasonable to call the SPPs in periodic structures reduced SPPs. The reduced SPPs were known since 1970s (Raether, 1988) and were revived as a type of resonances yielding extraordinary transmission in a perforated metallic film (Ebbesen et al., 1998).

When producing periodic array of nanoparticles, what is expected? Periodic structures are aggregation of monomers and dimers, and have photonic band structures. By structural control, it is expected to obtain desired photonic bands, for example, wave-number-independent, frequency-broad band, which is not obtained in dimers and so on. In terms of photovoltaic applications, light absorbers working at a wide energy and incident-angle (or wave-number) ranges are preferred. On the other hand, if one access highly enhanced electromagnetic fields, states of high quality factor, which are associated with narrow band, may be expected. Thus, the designs of plasmonic structures vary in accordance with needs. Main purpose of this chapter is to show some of concrete designs of plasmonic structures exhibiting collective oscillations of plasmons and broad-band plasmonic states, based on realistic and precise computations.

This chapter consists of 7 sections. Computational methods are described in section 2. One-dimensional (1D) and two-dimensional (2D) plasmonic structures are examined based on numerical results in sections 3 and 4, respectively. As for applications, light absorption management is examined in section 3 and polarization manipulators of subwavelength thickness are shown in section 4. Conclusion is given in section 5.

2. Computational methods

Before describing the results of 1D and 2D plasmonic structures, computational methods are noted in this section. In section 2.1, Fourier modal method or rigorously coupled-wave approximation (RCWA) is described, suitable to compute linear optical spectra such as reflection and transmission. In section 2.2, finite element method is explained, which is employed to evaluate electromagnetic field distributions. Although the two methods have been already established, the details in implementation are useful when researchers unfamiliar to plasmonics launch numerical study. Furthermore, the detailed settings are described.

Realistic simulations are intended here. As material parameters, constructive equations in Maxwell equations for homogeneous media have permittivity and permeability (Jackson, 1999). In the following computations, we took permittivity of metals from the literature compiling measured data (Rakić et al., 1998). The permittivity of transparent dielectric was set to be typical values: that of air is 1.00054 and that of SiO_2 is 2.1316. The permittivity of Si was also taken from literature (Palik, 1991). At optical wavelengths, it is widely believed that permeability is unity in solids (Landau et al., 1982); to date, any exception has not been found in solid materials.[1]

[1] Metamaterials were initially intended to realize materials of arbitrary permittivity and permeability by artificial subwavelength structures (Pendry & Smith, 2004). This strategy has been successful especially at microwaves.

2.1 Optical spectra

Linear optical responses from periodic structures are observed as reflection, transmission and diffraction. To calculate the linear optical responses, it is suitable to transform Maxwell equations into the Fourier representation. By conducting the transformation, the equation to be solved is expressed in the frequency domain; therefore, optical spectra are obtained in the computation with varying wavelength. In actual computations, it is crucial to incorporate algorithm which realizes fast convergence of the Fourier expansion. If one does not adopt it, Fourier expansion shows extremely slow convergence and practically one cannot reach the answer. The algorithm could not be found for a few decades in spite of many trials. The issue was finally resolved for 1D periodic systems in 1996 (Lalanne & Morris, 1996; Li, 1996b; Granet & Guizal, 1996) and succeedingly for 2D periodic systems in 1997 (Li, 1997). The Fourier-based method is often called RCWA. Commercial RCWA packages are now available. In this study, we prepared the code by ourselves incorporating the Fourier factorization rule (Li, 1997) and optimized it for the vector-oriented supercomputers.

In general, the periodic structures are not single-layered but are composed of stacked layers. Eigen modes in each layer expressed by Fourier-coefficient vectors are connected at the interfaces by matrix multiplication. The intuitive expression results in to derive transfer matrix (Markoš, 2008). Practically, transfer matrix method is not useful because it includes exponentially growing factors. To eliminate the ill-behaviour, scattering matrix method is employed. Transfer and scattering matrices are mathematically equivalent. In fact, scattering matrix was derived from transfer matrix by recurrent formula (Ko & Inkson, 1988; Li, 1996a). The derivations were independently conceived for different aims: the former was to solve electronic transport in quantum wells of semiconductors as an issue in quantum mechanics (Ko & Inkson, 1988) and the latter was to calculate light propagation in periodic media as an issue in classical electromagnetics (Li, 1996a).

In actual implementation, truncations of Fourier expansions are always inevitable as written in equation (1), which shows Lth-order truncation. Of course, the Fourier expansion is exact as $L \to \infty$.

$$E(x,y) = \sum_{m,n=0,\pm1,\pm2,\cdots,\pm L} E_{mn} \exp(ik_x x + ik_y y + 2\pi i m / d_x + 2\pi i n / d_y) \quad (1)$$

In equation (1), 2D periodic structure of the periodicities of d_x and d_y is assumed and incident wave vector has the components k_x and k_y. The term E_{mn} is Fourier coefficient of function $E(x, y)$. For 2D periodic structures shown later, the truncation order is set to be L= 20. Then, estimated numerical fluctuations were about 1%. For 1D periodic structures, one can assume that d_y is infinity in equation (1); as a result, requirements in numerical implementation become much less than 2D cases. It is therefore possible to set large order such as L=200 and to suppress numerical fluctuations less than 0.5%.

Optical spectra calculated numerically by the Fourier modal method were compared with measured spectra; good agreement was confirmed in stacked complementary 2D plasmonic crystal slabs, which have elaborate depth profiles (Iwanaga, 2010b, 2010d).

2.2 Electromagnetic-field distributions

Electromagnetic-field distributions were computed by employing finite element method (COMSOL Multiphysics, version 4.2). One of the features is to be able to divide constituents by grids of arbitrary dimensions.

To keep precision at a good level, transparent media were divided into the dimensions less than 1/30 effective wavelength. As for metals, much finer grids are needed. Skin depth of metals at optical wavelengths is a few tens of nm; therefore, grids of sides of a few nm or less were set in this study. Such fine grids result in the increase in required memory in implementation. Even for the unit domain in 2D periodic structures, which is minimum domain and becomes three-dimensional (3D) as shown in Fig. 7, the allocated memory easily exceeded 100 GB. As for 1D structures, the unit domain is 2D and requires much less memory in implementation. Accordingly, computation time is much shorter; in case of Fig. 3, it took about ten seconds to complete the simulation.

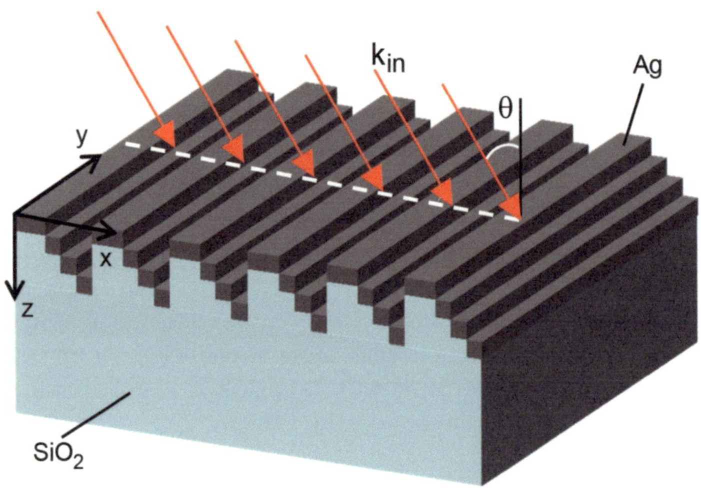

Fig. 1. Schematic drawing of an efficient 1D plasmonic light absorber of Ag nanorod array on SiO_2 substrate, which was found based on the search using genetic algorithm. Plane of incidence is set to be parallel to the *xz* plane.

The finite element method was applied for resolving the resonant states in the stacked complementary 2D plasmonic crystal slabs and revealed the eigen modes successfully (Iwanaga, 2010c, 2010d).

3. 1D periodic metallic nanorod array

Light absorbers of broad band both in energy and incident-angle ranges were numerically found (Iwanaga, 2009) by employing simple genetic algorithm (Goldberg, 1989). One of the efficient absorbers is a 1D metallic nanorod array as drawn in Fig. 1. The Ag nanorods (dark grey) are assumed to be placed on the step-like structure of SiO_2 (pale blue). Periodic direction was set to be parallel to the x axis, and the periodicity is 250 nm. The nanorods are parallel to the y axis and infinitely long. The nanorods in the top layer have the xz rectangular sections of 100×50 nm^2. The other nanorods have the xz square sections of 50×50 nm^2.

In 1D structures, it was found that depth profiles are crucial to achieve desired optical properties. Single-layered 1D structures have little degree of freedom to meet a designated optical property whereas 1D structures of stacked three layers have enough potentials to reach a given goal (Iwanaga, 2009).

In this section, we clarify the light-trapping mechanism by examining the optical and absorption properties, and electromagnetic field distributions. Collective electrodynamics between the nanorods plays a key role to realize the doubly broad-band absorber.

3.1 Optical responses and light absorption

Incident plane waves travel in the *xz* plane (that is, the wave vectors $\mathbf{k}_{in}$ are in the *xz* plane) as shown in Fig. 1, keeping the polarization to be *p* polarization, that is, incident electric-field vector $\mathbf{E}_{in}$ is in the *xz* plane. To excite plasmonic states in 1D periodic systems, the *p* polarization is essential. If one illuminates the 1D object by using *s*-polarized light (that is, $\mathbf{E}_{in}$ parallel to *y*), plasmonic states stemming from SPPs are not excited. Absorbance spectra under *p* polarization at incident angles θ of -40, 0, and 40 degrees are shown in Fig. 2(a) with solid line, dashed line, and crosses, respectively. The sign of incident angles θ is defined by the sign of *x*-component $k_{in,x}(=|\mathbf{k}_{in}|\sin\theta)$ of incident wave vector. Absorbance *A* in % is defined by

$$A = 100 - \sum\nolimits_{n=0,\pm1,\pm2,\cdots}(R_n + T_n) \qquad (2)$$

where R_n and T_n are *n*th-order reflective and transmissive diffractions, respectively. R_0 denotes reflectance and T_0 stands for transmittance. We computed linear optical responses R_n and T_n by the Fourier modal method described in section 2.1, and evaluated *A* by use of equation (2). The symbols R_0 and T_0 are respectively expressed simply as *R* and *T* from now on. In the 1D structure in Fig. 1, since the periodicity is 250 nm, T_n and R_n for $n \neq 0$ are zero for θ = 0° in Fig. 2(a) and zero for θ = -40° at more than 525 nm.

In Fig. 2(a), absorption significantly increases at θ = -40° in the wavelength range longer than 600 nm. It is to be stressed that absorption is more than 75% in a wide range from 600 to 1000 nm. Thus, the 1D structure in Fig. 1 works as a broad-band absorber in wavelengths from the visible to near-infrared ranges.

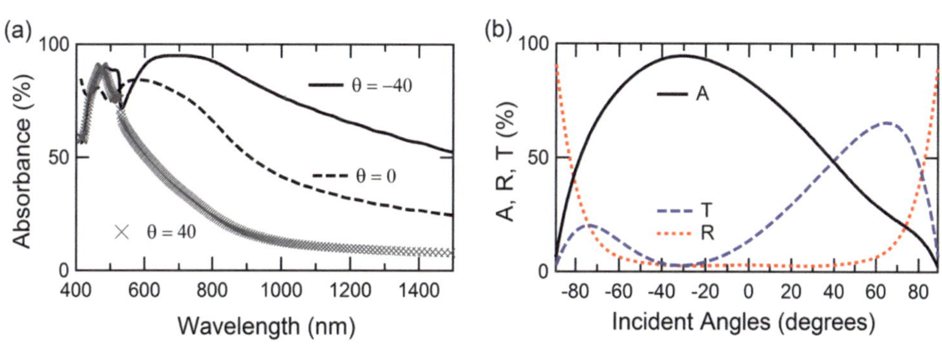

Fig. 2. (a) Absorption spectra at -40° (solid line), 0° (dashed line), and 40° (crosses) under *p* polarization. (b) Spectra of *A* (solid line), *R* (dotted line), and *T* (dashed line) at 620 nm dependent on incident angles.

In Fig. 2(b), The *A* spectrum at 620 nm dependent on incident angles is shown with solid line. The corresponding *T* and *R* spectra are shown with blue dashed and red dotted lines, respectively. Note that diffraction does not appear at this wavelength.

The *T* spectrum in Fig. 2(b) exhibits asymmetric distribution for incident angles θ, indicating that the structure in Fig. 1 is optically deeply asymmetric. In contrast, the *R* spectrum is symmetric for θ and the relation of $R(\theta)=R(-\theta)$ is satisfied; the property is independent of structural symmetry and is known as reciprocity (Potten, 2004; Iwanaga et al., 2007b). The *A* spectrum takes more than 80% at a wide incident-angle range from 5° to -60°. It is thus shown that the 1D periodic structure in Fig. 1 is a doubly broad-band light absorber.

3.2 Magnetic-field and power-flow distributions

To reveal the plasmonic state inducing the doubly broad-band absorption in Fig. 2, we examine here the electromagnetic field distributions at 620 nm and θ = -40°, evaluated by the finite element method. As described in section 3.1, incident plane waves are *p*-polarized and induce transverse magnetic (TM) modes in the 1D periodic structure. Therefore, magnetic-field distribution is suitable to examine the features of the plasmonic state.

In Fig. 3(a), magnetic-field distribution is presented; the magnetic field has only *y* component under *p* polarization and the *y*-component of magnetic field is shown with colour plot. Figure 3(a) shows a snapshot of the magnetic field, where the phase is defined by setting incident electric field $\mathbf{E}_{in}$=(sin(-40°), 0, cos(-40°)) at the left-top corner position. The propagation direction of incidence is indicated by arrows representing incident wave vectors $\mathbf{k}_{in}$. To show a wide view at the oblique incidence, the domain in the computation was set to include five unit cells. We assigned the *yz* boundaries (that is, the left and right edges) periodic boundary condition.

The magnetic field distribution in Fig. 3(a) forms spatially oscillating pairs indicated by the signs + and -. It is to be noted that the oscillating pairs are larger than each metallic nanorod and are supported by three or four nanorods. The distributions are enhanced at the vicinity of nanorod array and strongly suggest that collective oscillations take place, resulting in the broad absorption band. As for plasmonic states, resonant oscillations inside metallic nanorod have been observed in most cases, which are attributed to Mie-type resonances (Born & Wolf, 1999). The present resonance is distinct from Mie resonances and has not been found to our best knowledge.

In Fig. 3(b), time-averaged electromagnetic power-flow distribution is shown. The power flow is equivalent to Poynting flux at each point. The *z*-component of the power flow is shown with colour plot and the vectors of power flow are designated by arrows, which are shown in the logarithmic scale for clarity. Oblique incidence is seen at the top of the panel and the power flow successfully turns around the nanorods, going into SiO_2 substrate. In addition to this finding, let us remind that the sum of *R* and *T* are at most 10% as shown in Fig. 2(b), that the power flow in the substrate is not far-field component but mostly evanescent components, and that most of incident power is consumed at the vicinity of the nanorod array. Therefore, incident radiation is considered to be effectively trapped at the vicinity of the nanorod array, especially in the substrate. Management of electromagnetic power flow is a key to realize photovoltaic devices of high efficiency.

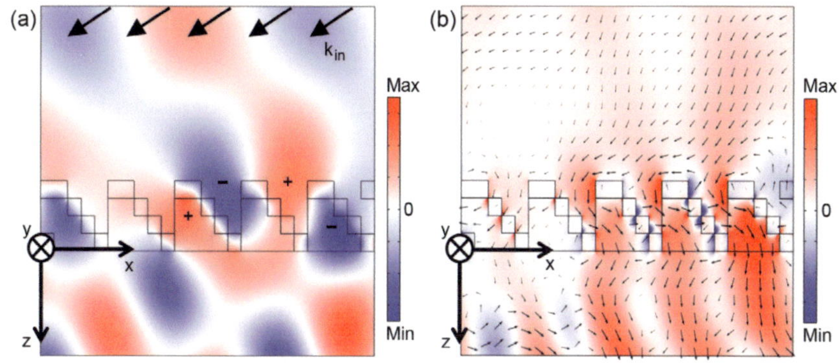

Fig. 3. (a) A snapshot of *y*-component of magnetic field (colour plot). Incident wave vectors are shown with arrows on the top. (b) Time-averaged electromagnetic power flow of *z* component (colour plot). Vectors (arrows) are represented in the logarithmic scale.

3.3 Management of incident light for photovoltaic applications

As is shown in sections 3.1 and 3.2, periodic structure of metallic nanorod array can be broad-band light absorber concerning both wavelengths and incident angles. Good light absorbers are preferred to realize more efficient photovoltaic devices. Possibility for the application is discussed here.

In considering producing efficient photovoltaic devices, it is crucial to exploit incident light fully. In the context, perfect light absorbers are usually preferred. However, light absorption and management of light have to be discriminated. If plasmonic absorbers consume incident light by the resonances resident inside metallic nanostructures such as Mie resonance, photovoltaic parts cannot use the incident light. Thus, it is not appropriate to optimize light absorption by metallic nanostructures when one tries to incorporate them into photovoltaic devices. Instead, one should manage to convert incident light to desired distributions by metallic nanostructures (Catchpole & Polman, 2008a, 2008b). In Fig. 3(b), we have shown that incident light effectively travels into substrate, in which photovoltaic parts will be made. Additionally, most of the light taken in is converted to enhanced evanescent waves. In comparison with the incident power, the power of the evanescent wave is more than a-few-fold enhanced. Such local enhancement of electromagnetic fields is preferable in photovoltaic applications.

As is widely known, management of incident light has been conducted in Si-based solar cells. At the surface, textured structures are usually introduced to increase the take-in amount of light (Bagnall & Boreland, 2008). The difference between the textured structures and the designed metallic nanostructures exists in the enhancement mechanism; the former has no enhancement while the latter can have local resonant enhancement as described above.

In actual fabrications of photovoltaic devices incorporating metallic nanostructures, plasmonic structures will be made on semiconductors. The structure in Fig. 1 is made on SiO_2 and has to be redesigned because the permittivity of SiO_2 and semiconductor such as Si is quite different at the visible range. In this section, we have shown actual potentials of plasmonic structures for light management through a concrete 1D periodic structure of nanorod array. Since genetic algorithm search is robust and applicable to issues one wants to find solutions (Goldberg, 1989), we positively think of finding plasmonic structures for photovoltaic applications.

In further search, 2D structures will be the targets, independent of incident polarizations. As for the actual fabrications, one may think that the step-like structure as shown in Fig. 1 are hard to produce by current top-down nanofabrication technique. In fact, there is hardly report that 90° etching is successfully executed. However, there is enough room to improve fabrication procedures; for example, if one could prepare hard mask and use calibrated aligner to conduct dry etching of semiconductors, it would be possible to etch down at almost 90° and even to produce step-like structures.

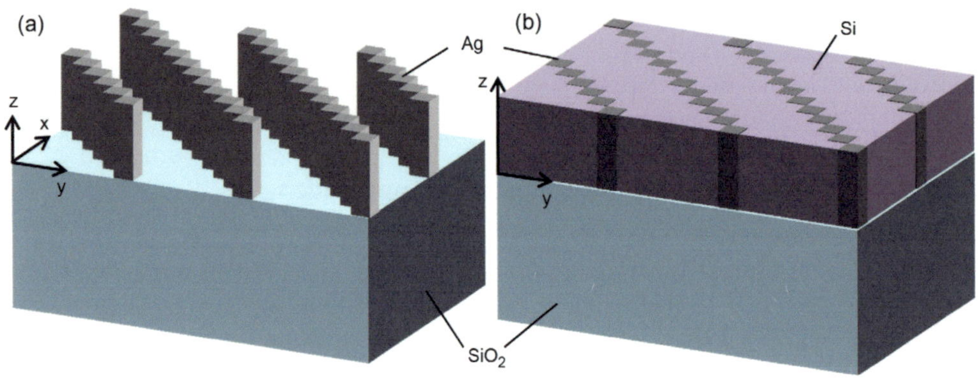

Fig. 4. Schematic drawing of 2D periodic Ag nanorod arrays on SiO_2 substrate. (a) Free standing in air. (b) Embedded in a Si layer.

4. 2D periodic metallic nanorod array

2D periodic nanorod array has much variety in design. In this section, we show how modification of unit cell drastically changes the optical properties. As concrete structures, we present the results on the rectangular nanorod array as shown in Fig. 4 and refer to those on circular nanorod array. In addition, it is shown that well-adjusted 2D nanorod arrays work as efficient polarizers of subwavelength thickness. As the application, circular dichroic devices are presented, which include 2D nanorod array as a component.

Before describing the numerical results on 2D metallic nanorod arrays, we mention how they can be fabricated. It is probably easier to produce the structure in Fig. 4(b) than that in Fig. 4(a). Since thin Si wafers can be fabricated in nm-precision as Si photonic crystal slabs are made (Akahane et al., 2003), the procedure of electron-beam patterning, development, metal deposition, and removal of resist results in the structure in Fig. 4(b) Free-standing metallic nanorods seem to be relatively hard to produce. Simple procedure described as for Fig. 4(b) is unlikely to be successful. Instead, other procedures have to be conceived. One of the ways is to modify the fabrication procedure to produce metallic nanopillars of about 300 nm height (Kubo & Fujikawa, 2011).

4.1 Optical properties

In Fig. 5, T and R spectra of free-standing Ag nanorod arrays are shown. Unit cell structures in the xy plane are drawn at the left-hand side. The periodicity is 250, 275, and 240 nm along both x and y axes in Figs. 5(a), 5(b), and 5(c), respectively. Grey denotes the xy section of Ag nanorods, which is 50×50 nm² in the xy plane. The height of the nanorods was set to be 340 nm. The gaps between nanorods were set to be 0, 5, and 10 nm along the x and y axes in Figs. 5(a), 5(b), and 5(c), respectively.

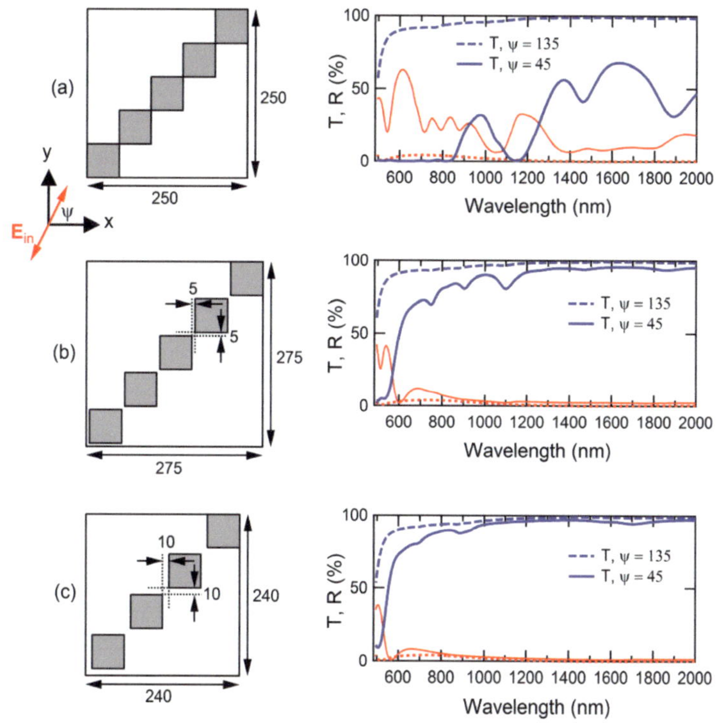

Fig. 5. Unit cell structures and the optical spectra of free-standing Ag nanorod array on SiO_2 substrate. Grey denotes Ag nanorod of 50×50 nm² in the xy plane. Gaps between each nanorod are set along the x and y axes: (a) 0 nm, (b) 5 nm, and (c) 10 nm. Dimensions are written in units of nm. T and R spectra at ψ = 45° are shown with solid and thin lines, respectively. T and R spectra at ψ = 135° are represented with dashed and dotted lines, respectively.

Incident plane waves illuminate the 2D structures at normal incidence. Incident polarization $\mathbf{E}_{in}$ was set to be linear, defined by azimuth angle ψ, that is, the angle between the x axis and the $\mathbf{E}_{in}$ vector, as drawn in Fig. 5(a). In accordance with the symmetry of the unit cell, two polarizations ψ = 45° and 135° were probed. T and R spectra at ψ = 45° are displayed with blue solid and red thin lines, respectively. T and R spectra at ψ = 135° are shown with blue dashed and red dotted lines, respectively.

T spectra at ψ = 45° are sensitive to the gaps. In Fig. 5(a), at the visible range of wavelength less than 800 nm, definite contrast of T at ψ = 45° and 135° is observed. As gaps becomes larger, the contrast of T rapidly diminishes. Actually, in Fig. 5(c) where the gap is 10 nm, T spectra at ψ = 45° and 135° become quite similar in spectral shapes and lose the difference seen in Fig. 5(a). The gap dependence of T spectra implies that there exists resonant state in the structure of Fig. 5(a) at $\psi = 45°$ and less than 800 nm and that the resonant state is lost by the nm-order gaps between nanorods.

The strong contrast of T in Fig. 5(a) indicates that the 2D nanorod arrays serves as a good polarizer of subwavelength thickness, which is employed in section 4.3.

Dimers or aggregations of rectangular and circular metallic nanostructures have attracted great interest in terms of so-called gap plasmons in terms of enhanced Raman scattering (Futamata et al., 2003; Kneipp, 2007). T spectra in Fig. 5 suggest that gap plasmons rapidly disappear as the gap increases and are lost even with a small gap of 10 nm.

In Fig. 5, we show the results on rectangular Ag nanorod array; similar spectral examinations were conducted for circular Ag nanorod arrays though the spectra are not shown here. The qualitative tendency is similar and the contrast of T is rapidly lost as the gaps between the circular nanorods increases in nm order.

In Fig. 6, we show T spectra of Ag nanorod arrays embedded in a Si layer of 340 nm height along the z axis. Incident polarizations were $\psi = 45°$ and 135°. T spectra at $\psi = 45°$ and 135° are shown with blue solid and blue dashed lines, respectively. It is first to be noted that T spectra at $\psi = 135°$ are almost independent of the gaps between Ag nanorods; T's at the wavelength range more than 1000 nm are several tens of % and exhibit Fabry-Perot-like oscillations coming from the finite thickness of the periodic structure, suggesting that the 2D structure for $\psi = 135°$ is transparent due to off resonance. In contrast, T spectra at $\psi = 45°$ vary the shape significantly with changing the gaps and are very sensitive to the gaps. At the 0 nm gap in Fig. 6(a), contrast of T is observed at the wavelength range longer than 1500 nm, indicating that the 2D structure in Fig. 4(b) also works as an efficient polarizer. The states at 1770 nm (arrow in Fig. 6(a)) are examined by electric field distributions in Fig. 7.

4.2 Electromagnetic-field distributions on resonances

In Fig. 7, electric-field distributions are shown which correspond to the 2D periodic structure of the unit cell in Fig. 6(a). Incident wavelength is 1770 nm; the wavelength is indicated by an arrow in Fig. 6(a). Colour plots denote intensity of electric field $|\mathbf{E}|$ and arrows stand for 3D electric-field vector. The unit domain used in the computations by the finite element method is displayed. Periodic boundary conditions are assigned to the xz and yz boundaries.

Figures 7(a) and 7(b) present the electric-field distributions at incident azimuth angle ψ = 45° and 135°, respectively. Incident plane wave travels from the left xy port to the right xy port. The phase of incident wave at the input xy port was defined by $\mathbf{E}_{in}$ = -(sin(45°), cos(45°), 0) in Fig. 7(a) and by $\mathbf{E}_{in}$ = (sin(135°), cos(135°), 0) in Fig. 7(b). The left panels show 3D view and the right panels shows the xy section indicated by cones in the left panels.

Electric-field distributions at $\psi = 45°$ in Fig. 7(a) are prominently enhanced at the vicinity of the connecting points of Ag nanorods. The enhanced fields are mostly induced outside the Ag nanorods and oscillate in-phase (or coherently), suggesting that the resonant states are not Mie type. On the other hand, electric-field distributions at $\psi = 135°$ in Fig. 7(b) have local hot spots at the corners of the Ag nanorods. It is usually observed at off resonant conditions. The electromagnetic wave propagates dominantly in the Si part.

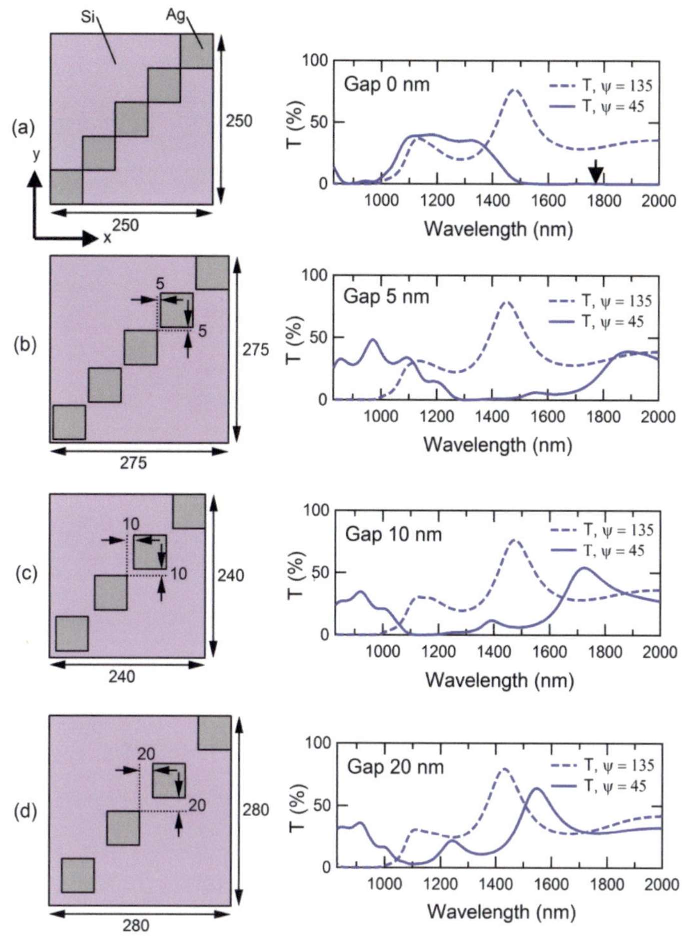

Fig. 6. Unit cell structures and the T spectra of Ag nanorod array embedded in a Si layer on SiO_2 substrate. Gaps of Ag nanorods along the x and y axes are (a) 0 nm, (b) 5 nm, (c) 10 nm, and (d) 20 nm, respectively. Unit cells were set similarly to Fig. 5. T spectra at $\psi = 45°$ and 135° are shown with solid and dashed lines, respectively.

Incident power was set to be 2.56×10^2 W/m^2 at the input xy port and the corresponding electric-field intensity was 4.39×10^2 V/m. The resonant electric field at the vicinity of nanorod array reaches 4.6×10^3 V/m at the maximum and shows about tenfold enhancement; the scale bar has the maximum of 8.2×10^2 V/m and the distributions are displayed in a saturated way to clearly present them near the connecting points. In air, incident and reflected waves are superimposed in phase and consequently the electric-field intensity takes larger values than the incident power.

4.3 Application for subwavelength circular dichroic devices

As shown in section 4.1, 2D periodic metallic nanorod arrays can serve as polarizers of subwavelength thickness. In this section, we make use of such an efficient polarizer and introduce subwavelength optical devices of circular dichroism.

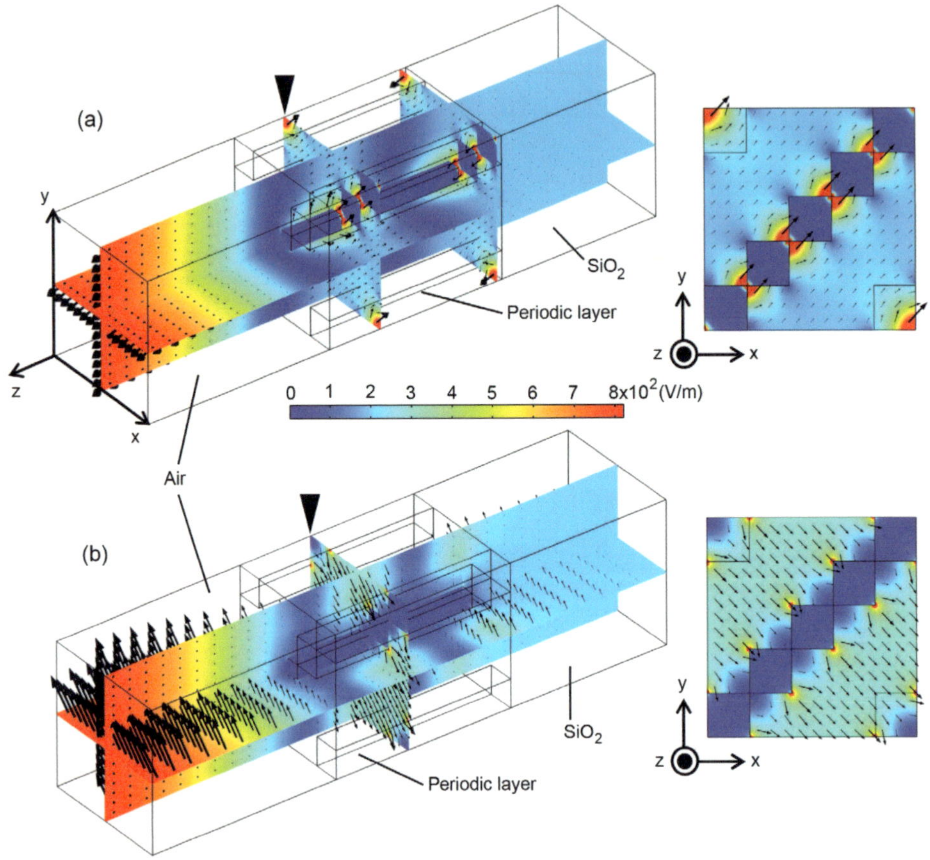

Fig. 7. Electric field distributions of the 2D periodic structure of the unit cell of Fig. 6(a): (a) 1770 nm and ψ = 45°; (b) 1770 nm and ψ = 135°. Left: 3D views of the unit domain. Right: the *xy* sections, indicated by cones in the left panels.

Efficient polarizers selecting polarization vectors are one of the key elements to realize various subwavelength optical devices. Another key element is wave plates which manipulate the phase of electromagnetic waves. In Fig. 8(a), a concrete design of wave plate of subwavelength thickness is presented, which is multilayer structure composed of Ag and SiO_2 and made thin along the layers. Thickness of each Ag and SiO_2 layer along the *x* axis is assumed to be 30 and 270 nm, respectively. Incident light sheds on the side or the *xy* plane. Azimuth angle ψ is defined similarly to Fig. 5(a).

To clarify the basic optical properties of the multilayer structure, *R* spectra at ψ = 0° and 90° under normal incidence are shown in Fig. 8(b). For simplicity, the thickness is assumed to be sufficiently thick to eliminate interference pattern in *R* spectra. For ψ = 0° (that is, $\mathbf{E}_{in}$ is parallel to the x axis), the structure shows small *R* (red dashed line) and is transparent. For ψ = 90° (that is, $\mathbf{E}_{in}$ is parallel to the y axis), it shows *R* spectrum (red solid line) just as a typical Drude metal (Ashcroft & Mermin, 1976). At longer wavelength range than 1000 nm, the structure serves as wire-grid polarizer whereas, at shorter wavelength range than 1000 nm, the structure becomes transparent for any polarization and can work as an efficient wave plate (Iwanaga, 2008).

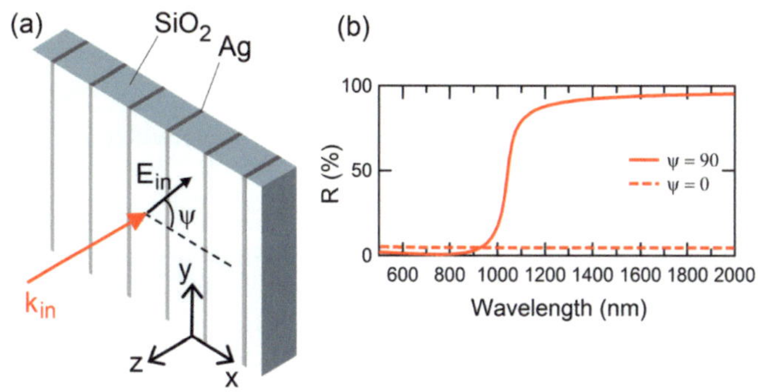

Fig. 8. Wave plate of subwavelength thickness. (a) Schematic drawing of structure and optical configuration. (b) Reflectance spectra under normal incidence at $\psi = 0°$ (dashed line) and $\psi = 90°$ (solid line).

The strong anisotropy of the multilayer structure is quantitatively expressed by using effective refractive index (Iwanaga, 2007a); it was found that, as a wave plate, the x axis is fast axis of effective refractive index of about 2 and the y axis is very slow axis of effective refractive index less than 1. It is to be emphasized that such strong anisotropy has not been found in solid material and makes it possible for the wave plates of multilayer structure to be extremely thin. In artificial structures, working wavelength can be tuned at desired wavelength by structural modifications; in contrast, it is very hard to change the working wavelength in solid materials because resonance is intrinsic property of materials. The feasibility in tuning is another advantage in subwavelength artificial structures.

Once key elements are found, the combinations are naturally derived and usually realized by producing stacked structures. A recent example is orthogonal polarization rotator of subwavelength thickness, which was numerically substantiated by designing a skew stacked structure of wave plates (Iwanaga, 2010a). By introducing stacked structures, potentials of subwavelength optical devices are greatly extended.

Figure 9(a) shows a concrete design of circular dichroic device (named I), which transforms incident circular polarization to transmitted linear polarization. The circular dichroic device has stacked structure of a wave plate of multilayer structure and a polarizer of 2D Ag nanorod array. The wave plate basically plays a role as a quarter wave plate at the wavelength range of the present interest. Each unit cell of the stacked layers is drawn in detail at the bottom; the dimensions are written in units of nm. Grey denotes Ag and pale blue SiO_2. The thickness of the wave plate and the polarizer was set to be 284 and 210 nm, respectively; therefore, the total thickness of the circular dichroic device was 494 nm.

In Fig. 9(b), T spectra under right-handed circular (RHC) and left-handed circular (LHC) polarizations are shown with blue solid and blue dashed lines, respectively. Obviously, definite contrast of T appears at about 850 nm. The degree of circular dichroism σ is defined by the following equation.

$$\sigma = \frac{T_{\mathrm{RHC}} - T_{\mathrm{LHC}}}{T_{\mathrm{RHC}} + T_{\mathrm{LHC}}} \tag{3}$$

When σ = 1, the optical device is optimized for RHC and when σ = -1, it is done for LHC. In Fig. 9(c), the σ is shown and takes the value almost equal to unity at 855 nm, which is indicated by a red arrow in Fig. 9(b). Since the thickness of the device is 494 nm, it is confirmed that the circular dichroic device I is certainly subwavelength thickness, which is 58% length for the working wavelength.

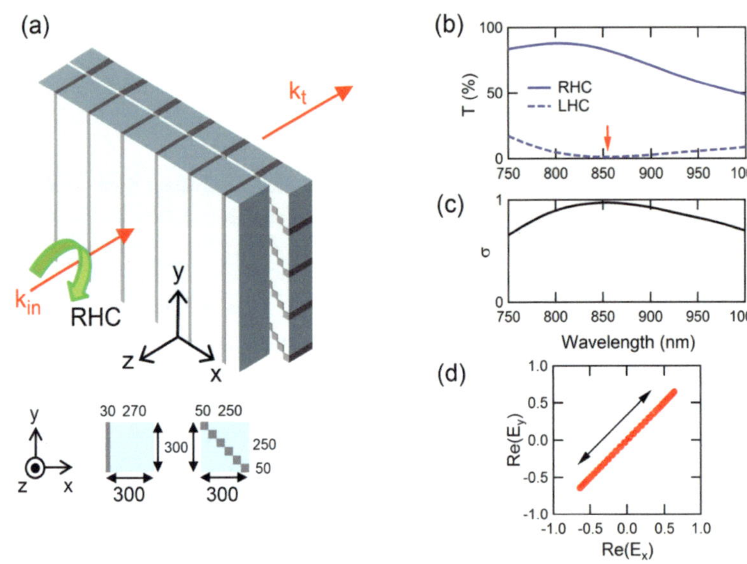

Fig. 9. Circular dichroic device I of subwavelength thickness. (a) Stacked subwavelength structure transforming circular into linear polarizations. (b) Transmittance spectra under RHC (solid line) and LHC (dashed line). (c) Degree of circular polarization σ, defined by equation (3). (d) Polarization of transmitted light.

Figure 9(d) presents polarization of transmitted light (red closed circles). Electric field was recorded for one wavelength and projected onto the *xy* plane. Clearly, the transmitted light is linearly polarized, characterised as ψ = 45° by using azimuth angle ψ.

Figure 10(a) shows schematic drawing of circular dichroic device II, transforming incident circular polarization into counter-circular polarization. The device is composed of stacked structure of wave plate, polarizer, and wave plate and has subwavelength thickness for the working wavelength. Each unit cell of the components is drawn at the bottom and is specified; the first layer means that incident plane waves illuminate first, and succeedingly the incident waves travel through the second and third layers. Grey denotes Ag and pale blue SiO_2. The thickness of the first, second, and third layers is 284, 210, and 255 nm, respectively; the total thickness is 749 nm. Although the basic design of the circular dichroic device I is similar to the device II in Fig. 9, the thickness of the third layer was finely adjusted to obtain ideally circular polarization of transmitted light because evanescent components contribute at the interface of the second and third layers, and modify the electrodynamics in the device II from that at homogeneous interface by plane waves.

Figure 10(b) shows *T* spectra under RHC (blue solid line) and LHC (blue dashed line) incidence. Definite contrast of *T* is observed at 820-900 nm; *T* under LHC incidence is well suppressed. In Fig. 10(c), the degree of circular dichroism σ is shown, evaluated by equation (3). Almost ideal circular dichroism is realized at 855 nm, indicated by a red arrow in Fig. 10(b).

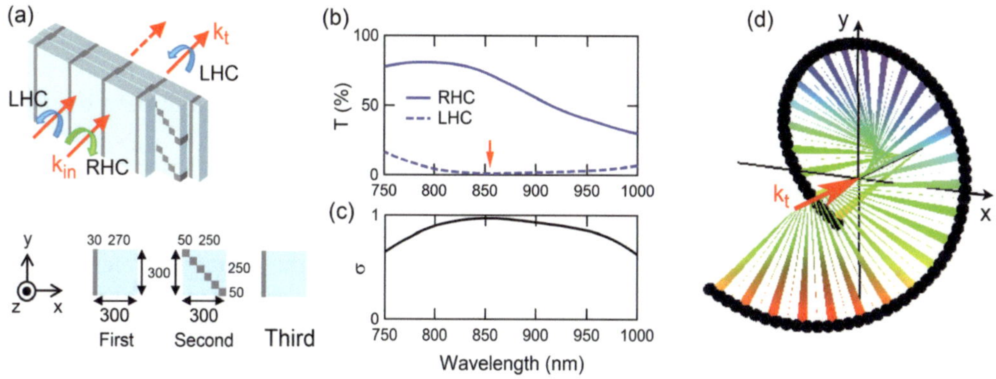

Fig. 10. Circular dichroic device II of subwavelength thickness. (a) Stacked subwavelength structure transforming circular into counter-circular polarizations. (b) Transmittance spectra under RHC (solid line) and LHC (dashed line). (c) Degree of circular polarization σ, defined by equation (3). (d) 3D plot of the trajectory of polarization of transmitted light with wave vector $\mathbf{k}_t$; the wavelength is 855 nm and indicated by an arrow in (b).

Figure 10(d) presents 3D plot of trajectory of the transmitted polarization (or electric field) at 855 nm. The polarization circularly rotates in the left-handed direction along the transmitted wave vector $\mathbf{k}_t$. Thus, it is shown that a unique circular dichroic device transforming incident circular polarization into the counter-circular polarization can be realized with subwavelength thickness by employing metallic nanostructures such as thin line and nanorods. Miniature optical devices shown in this section can serve as key elements in micro-optics circuits in the future.

In considering circular dichroism, it is often preferred to use helical structures. Probably, helical structures are connected unconsciously in mind to helical distributions of polarization such as Fig. 10(d), that is, circular polarizations. In principal, circular dichroism originates from simultaneous manipulations of polarization vector and the phase as proved in Figs. 9 and 10.

Resonances inducing simultaneous change of polarization vector and phase are possibly to be resident in helical structures; an example is periodic array of gold helical structures which serves as a circular dichroic device in infrared range (Gansel et al., 2009). Note that it is impossible to move the working wavelength to the visible range by simply making the smaller structures because scaling law does not hold in the periodic structures including metals which have wavelength-dependent permittivity.

To find new structures serving as circular dichroic devices, nature can provide clues. Jewelled beetles are rather widely known as circular dichroic insects. It turned out that the wings have helically stacked structures (Sharma et al., 2009). Biomimetics thus reminds us to learn from nature.

5. Conclusion

Periodic metallic nanorod arrays have been investigated based on the numerical methods. It was shown that the nanorod arrays have collective resonant states or coherent oscillations outside each nanorod; the field distributions suggest that the resonant states are distinct from Mie resonance. The resonant states form broad band in wavelength, indicating that they are continuum. In the 1D periodic structure of stacked layers, the resonant states are also broad band in incident angles. The doubly broad-band features have not been reported in metallic nanostructures. It was also shown in 2D structures of nanorod array that the structures are crucial to realize the broad-band plasmonic states and that geometrical modifications in the order of 5 nm significantly affect the collective states. Further development of nanofabrications will lead us to the novel plasmonic states. In terms of applications, the 1D nanorod array was discussed as a light managing element for photovoltaic devices and the 2D nanorod arrays were incorporated in highly efficient subwavelength circular-dichroic devices which were concretely designed to work at optical wavelengths, transforming incident circular polarizations to desired polarizations.

6. Acknowledgment

This study was partially supported by Cyberscience Centre, Tohoku University, by JST, PRESTO, and by JSPS KAKENHI (Grant No. 22760047).

7. References

Akahane, Y., Asano, T., Song, B.-S., & Noda, S. (2003). High-Q photonic nanocavity in a two-dimensional photonic crystal. *Nature*, Vol. 425, No. 6961, (October 2003), pp. 944-947, ISSN 0028-0836

Ashcroft, N. W. & Mermin, N. D. (1976). *Solid State Physics* (International ed.), Saunders College Publishing, ISBN 0-03-049346-3, Orlando, United States of America

Bagnall, D. M. &, Boreland, M. (2008). Photovotaic technologies. *Energy Policy*, Vol. 36, No. 12, (December 2008), pp. 4390-4396, ISSN 0301-4215

Born, M. & Wolf, E. (1999). *Principals of Optics* (seventh extended ed.), Cambridge Univ. Press, ISBN 0-521-642221, Cambridge, United Kingdom

Catchpole, K. R. & Polman, A. (2008a). Design principles for particle plasmon enhanced solar cells. *Applied Physics Letters*, Vol. 93, No. 19, (November 2008), pp. 191113-1–191113-3, ISSN 0003-6951

Catchpole, K. R. & Polman, A. (2008b). Plasmonic solar cells. *Optics Express*, Vol. 16, No. 26, (December 2008), pp. 21793-21800, ISSN 1094-4087

Ebbesen, T. W., Lezec, H. J., Ghaemi, H. F., Thio, T., & Wolff, P. A. (1998). Extraordinary optical transmission through sub-wavelength hole arrays. *Nature*, Vol. 391, No. 6668, (February 1998), pp. 667-669

Futamata, M., Maruyama, Y., & Ishikawa, M. (2003). Local Electric Field and Scattering Cross Section of Ag Nanoparticles under Surface Plasmon Resonance by Finite Difference Time Domain Method. *The Journal of Physical Chemistry B*, Vol. 107, No. 31, (August 2003), pp. 7607-7617, ISSN 1520-6106

Gansel, J. K., Thiel, M., Rill, M. S., Decker M., Bade, K., Saile, V., von Freymann, G., Linden, S., & Wegener, M. (2009). Gold Helix Photonic Metamaterial as Broadband Circular

Polarizer. *Science,* Vol. 325, No. 5947, (September 2009), pp. 1513-1515, ISSN 0036-8075

García de Abajo, F. J. (2008). Nonlocal Effects in the Plasmons of Strongly Interacting Nanoparticles, Dimers, and Waveguides. *The Journal of Physical Chemistry C,* Vol. 112, No. 46, (October 2008), pp. 17983-17987, ISSN 1932-7447

Goldberg, D. E. (1989). *Genetic Algorithms in Search, Optimization & Machine Learning,* Addison-Wesley, ISBN 0-201-15767-5, Boston, United States of America

Granet, G. & Guizal, B. (1996). Efficient implementation of the coupled-wave method for metallic lamellar gratings in TM polarization. *Journal of the Optical Society of America A,* Vol. 13, No. 5, (May 1996), pp. 1019-1023, ISSN 0740-3232

Iwanaga, M. (2007a). Effective optical constants in stratified metal-dielectric metamaterial. *Optics Letters,* Vol. 32, No. 10, (May 2007), pp. 1314-1316, ISSN 0146-9592

Iwanaga, M., Vengurlekar, A. S., Hatano, T., & Ishihara, T. (2007b). Reciprocal transmittances and reflectances: An elementary proof. *American Journal of Physics,* Vol. 75, No. 10, (October 2007), pp. 899-902, ISSN 0002-9505

Iwanaga, M. (2008). Ultracompact waveplates: Approach from metamaterials. *Applied Physics Letters,* Vol. 92, No. 15, (April 2008), pp. 153102-1 – 153102-3

Iwanaga, M. (2009). Optically deep asymmetric one-dimensional plasmonic crystal slabs: Genetic algorithm approach. *Journal of the Optical Society of America B,* Vol. 26, No. 5, (May 2009), pp. 1111-1118, ISSN 0740-3224

Iwanaga, M. (2010a). Subwavelength orthogonal polarization rotator. *Optics Letters,* Vol. 35, No. 2, (January 2010), pp. 109-111

Iwanaga, M. (2010b). Polarization-selective transmission in stacked two-dimensional complementary plasmonic crystal slabs. *Applied Physics Letters,* Vol. 96, No. 8, (February 2010), pp. 083106-1 – 083106-3

Iwanaga, M. (2010c). Subwavelength electromagnetic dynamics in stacked complementary plasmonic crystal slabs. *Optics Express,* Vol. 18, No. 15, (July 2010), pp. 15389-15398

Iwanaga, M. (2010d). Electromagnetic eigenmodes in a stacked complementary plasmonic crystal slab. *Physical Review B,* Vol. 82, No. 15, (October 2010), pp. 155402-1 – 155402-8, ISSN 1098-0121

Jackson, J. D. (1999). *Classical Electrodynamics* (third ed.), Wiley, ISBN 0-471-30932-X, Hoboken, United States of America

Kneipp, K. (2007). Surface-enhanced Raman scattering. *Physics Today,* Vol. 60, No. 11, (November 2007), pp. 40-46, ISSN 0031-9228

Ko, D. Y. K. & Inkson, J. C. (1988). Matrix method for tunneling in heterostructures: Resonant tunneling multilayer systems. *Physical Review B,* Vol. 38, No. 14, (November 1988), pp. 9945-9951

Kubo, W. & Fujikawa, S. (2011) Au Double Nanopollars with Nanogap for Plasmonic Sensor. *Nano Letters,* Vol. 11, No. 1, (January 2011), pp. 8-15, ISSN 1530-6984

Kuwata, H., Tamaru, H., Esumi, K., & Miyano, K. (2003). Resonant light scattering from metal nanoparticles: Practical analysis beyond Rayleigh approximation. *Applied Physics Letters,* Vol. 83, No. 22, (December 2003), pp. 4625-4627

Lalanne, P. & Morris, G. M. (1996). Highly improved convergence of the coupled-wave method for TM polarization. *Journal of the Optical Society of America A,* Vol. 13, No. 4, (April 1996), pp. 779-784

Landau, L. D., Lifshitz, E. M., & Pitaevskii, L. P. (1982). *Electrodynamics of Continuous Media* (second ed. revised and extended), Elsevier, ISBN 0-7506-2634-8, Oxford, United Kingdom

Li, L. (1996a). Formulation and comparison of two recursive matrix algorithm for modelling layered diffraction gratings. *Journal of the Optical Society of America A*, Vol. 13, No. 5, (May 1996), pp. 1024-1035

Li, L. (1996b). Use of Fourier series in the analysis of discontinuous periodic structures. *Journal of the Optical Society of America A*, Vol. 13, No. 9, (September 1996), pp. 1870-1876

Li, L. (1997). New formulation of the Fourier modal method for cross surface-relief gratings. *Journal of the Optical Society of America A*, Vol. 14, No. 10, (October 1997), pp. 2758-2767

Liu, N., Guo, H., Fu, L., Kaiser, S., Schwerizer, H., & Giessen, H. (2007). Plasmon Hybridization in Stacked Cut-Wire Metamaterials. *Advanced Materials*, Vol. 19, No. 21, (November 2007), pp. 3628-3632, ISSN 1521-4095

Liu, N., Liu, H., Zhu, S., & Giessen, H. (2009). Stereometamaterials. *Nature Photonics*, Vol. 3, No. 3, (March 2009), pp. 157-162, ISSN 1749-4885

Markoš, P. & Soukoulis, C. M. (2008). *Wave Propagation from Electrons to Photonic Crystals and Left-handed Materials* (first ed.), Princeton Univ. Press, ISBN 978-0-691-13003-3, New Jersey, United States of America

Maxwell-Garnett, J. C. (1904). Colours in Metal Glasses in Metallic Films. *Philosophical Transactions of the Royal Society of London, Series A*, Vol. 203, pp. 385-420, ISSN 0080-4614

Maxwell-Garnett, J. C. (1906). Colours in Metal Glasses in Metallic Films, and in Metallic Solutions. *Philosophical Transactions of the Royal Society of London, Series A*, Vol. 393, pp. 237-288

McMahon, J. M., Gray, S. K., & Schatz, G. C. (2010). Optical Properties of Nanowire Dimers with a Spatially Nonlocal Dielectric Function. *Nano Letters*, Vol. 19, No. 22, (November 2007), pp. 3473-3481

Murray, A., & Barnes, W. L. (2007). Plasmonic Materials. *Advanced Materials*, Vol. 19, No. 22, (November 2007), pp. 3771-3782

Palik, E. D. (1991). *Handbook of Optical Constants of Solids II*, Academic Press, ISBN 978-0-125-44422-4, San Diego, United States of America

Pendry, J. B. & Smith, D. R. (2004). Reversing Light With Negative Refraction. *Physics Today*, Vol. 57, No. 6, (June 2004), pp. 37-43

Potten, R. J. (2004). Reciprocity in optics. *Reports on Progress in Physics*, Vol. 67, No. 5, (May 2004), pp. 717-754, ISSN 0034-4885

Prodan, E., Radloff, C., Halas, N. J., & Nordlander, P. (2003). A Hybridazation Model for the Plasmon Response of Complex Nanostructures. *Science*, Vol. 302, No. 5644, (October 2003), pp. 419-422

Raether, H. (1998). *Surface Plasmons on Smooth and Rough Surfaces and Gratings*, Springer, ISBN 0387173633, Berlin, Germany

Rakić, A. D., Djurišić, A. B., Elazar, J. M., & Majewski, M. L. (1998). Optical properties of metallic films for vertical-cavity optoelectronic devices. *Applied Optics*, Vol. 37, No. 22, (August 1998), pp. 5271-5283, ISSN 0003-6935

Sharma, V., Crne, M., Park, J. O., & Srinivasarao, M. (2009). Structural Origin of Circular Polarized Iridescence in Jewelled Beetles. *Science*, Vol. 325, No. 5939, (July 2009), pp. 449-451

5

ZnO Nanorod Arrays Synthesised Using Ultrasonic-Assisted Sol-Gel and Immersion Methods for Ultraviolet Photoconductive Sensor Applications

Mohamad Hafiz Mamat[1], Zuraida Khusaimi[2],
Musa Mohamed Zahidi[1] and Mohamad Rusop Mahmood[1,2]
[1]NANO-ElecTronic Centre (NET), Faculty of Electrical Engineering;
[2]NANO-SciTech Centre (NST), Institute of Science (IOS);
Universiti Teknologi MARA (UiTM), Shah Alam, Selangor,
Malaysia

1. Introduction

Zinc oxide (ZnO) nanomaterials have emerged as one of the most promising materials for electronic devices such as solar cells, light-emitting devices, transistors, and sensors. The diverse structures of ZnO nanomaterials produce unique, useful, and novel characteristics that are applicable for high-performance devices. The ZnO nanorod array is a beneficial structure that has become extremely important in many applications due to its porosity, large surface area, high electron mobility, and variety of feasible techniques. The chemistry and physical tuning of its surface state, including processes such as annealing and chemical treatments, enhance its functionality and sensitivity and consequently improve the device performance. These useful characteristics of ZnO nanorod arrays enable the fabrication of ultraviolet (UV) photoconductive sensors with high responsivity and reliability. Although there are many techniques available to synthesise the ZnO nanorod arrays, solution-based methods offer many advantages, including the capacity for low-temperature processing, large-scale deposition, low cost, and excellent ZnO crystalline properties. In this chapter, the synthesis of ZnO nanorod arrays via ultrasonic-assisted sol-gel and immersion methods will be discussed for application to UV photoconductive sensors. The optical, structural, and electrical properties of deposited ZnO nanorod arrays will be reviewed, and the performance of the synthesised ZnO nanorod array-based UV photoconductive sensors will be discussed.

2. Ultraviolet photoconductive sensor using ZnO nanomaterials

Recently, ZnO nanostructures have received much attention due to their promising characteristics for electronic, optical, and photonic devices. Generally, ZnO exhibits semiconducting properties with a wide band gap of 3.3 eV at room temperature and a strong binding energy of 60 meV, which is much larger than that of gallium nitride (GaN, 25 meV) or the thermal energy at room temperature (26 meV). ZnO is naturally an n-type semiconductor material that is very transparent in the visible region, especially as a thin film, and has good UV absorption. ZnO is a biosafe and biocompatible material that has many applications, such as in electronics and biomedical and coating technologies. A reduction in size of the ZnO particle to the nanoscale level produces novel and attractive electrical, optical, mechanical, chemical, and physical properties due to quantum confinement effects. Moreover, ZnO nanostructures have a high aspect ratio, or a large surface-to-volume ratio and high porosity, which can fulfil the demand for high performance and efficiency in numerous applications (Lee et al., 2009, Galoppini et al., 2006, Park et al., 2011, Hullavarad et al., 2007).

UV photoconductivity, where the electrical conductivity changes due to the incident UV radiation, is characteristic of few semiconductors (wide band gap) or materials. This characteristic involves a number of mechanisms, including the absorption of light, carrier photogeneration, and carrier transport (Soci et al., 2010). Generally, a change in conductivity is related to the number of photogenerated carriers per absorbed photon or quantum yield and the mobility of the photogenerated carriers. The photoresponse time usually involves factors such as carrier lifetime and the defects state of the material. In other words, the UV photoconductivity represents important electrical properties that are related to carrier mobility, carrier lifetime, and defects in the materials.

There are various reports regarding UV photoconductive sensors that utilise ZnO nanostructures as the sensing elements. For example, Pimentel et al. developed ZnO thin-film-based UV sensors using radio frequency (RF) magnetron sputtering (Pimentel et al., 2006). They produced ZnO thin films with resistivities from 5 $x10^4$ to 1 $x10^9$ Ω cm and revealed that the preparation of ZnO films without oxygen exposure in an RF sputtering chamber produced a UV detector with higher sensitivity at thicknesses below 250 nm than ZnO films with oxygen exposure. They theorised that the result might have been due to the smaller grain size of the ZnO films without oxygen exposure, which increased the sensor active areas for UV detection.

Additionally, Xu et al. developed an Al-doped ZnO thin-film-based UV sensor using the sol-gel method (Xu et al., 2006). They produced a 5 mol % Al-doped ZnO film that was highly oriented along the c-axis of a Si (111) substrate. Their study detailed the suitability of Al-doped ZnO thin films for UV detection, where a high photocurrent value was obtained when the film was irradiated with UV light between 300 nm and 400 nm. However, their study revealed that the cut-off wavelength of Al-doped ZnO was blue-shifted to a shorter wavelength compared with the undoped film. They also observed that the photocurrent value of the Al-doped ZnO film in the visible region was reduced slightly compared to the undoped ZnO film, which improved the UV sensor sensitivity.

Zheng et al. developed a photoconductive ultraviolet detector based on ZnO films (Zheng et al., 2006). The ZnO thin films were deposited by pulsed laser deposition (PLD) at a thickness of 300 nm on glass substrates. Al metal contacts with 0.1 mm separation were deposited onto the ZnO films to complete the UV photoconductive sensor configuration. The crystallite size of the PLD-deposited ZnO film was around 23 nm, and the ZnO films grew along the c-axis, or perpendicular to the substrate. They found that the crystallite boundaries that were induced by the small crystallite size of the ZnO nanoparticles contributed to the oxygen adsorption at the interfaces of the ZnO crystallites. This condition also resulted in carrier scattering, which decreased the carrier mobility. They also observed that the ZnO-nanoparticles based UV detector from their method showed a large dark current of approximately 0.2 mA at a bias voltage at 5 V, which was due to intrinsic defects, such as oxygen vacancies and zinc interstitials.

Jun et al. fabricated ultraviolet photodetectors based on ZnO nanoparticles with a diameter size of 70 nm using a paint method on thermally oxidised Si substrate (Jun et al., 2009). They used gold as the metal contacts with a gap of 20 µm. They addressed the surface defect problem experienced by nanoparticle-based UV detectors. Surface defects cause a rise time delay during UV illumination and irradiative recombination between the holes and electrons, which lowers the performance of ZnO nanoparticle-based devices.

Liu et al. fabricated a ZnO/diamond-film-based UV photodetector on a Si substrate (Liu et al., 2007). The ZnO films were deposited on a freestanding diamond-coated Si substrate by RF magnetron sputtering. They used gold as the metal contacts, which were deposited onto the film by DC magnetron sputtering with 2 mm of electrode separation. They found that the dark current of their UV sensor decreased with the grain size, which was due to the reduction of the ZnO grain boundaries. It was also mentioned that the ZnO-film-based UV photodetector showed a slow photoresponse due to a carrier-trapping or polarisation effect.

Hullavarad et al. developed UV sensors based on nanostructured ZnO spheres in a network of nanowires (Hullavarad et al., 2007). They produced the nanostructured ZnO using a direct vapour phase (DVP) technique. The sizes of the microspheres varied from 600 nm-2 µm, while the nanowire diameters were 30-65 nm. Based on their analysis, the dark current value of their sensor was 1×10^{-10} A at 1 V, which is less than the dark current of a ZnO thin film-based sensor reported by Yang et al. (Yang et al., 2003) and is a result of the low surface-defect properties of their ZnO nanostructures, as observed in the photoluminescence (PL) spectra.

Another interesting study that utilised a single nanobelt as a UV photoconductive sensor was conducted by Yuan et al. (Yuan et al., 2011). The nanobelt has a very similar structure as the nanorod, except the nanobelt exists in a box-like dimension where it has height (nanobelt thickness), width and length. In this case, the prepared nanobelt had a thickness of 120 nm and a width of 600 nm. With this structure, a sensor was constructed with a photocurrent value that was four orders of magnitude higher than the dark current. The sensor also possessed other excellent performance features, such as a high photosensitivity of 10^4, a low dark current of 10^{-3} µA, a low power consumption of 2.45 µW, a typical rise time of 0.12 s, and a decay time of 0.15 s. They explained that the high surface-to-volume ratio and the high coverage-area-to-total-area ratio contributed to the superior performance of their device.

A UV photoconductive sensor using a film of ZnO nanowall networks has been fabricated by Jiang et al. (Jiang et al., 2011). The films were prepared on a Si (111) substrate using plasma-assisted molecular beam epitaxy, with the inner diameters of the nanowalls ranging from 100 to 500 nm. In their sensor configuration, 200 nm-thick Au metal contacts were deposited in an interdigitated electrode design with electrode fingers that were 5 µm wide, 500 µm long, and on a pitch of 2 µm. The sensor showed a huge response to 352 nm UV light, with a responsivity of 24.65 A/W under a biased voltage of 5 V. The cut-off wavelength of the sensor was approximately 360 nm. They showed that the nanostructure-based device had a high photoconductive gain due to the presence of oxygen-related hole-trap states on the nanowall surface.

Based on these previous studies, the use of nanostructure materials for UV photoconductive sensor applications have many advantages over bulk structures, including high gain, low power consumption, high sensitivity, reduced dimensionality, and the use of an extremely small fraction of the device's active materials. There are a number of factors that contribute to the high photosensitivity of nanostructure-based devices, including the surface-to-volume ratio, surface defects, light trapping, and porosity (Soci et al., 2007). Current research has mainly focused on the fabrication of UV photoconductive sensors using ultra-small nanostructures that contribute to the large surface area of the sensing element. Research has also emphasised prolonging the carrier lifetime of the device during UV illumination to lower the charge-carrier recombination. The carrier transit time also plays an important role in the device performance; thus, high mobility nanostructures are needed for good device performance.

3. ZnO nanorod arrays in ultraviolet photoconductive sensor

Currently, ZnO nanorods are receiving considerable attention for UV photoconductive sensor applications due to their unique characteristics and quantum confinement properties. The nanorod structure shows good surface area availability with excellence carrier transport characteristics that are very suitable for UV sensor applications. Depending on the method and experimental parameters used, the nanorod sizes (i.e., diameter and length) are tuneable, which may give different sensor performances. Additionally, by modifying the surface, the performance of the sensor can also be improved because of the relationship between surface defects and surface adsorption of gas molecules from the atmosphere, which tremendously influence the sensor characteristics. According to Soci et al., one-dimensional (1D) structures have several advantages over bulk or thin films in UV sensor applications, including light scattering enhancements that reduce optical losses, improved light absorption, large photosensitivity due to the high gain, and the possibility to integrate functionalities within single 1D devices (Soci et al., 2010). The prolonged photocarrier lifetime, which is due to charge separation promoted by surface states, and the reduction in carrier transit time, which can be achieved in high-quality, low-defect ZnO nanorod together with small gap of metal contacts, both contribute to the high gain in the nanorod-based devices.

Surface area plays a very important role in the UV sensing mechanism, as the sensing mechanism involves the surface reactions between free carriers and the surrounding environment, such as oxygen molecules and humidity (Mamat et al., 2011). The nanorod area possesses a high surface area on the film surface that is suitable for UV photoconductive sensor applications. Moreover, these nanorods exhibit higher carrier mobility than that of ZnO nanoparticles, which works effectively during the surface reaction process. Generally, the photoresponse of a UV photoconductive sensor is influenced by the adsorption and desorption of oxygen on its surface during UV illumination. Oxygen molecules from the surrounding are adsorbed onto the nanorod surface by capturing free electrons from ZnO, as shown in the following equation (Su et al., 2009, Zheng et al., Lupan et al., 2010):

$$O_2 + e^- \rightarrow O_2^- \tag{1}$$

where O_2 is an oxygen molecule, e^- is a free electron, and O_2^- is an adsorbed oxygen on the nanorod surface. When the UV light is incident on the nanostructure surface, electron-hole pairs are photogenerated according to the following equation:

$$hv \rightarrow h^+ + e^- \tag{2}$$

where hv is the photon energy of UV light, h^+ is a photogenerated hole in the valence band and e^- is a photogenerated electron in the conduction band. A large surface area availability in the nanorod film facilitates a fast surface reaction process as the photogenerated hole reacts with a negatively charged adsorbed oxygen, as shown by:

$$O_2^- + h^+ \rightarrow O_2 \tag{3}$$

This condition leaves behind the electron of the pair, which increases the conductivity of the nanostructures. When the illumination is turned off, the oxygen molecule recombines with the electron, leading to a decrease in film conductivity.

This sensor behaviour that is related to the adsorption and desorption of oxygen was studied by Jun et al. (Jun et al., 2009). They measured their fabricated UV sensor under different atmospheric pressures (0.1-1 atm), with different oxygen levels. They found that the photoresponse decay time constant of the sensor increased with decreasing atmospheric pressure. Because the lower atmospheric pressure had a lower oxygen content, it reduced the ability of the sensor to return to its initial state (dark current) due to a reduction of oxygen adsorption onto the ZnO surface. This condition increased the decay time constant of the device as the atmospheric pressure was lowered.

Basically, the nanorod-based UV photoconductive sensor represents the simplest configuration of the UV sensor. It consists of just the nanorods and metal contacts for the photogenerated carrier transport to the outer circuit. In this UV photoconductive sensor configuration, ZnO nanorods are used either vertically or horizontally with the substrates. The vertical standing nanorod is commonly used in an array form or a film-based sensor, while the horizontal nanorod is used in single-nanorod-based sensors. However, a single-nanorod-based UV sensor is very complicated and involves a very challenging fabrication process using high-cost instruments. The realisation of single-nanorod-based UV sensors might reduce the size and the power consumption of the UV sensor. For example, a single-ZnO-nanorod-based UV sensor has been fabricated by Chai et al. (Chai et al., 2011). They used chemical vapour deposition (CVD) method to synthesise a ZnO nanorod with a diameter approximately 1-3 μm and a length of 20-200 μm. To fabricate the UV sensor, a focused ion beam (FIB) *in situ* lift-out technique was used. In their sensor configuration, Au/Ti metal electrodes separated by 20 μm were used. They showed that the single nanowire had a good response to UV light, where the resistance decreased from 52.4 to 48.0 kΩ during 365 nm UV illumination with an optical power of 0.1 mW.

A nanorod array-based UV photoconductive sensor is a promising device structure that has an easier fabrication process compared to a single nanorod-based UV sensor. Moreover, it produces large photocurrent signals due to the large surface coverage and nanorod density.

Various techniques are available to fabricate the ZnO nanorod array, including metal-organic chemical vapour deposition (MOCVD), CVD, sputtering, and solution-based synthesis. Solution-based synthesis has shown promising results for producing aligned ZnO nanorod arrays. This technique is simple, versatile, low-temperature and can be used for large-scale depositions. Another advantage of this technique is that it is a vacuum and gas-free deposition method in which the chemical reactions completely depend on the prepared solution. The biggest advantage of this method is its low-temperature processing, which could even be used to deposit nanostructures on polymer substrates. Unlike other methods that require high temperature for nanostructure growth, this hydrothermal synthesis can be operated at temperatures as low as 50°C for the deposition of ZnO nanostructures (Niarchos et al., 2010).

4. Synthesis of ZnO nanorod arrays via ultrasonic-assisted sol-gel and immersion methods

Recently, ultrasonic irradiation has been applied in hydrothermal processes to prepare ZnO nanostructures. This sonochemical methods use ultrasound irradiation at ranges between 20 kHz to 10 MHz (Suslick et al., 1991). For example, Mishra et al. have synthesised flower-like ZnO nanostructures using a starch-assisted sonochemical method (Mishra et al., 2010). Jia et al. also produced ZnO nanostructures using a sonochemical method (Jia et al., 2010). Using ultrasonic irradiation, they produced hollow ZnO microspheres during hydrothermal synthesis. Another example of ultrasonic-assisted hydrothermal synthesis is a study performed by Mazloumi et al. They produced cauliflower-like ZnO nanostructures using a sonochemical method (Mazloumi et al., 2009). The products that were synthesised by the three sonication methods were similar in that they consisted of powder-form nanostructures. Unfortunately, the powder-forms structures require a seperate process to deposit them onto the substrate for electronic device applications.

In our process, we apply ultrasonic irradiation to the precursor solution, which is used to grow ZnO nanorod arrays on a seed-layer-coated substrate using an immersion process. The sonication process uses powerful ultrasound radiation that can induce molecules to undergo chemical reactions. Sonication is usually used in cleaning processes to remove contaminations, such as substrate and glass wear, from solid surface. Ultrasound radiation involves the creation, growth and collapse of bubbles that can break the chemical bonds of materials in a liquid medium. Generally, growing nanomaterials using a chemical solution method requires a precursor, stabiliser, and solvent. The precursor material supplies the main atoms or ions of the nanomaterials, while the stabiliser material is used to ensure that the growth of the nanomaterial is controlled to a specific rate or structure. The reaction process between the stabiliser and the precursor material prevents the nanomaterials from growing too fast in a certain direction or plane. However, if the reaction process between the precursor and stabiliser does not occur uniformly throughout the solution (e.g., due to agglomerated precursor materials at the beginning or early stages), the size of the end product materials will be large, consequently reducing the surface area. This condition reduces the quantum confinement effect in the produced nanomaterials. In our case, we apply sonication to rupture agglomerated precursor and stabiliser materials and, at the same time, ensure a highly homogenous and uniform reaction process between the precursor and stabiliser.

ZnO nanorod array films were fabricated using an ultrasonic-assisted sol-gel and immersion method using zinc nitrate hexahydrate ($Zn(NO_3)_2 \cdot 6H_2O$) as a precursor, hexamethylenetetramine (HMT, $C_6H_{12}N_4$) as a stabiliser, and aluminium nitrate nonahydrate ($Al(NO_3)_3 \cdot 9H_2O$, 98 %, Analar) as a dopant (Mamat et al., 2010). Aluminium (Al) doping is especially attractive because it contributes to the higher conductivity of the film without deteriorating the optical and crystalline properties of the ZnO. The precursor, stabiliser, and dopant were dissolved in deionised (DI) water before being subjected to the sonication process using an ultrasonic water bath (Hwasin Technology Powersonic 405, 40 kHz) for 30 min at 50°C. Subsequently, the solution was stirred and stored at room temperature for 3 h.

Next, the solution was poured into a vessel, where the seed-layer–coated glass substrate was positioned at the bottom of the vessel. The seed layer, or Al-doped ZnO nanoparticle layer, was coated onto the substrate with a thickness of approximately 200 nm using sol-gel spin-coating (Mamat et al., 2010). The existence of the seed layer on the glass substrate reduced the formation energy for the crystallisation of the ZnO and, thus, helped the nanorod grow more easily on the glass substrate. The vessel was then sealed before being immersed into a water bath for 4 h at 95°C. After the immersion process, the sample was removed from the vessel and rinsed with DI water. The sample was then dried at 150°C for 10 min and annealed at 500°C for 1 h in a furnace. Next, 60-nm-thick Al metal contacts were deposited onto the nanorod array using thermal evaporation to complete the sensor structure. The distance between the electrodes was approximately 2 mm.

The surface morphologies of the ZnO nanorod array films were observed by field-emission scanning electron microscopy (FESEM, ZEISS Supra 40VP and JEOL JSM-7600F). The surface topology of the nanorod arrays was characterised using atomic force microscopy (AFM, Park System). The crystallinity of the samples was investigated using X-ray diffraction (XRD, Rigaku Ultima IV). The transmittance and absorbance characteristics of the seed layer and the thin film were characterised using an ultraviolet-visible (UV-Vis) spectrophotometer (Perkin Elmer Lambda 750). The photoluminescence (PL) properties of the synthesised nanorods were investigated using a PL spectrophotometer with a helium-cadmium (He-Cd) excitation laser operating at 325 nm (PL, Horiba Jobin Yvon-79 DU420A-OE-325). The UV photoresponse measurements of the fabricated sensor were conducted using a spectral sensitivity analysis system (Bunko-Keiki, CEP 2000) with a monochromatic xenon (Xe) lamp operating at 365 nm and a power intensity of 5 mW/cm^2 as well as photocurrent measurement system operating at 365 nm and a power density of 750 $\mu W/cm^2$. The thicknesses of the samples were measured using a surface profiler (VEECO/D 150+). The fabrication process of the ZnO-nanorod-based UV photoconductive sensor is shown in Fig. 1.

5. Performance of synthesised ZnO-nanorod-array-based ultraviolet photoconductive sensor

We have investigated the performance of the ZnO-nanorod-array-based UV photoconductive sensor prepared via ultrasonic-assisted sol-gel and immersion methods. There are numerous factors that influence the sensor performance, such as nanorod size, surface area, surface defects, film thickness, metal contacts, and doping. In this subchapter, we will highlight the effects of surface modifications, film thickness, and Al ions doping on the performances of the fabricated ZnO-nanorod-array-based UV sensor.

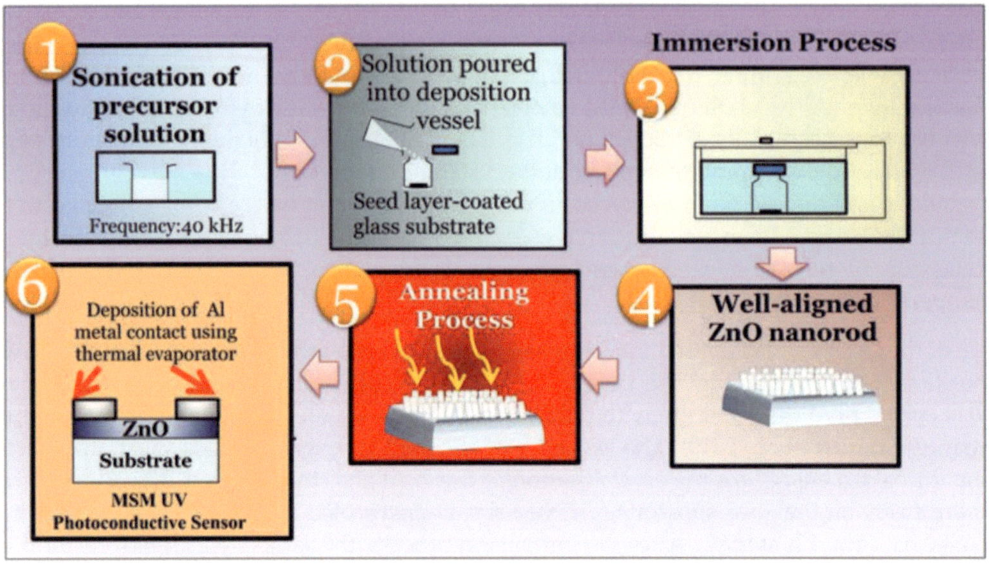

Fig. 1. Fabrication process of aligned ZnO nanorod array-based UV photoconductive sensor via sonicated sol-gel and immersion methods.

5.1 Surface modification

A ZnO nanorod array was prepared on glass substrate utilising an Al-doped ZnO nanoparticle thin film catalytic seed layer. Figure 2(a) shows a field-emission scanning electron microscopy (FESEM) image of the seed layer that was prepared using the sol-gel spin-coating technique. The particle sizes of the Al-doped ZnO nanoparticles were estimated to range from 10 to 40 nm. From the FESEM figure, synthesised Al-doped ZnO nanoparticles exhibited some edges rather than perfect curved surfaces due to the differenced in the surface energy of crystallographic directions of the ZnO growth. In the wurtzite structure, the relative growth rate of each crystallographic plane differed somewhat according to the crystal orientation, so it was difficult for crystalline ZnO to grow symmetrically into spherical particles (Lee et al., 2007). The FESEM figure also indicates that the particles were well connected to each other and that it was very important to develop a continuous transport pathway in the granular film for electron movement in the UV sensor application. Figure 2(b) shows an AFM image of the seed catalyst layer. Based on the AFM image, the root mean square (RMS) roughness of the Al-doped ZnO nanoparticle thin film was 17.51 nm over an area of 100 μm^2.

In the UV photoconductive sensor, this seed layer plays a very important role in increasing the sensor performance. Generally, one of the factors that degrades the sensor performance is the strain of the film or the material, which influences the density of the defects and the photoelectric activity of the sensor (Shinde & Rajpure, 2011). This seed layer facilitates the homogenous growth of the compressive-strained layer, i.e., the high quality ZnO nanorod material, which has a low defect density and allows for a smooth charge transfer process during UV photo-illumination. As a result, the seed layer results in a higher responsivity of the ZnO-nanorod-array-based UV photoconductive sensor.

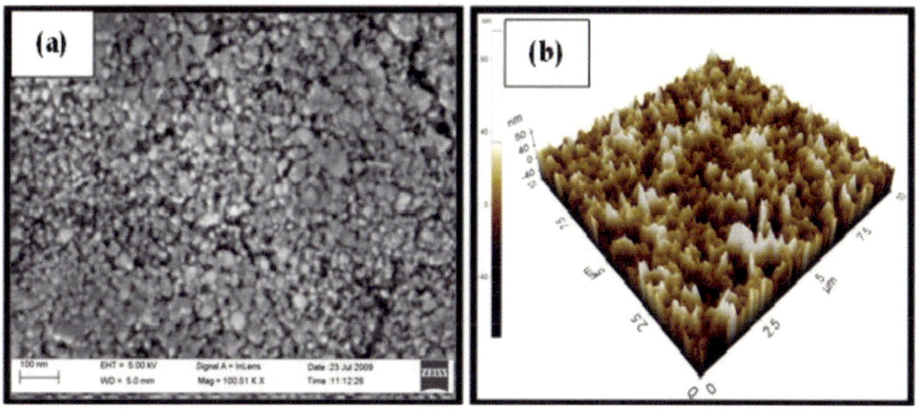

Fig. 2. (a) FESEM image of Al-doped ZnO nanoparticle thin film seed layer. (b) AFM topography image of the seed layer.

Figure 3 shows an FESEM image of the Al-doped ZnO nanorod array before (Fig. 3(a)) and after (Fig. 3(b) and 3(c)) the annealing process at 500°C. The nanorods were prepared using a 1000 ml solution that was sonicated in a beaker. The images show that well-oriented, hexagonal-shaped ZnO nanorod arrays were deposited onto the seed-layer-coated glass substrate with good uniformity and high density. The diameters and lengths of the nanorods were not strongly affected by the annealing process, as the diameter of the nanorods ranged between 40 to 150 nm and the nanorods were 1.1 μm long. The nanorods were aligned well, which indicates that this low-temperature ultrasonic-assisted sol-gel and immersion processes produce high-quality ZnO nanorod arrays. We believe that the excellent alignment of the nanorod arrays is due to the seed layer films, which act as nucleation centres that provide an almost mismatch-free interfacial layer between the nanorods and the seed layers. This layer assists an epitaxial nanorod growth process on the seed-layer-coated glass substrates. Figure 3(d) shows an AFM topography image of an annealed ZnO nanorod array measured in a 1 μm^2 area. Based on this topography image, the root mean square (RMS) roughness of the nanorod array was approximately 21.95 nm.

As shown in Fig. 3(b) and 3(c), nanoholes appeared on the surfaces of the nanorods after the annealing process. A closer look at the cross-sectional images indicates that nanoholes exist on nearly the entire nanorod surface. We suspect that these nanoholes are the result of the evaporation of impurities, such as hexamethylenetetramine (HMT), during the annealing process at high temperature. Interestingly, these nanoholes facilitate a larger surface area availability of the single nanorod and facilitate effective sites for the oxygen adsorption process. Thus, the existence of these nanoholes on the nanorod surface could improve the performance of the UV photoconductive sensor because of the increased surface area and surface photochemistry. The condition of the Al-doped ZnO nanorod with nanoholes after annealing process is shown in Fig. 4.

XRD spectra of the as-grown nanorods and the 500°C annealed nanorods are shown in Figure 5. The spectra confirmed that the synthesised nanorods belong to the ZnO hexagonal wurtzite structure (joint committee on powder diffraction standards (JCPDS) PDF no. 36-1451). Both the as-grown and annealed samples contained a dominant XRD peak at the

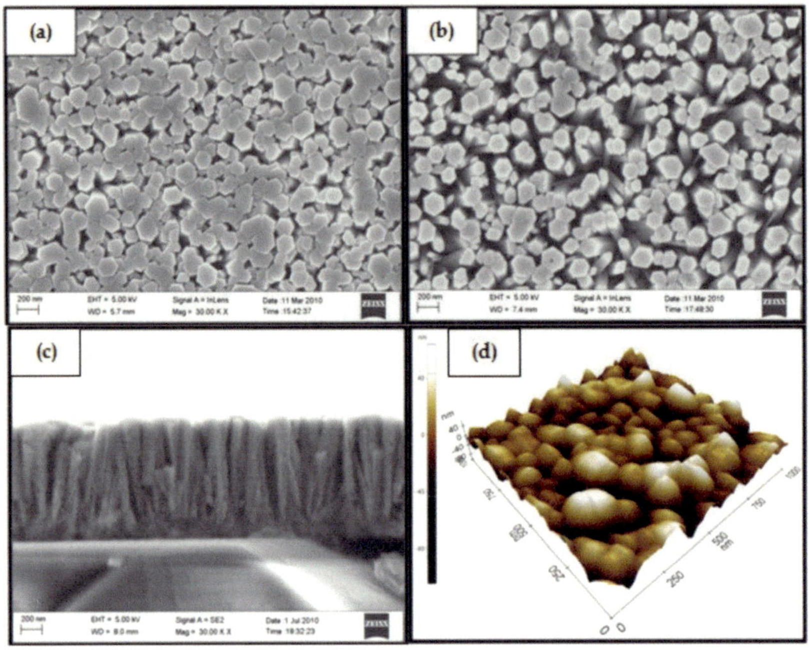

Fig. 3. FESEM image of (a) as-grown and (b) 500°C annealed Al-doped ZnO nanorod arrays. (c) Cross-sectional image of the annealed Al-doped ZnO nanorod arrays, clearly showing nanoholes on the ZnO surface. (d) AFM image of the annealed Al-doped ZnO nanorod array.

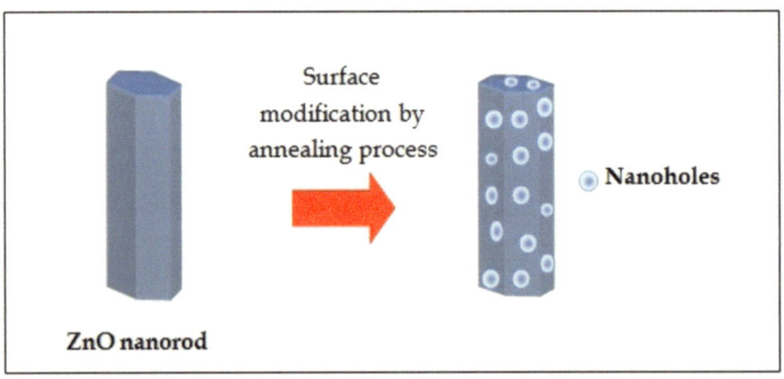

Fig. 4. Nanoholes produced on the surface of the nanorod after the annealing process as a result of the evaporation of impurities.

(002)-plane, implying that the nanorods were grown along the c-axis, or perpendicular to the substrates. This result indicates that the Al-doped ZnO nanorod arrays prepared in this work had a very good ZnO crystal quality. The weak peaks intensities of the other orientations might be due to the vertical alignment imperfections of the nanorods (Qiu et al., 2009). Based on the spectra, the peak intensities of the annealed nanorods were higher than the as-grown sample, indicating an improvement in the nanorod crystallinity after the annealing treatment. Based on these results, we predicted a possible growth mechanism for the formation of the well-aligned ZnO nanorod arrays on the seed-layer-coated glass substrate using the ultrasonic-assisted sol-gel and immersion methods (Mamat et al., 2011). We suspect that the growth along the c-axis might be caused by the effects of the sonication process on the precursor solution, which disperses and mixes the zinc nitrate (i.e., the precursor), aluminium nitrate (i.e., the dopant), and the HMT (i.e., the stabiliser) very well. The sonication process also helps dissolve the agglomerated zinc nitrate and HMT particles, which hasten the physical and chemical reaction activity in the solution. This process enables the Zn^{2+} ion to react effectively with the HMT, as shown by (Khusaimi et al., 2010)

$$Zn(NO_3)_2 + C_6H_{12}N_4 \rightarrow [Zn(C_6H_{12}N_4)]^{2+} + 2NO_3^- \quad (4)$$

The HMT plays a very important role in controlling the growth of the aligned ZnO nanorods. When it attaches to the Zn^{2+} ions, particle agglomeration is reduced and ZnO formation in the solution is slowed.

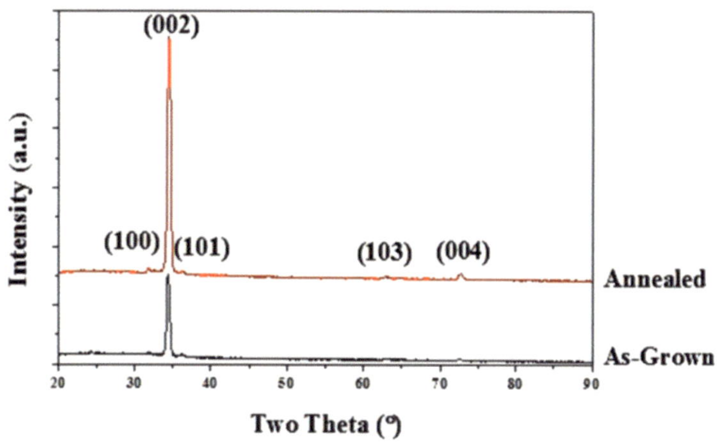

Fig. 5. XRD patterns of Al-doped ZnO nanorod array before (as-grown) and after annealing.

During the hydrothermal process, the seed catalyst layer provides a base that initiates the growth of the nanorod arrays through heterogeneous nucleation. It is generally accepted that heterogeneous nucleation on a seed layer surface occurs more easily than does homogenous nucleation (Guo et al., 2005). The seed layer provides the lowest energy barrier for heterogeneous nucleation, which produces almost negligible lattice mismatch between the nanorods (Chen et al., 2009, Giri et al., 2010). This condition results in the growth of high-quality aligned nanorod arrays on the substrates. The general reactions occurring during the hydrothermal process can be described by the following equations (Khusaimi et al., 2010, Lupan et al., 2007):

$$C_6H_{12}N_4 + 4H_2O \rightleftharpoons C_6H_{12}N_4H_4^+ + 4OH^- \quad (5)$$

$$[Zn(C_6H_{12}N_4)]^{2+} + 4OH^- \rightarrow Zn(OH)_4^{2-} + C_6H_{12}N_4 \quad (6)$$

$$Zn(OH)_4^{2-} \rightarrow Zn^{2+} + 4OH^- \quad (7)$$

$$Zn^{2+} + 2OH^- \rightleftharpoons ZnO + H_2O \text{ or } Zn^{2+} + 2OH^- \rightleftharpoons Zn(OH)_2 \rightleftharpoons ZnO + H_2O \quad (8)$$

Initially, when the Zn^{2+} and OH^- ion concentrations exceed the boundaries of supersaturation, ZnO nuclei form on the seed layer surface, initiating the growth of aligned ZnO nanorods. It has been suggested that the HMT also acts as a chelating agent that attaches to the nonpolar facets of ZnO nanorods (Sugunan et al., 2006). Because of this attachment behaviour, epitaxial growth along the c-axis is facilitated because only the (0001) plane is exposed during the growth process. Because the growth rate in the (0001) plane proceeds the fastest in the hydrothermal system (Laudise & Ballman, 1960, Laudise et al., 1965), the ZnO nanorods grow preferentially in the (0001) plane, which is vertically aligned with the substrate.

Good dispersion and mixing processes between the precursor and the stabiliser through sonication help control the diameter sizes of the nanorods because the HMT can immediately attach onto the nonpolar facets (i.e., six prismatic side-planes) after the ZnO nanorod nucleation process is initiated on the seed layer. The HMT acts as a buffer layer at the nonpolar surfaces, which disturbs the ZnO deposition onto these surfaces. This condition disables rapid growth on the side walls, or the nonpolar surface of the nanorod, which served to maintain an almost constant diameter size throughout the length. The rapid attachment of the HMT onto the ZnO surface prevents any nanorod growth in the direction of the nonpolar facets, thus hindering an enlargement in the nanorod diameter. Therefore, because the ZnO nanorods are confined by the HMT molecules on their nonpolar surfaces and only grow from the polar surface for further growth, directional growth along the c-axis is achieved.

The photoluminescence (PL) spectra of the as-grown nanorod and the annealed nanorod with nanoholes are depicted in Fig. 6. The main peaks that were observed in the spectrum of the annealed nanorods are located at 380 and 580 nm, and the main peaks in the spectrum of the as-grown nanorods is located at 380 and 590 nm. The UV emission at 380 nm corresponds to free exciton recombination, while the orange emission at 580 and 590 nm are related to the emission from defects, such as oxygen deficiencies and zinc interstitials. According to Rosa et al., this orange emission is due to zinc interstitials that occur close to or on the surface of the ZnO structure (De la Rosa et al., 2007). The shift in the visible emission peak from 590 to 580 nm for annealed nanorods might be due to the desorption of the OH groups from the nanorod surface (Lee et al., 2010). The as-grown nanorods exhibited a very weak UV emission peak intensity compared to the annealed nanorods. The annealing process increased the UV emission peak intensity, which indicates that the crystallinity of the sample improved at the higher annealing temperatures.

During the annealing treatment, the concentrations of defects in the nanorods were reduced according to equations (9-10) shown below (Lin et al., 2001):

$$V_O + \frac{1}{2}O_2 = O_O \tag{9}$$

$$Zn_i + \frac{1}{2}O_2 = Zn_{Zn} + O_O \tag{10}$$

where V_O is an oxygen vacancy, Zn_i is a zinc interstitial, and Zn_{Zn} and O_O represent zinc and oxygen at a lattice site, respectively. Because the supplied thermal energy induces an oxygenation process during the annealing treatment, oxygen from the atmosphere occupies the vacant sites of the ZnO lattice, which eventually reduces both the zinc interstitial and oxygen vacancy concentrations. To investigate this phenomenon, we calculated the ratio of the UV emission intensity over the visible emission intensity to be 0.6 and 7.1 for the as-grown and the 500°C annealed nanorods, respectively. It is generally accepted that this ratio value is a good way to evaluate the optical quality and stoichiometric properties of ZnO. An increase of this ratio indicates that the stoichiometric properties of ZnO and its optical properties were improved after the annealing process. Lee et al. reported that the ratio of the UV peak intensity over the visible peak intensity is directly related to the oxygen deficiency in the ZnO nanorods (Lee et al., 2010). Based on this ratio analysis, the defect concentrations from oxygen vacancies and zinc interstitials were suppressed during the annealing treatment, as indicated by in equations (9-10). The reduction of defect concentrations in the nanorods strengthened the UV emission and reduced the visible emission of the nanorods after the annealing process.

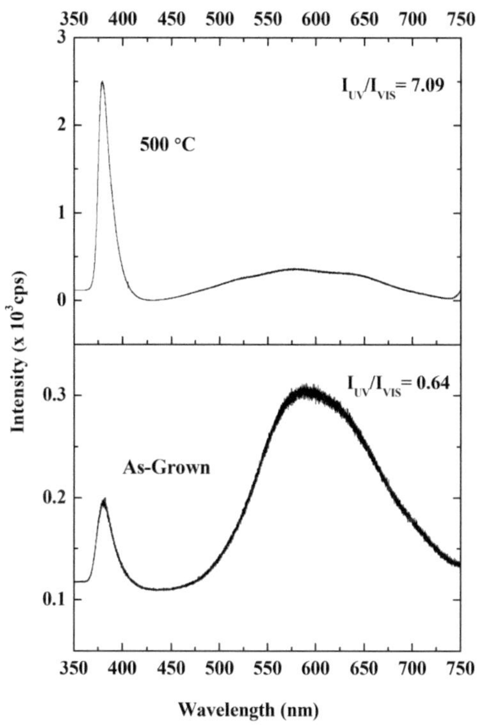

Fig. 6. Room temperature PL spectra of the as-grown and annealed ZnO nanorod arrays.

Figure 7 shows the time-dependent photocurrent properties of the fabricated UV photoconductive sensors under a bias voltage of 10 V. The measurements were conducted using 365 nm UV illumination with an optical power density of 5 mW/cm^2. Both the as-grown and annealed nanorods showed a response to the UV light but with different sensing characteristics. The dark currents/photocurrents of the fabricated devices were 3.49 x 10^{-6} A/4.08 x 10^{-4} A and 1.78 x 10^{-6} A/1.35 x 10^{-4} A for the annealed and as-grown nanorods, respectively. The responsivity of the fabricated devices was calculated according to equation (11) (Jun et al., 2009):

$$R = \frac{I_{ph} - I_{dark}}{P_{op}} \tag{11}$$

where I_{ph} is the photocurrent, I_{dark} is the dark current, and P_{op} is the optical power of the UV source. For the annealed nanorod-array-based UV sensor, the calculated responsivity was 1.35 A/W. For the as-grown nanorod-array-based UV sensor, the responsivity was 0.44 A/W. The desirable crystallinity properties and low defect concentrations of the annealed nanorods may have contributed to the high responsivity of this sensor. The rise and decay time constants for the fabricated sensors were estimated using equations (12-13) (Jun et al., 2009, Li et al., 2009):

Rise process upon UV illumination ON:

$$I = I_0\left(1 - e^{\frac{-t}{\tau_r}}\right) \tag{12}$$

Decay process after UV illumination OFF:

$$I = I_0 e^{\frac{-t}{\tau_d}} \tag{13}$$

where I is the magnitude of the current, I_0 is the saturated photocurrent, t is time, τ_r is the rise time constant and τ_d is the decay time constant. The calculations show that the annealed Al-doped ZnO nanorod array-based UV sensor exhibited small rise and decay time constants at 20 s and 22 s, respectively. For the as-grown sample, the rise time constant was 280 s, while the decay time constant was estimated to be 300 s. This result indicates that the annealing process greatly improved the performance of the UV sensor.

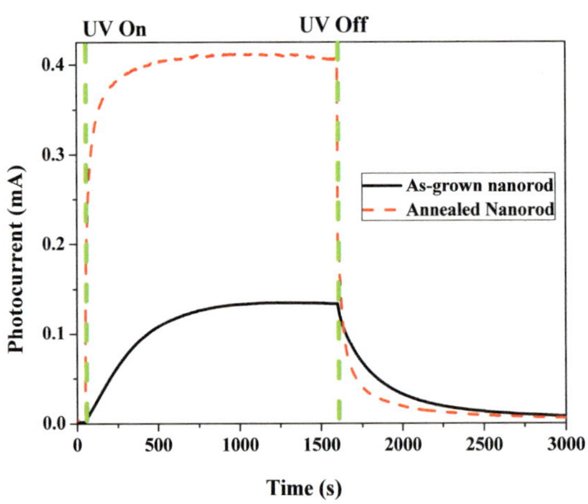

Fig. 7. Plot of the growth and decay of the photocurrent measured at 10 V of the as-grown and annealed Al-doped ZnO nanorod array-based UV photoconductive sensors under 365 nm, 5 mW/cm² UV illumination.

The annealing process plays an important role in improving the sensor performance and is considered a surface treatment or surface modification. For example, Kim et al. reported that water molecules and residual carbon from the fabrication process could be effectively desorbed from the surface of a ZnO nanowire during the annealing process (Kim et al., 2011). As a consequence, oxygen molecules from the air could occupy the existing defect sites more easily, which contributed to a faster photocurrent decay, higher sensitivity, and faster response when the UV light was turned off. Because their fabricated sensor consisted of a ZnO nanowire network that depended on the interconnections between the nanowires for the carrier transportation, the annealing process improved the contact between the nanowire interfaces. As a result, the contact resistance was reduced, and the potential barrier was lowered. Other important effects of the annealing process included an improvement of the nanowire crystallinity and a reduction in defects, which also significantly improved the photocurrent properties of the sensor.

Improvements of UV sensor performances from surface modifications have also been reported by other groups. For instance, Park et al. showed that a larger photocurrent value and faster photoresponse time were achieved for roughened ZnO nanorods in a UV photodetector (Park et al., 2011). They demonstrated that the surfaces of the nanorods could be roughened by immersing the nanorods in isopropyl alcohol (IPA) for 30 days. The etched areas of the nanorod surfaces from IPA contained defects state that enhanced the adsorption of oxygen molecules onto the surface. This condition resulted in a large quantity of oxygen molecules that adsorbed onto the nanorod surfaces.

Another method for increasing the sensor responsivity involved the surface passivation of ZnO nanorods by thin layer coating (e.g., with polyvinyl alcohol (PVA)). The idea is to increase the photoluminescence (PL) UV emission of the ZnO while reducing the green emission that is related to ZnO surface defects. Recent research on coating ZnO nanoparticles with PVA has shown a suppression in the number of defects evidenced by a reduction in the parasitic green emission (Qin et al., 2011). This characteristic increased the ratio of the photocurrent over the dark current compared to the uncoated ZnO particles. The coatings effectively decreased the number of holes in the deep level, which helped the UV-excited electrons to recombine with the holes in the valence band without being trapped in the deep-level defects of ZnO.

5.2 Role of nanorod array thickness

We investigated the performance of the UV sensor at different nanorod array thicknesses. In this study, we prepared the nanorod array films at different thicknesses by varying the immersion times from 1 to 5 h. In this case, we used a 100 ml solution in the vessel to grow the Al-doped ZnO nanorod arrays. We have previously shown that the volume of solution in the vessel affects the nanorod length or film thickness during the deposition process (Mamat et al., 2011). Therefore, it was expected that the film prepared for this study would be thinner than the nanorod array discussed in section 5.1. Figure 8 depicts FESEM images of the Al-doped ZnO nanorod arrays at different immersion times. The sizes of the nanorod diameters in the sample prepared after a 1 h immersion (Fig. 8(a)) ranged from 40 to 150 nm. Notably, the diameters of the nanorods remained almost unchanged after increasing the immersion time to 2, 4, and 5 h, as observed in Fig. 8(b), 8(c), and 8(d).

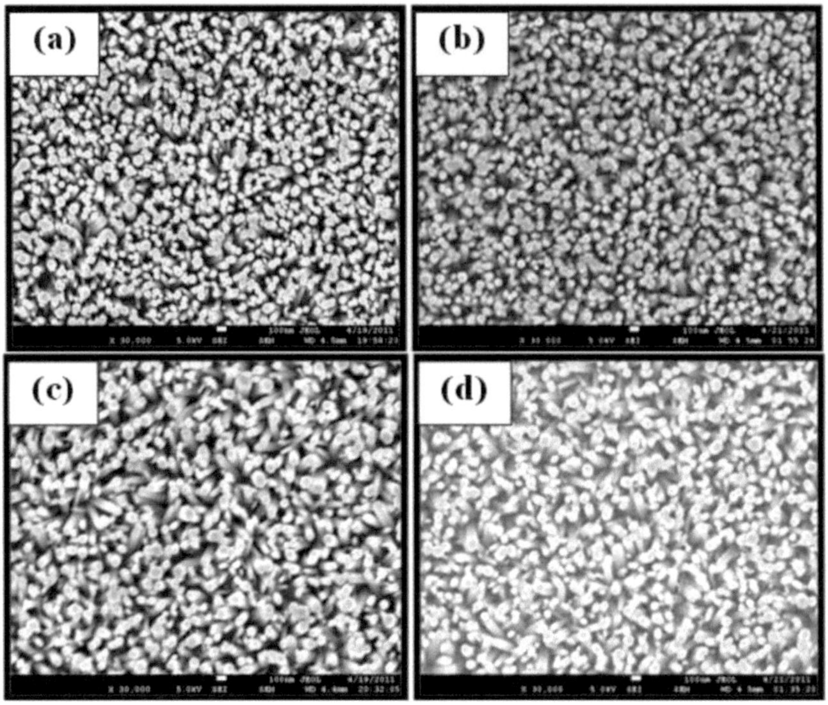

Fig. 8. Top-view FESEM images of Al-doped ZnO nanorod arrays prepared at immersion times of (a) 1, (b) 2, (c) 4, and (d) 5 h.

To investigate the growth behaviour, we performed thickness measurements to characterise the lengths of the nanorods, i.e., the film thicknesses that were grown with at different immersion times. The thicknesses of the nanorods were 629, 677, 727, 768, and 834 nm after being immersed for 1,2,3,4, and 5 hours, respectively. From this result, we concluded that the nanorod growth behaviour occurs primarily along the c-axis when the immersion process is carried out at longer times. It is interesting to note that the nanorod dimension increased only in length when immersed for longer times, without significantly affecting the size of the nanorod diameter. This result indicates that the controllable growth of the nanorods along the c-axis could be achieved under different immersion times using our sonicated sol-gel immersion method while maintaining the diameter sizes of the nanorods. It demonstrated that the sonication process provides a good dispersion process of the starting materials to produce a mixture of zinc-aluminium-hexamethylenetetramine (Zn-Al-HMT) complexes, which inhibit growth at the nonpolar surfaces of ZnO while promoting growth along the c-axis, or polar surface, at longer immersion times.

Figure 9 shows the photoresponse spectra of the Al-doped ZnO nanorod array-based UV sensor at different thicknesses. The photocurrent value of the UV sensor decreased with increasing film thickness up to 834 nm. The decrease in photocurrent value with film thickness also influenced the responsivity of the device, as it exhibited a lower value with thicker films. The responsivity of the device was calculated to be 2.13, 1.75, 1.21, 0.94, and 0.83 A/W for sensors with thicknesses of 629, 677, 727, 768, and 834 nm, respectively. Thicker films also increased the rise (decay) time constant of the devices. The rise (decay) process time constants of the device were calculated to be 3(12), 4(13), 6(15), 8(16), and 9(20) s for 629, 677, 727, 768, and 834 nm-thick nanorod array-based UV sensors, respectively.

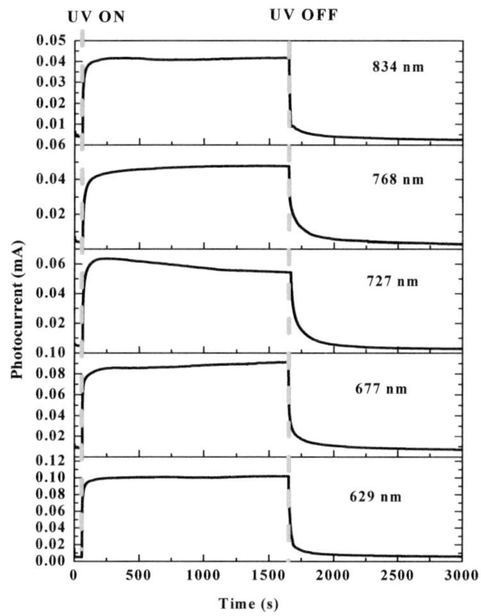

Fig. 9. UV photoresponse properties of Al-doped ZnO nanorod arrays at different thicknesses under a 365 nm, 750 μW/cm^2 UV light at 10 V bias voltage.

In the UV sensor, the film has an optimum thickness for effectively detecting UV irradiation. If the film is too thin, the UV light can pass through the film and not be fully absorbed because the transmittance depth of light is proportional to its wavelength. Therefore, the photocurrent value would be weak due to the low carrier photogeneration within the limited absorbed UV light. A thicker film should increase the UV absorption, which would result in an enhancement of the photocurrent value and responsivity of the UV sensor. However, an excessively thick film would saturate the UV absorption and would not contribute to an increase in photogenerated carriers. In this study, a reduction of responsivity of the sensor with increasing film thickness might be influenced by the extension of the diffusion lengths of the carriers to the metal contacts (Fu & Cao, 2006). An increase in film thickness should increase the recombination probability of the electron-hole pairs, leading to a decrease in photocurrent, and thus reducing the responsivity value of the UV sensor. This phenomenon might also contribute to a larger rise (decay) time constant of the device.

A similar observation of thickness-dependent UV photoconductive sensor was also reported by Shinde et al. when they produced a gallium (Ga)-doped ZnO thin-film-based UV photoconductive sensor fabricated using a spray pyrolysis method (Shinde & Rajpure, 2011, 2011). They found that by decreasing the Ga-doped ZnO thin film from 225 nm to 139 nm, the responsivity of the sensor at a 5 V bias voltage at 365 nm improved from 1125 A/W to 1187 A/W.

5.3 Aluminium doping effects

The doping process also plays an important role in the performance of the UV sensor. There are several elements that could be used as a dopant in the nanorods, such as gallium (Ga), indium (In), and aluminium (Al). In our study, we used Al ions doping because it could be easily incorporated into the ZnO lattice and because it enhanced some of the ZnO nanorod properties, such as optical transmittance and electrical conductivity. Furthermore, Al can serve as a donor and induce chemical defects, which tremendously improve the optical and electrical properties of ZnO (Yun & Lim, 2011).

Figure 10 shows the FESEM morphologies and cross-sectional images of undoped (Fig 10(a-c)) and Al-doped ZnO nanorod arrays (Fig. 10(d-e)). In this experiment, we sonicated 500 ml of the precursor solution in a smaller beaker to increase the ultrasonic power density applied to the solution. We observed that the diameter sizes of nanorods decreased after Al ions doping. The sizes of the undoped ZnO nanorods varied from 80 to 120 nm, while the sizes of the Al-doped ZnO nanorods ranged from 30 to 70 nm. The reduction in size may have originated from the different radii of Zn^{2+} and Al^{3+} ions, which are 0.074 nm and 0.054 nm, respectively. The existence of Al in the ZnO lattice may influence the attractive forces between the atoms and thus reduce the diameter sizes of the ZnO nanorod. A similar behaviour of the decrease in diameter was also reported by Hsu et al. (Hsu & Chen, 2010). We also observed that the diameter sizes of Al-doped ZnO nanorods were smaller than those of nanorods produced using a solution that was sonicated in a volume of 1000 ml, as discussed in section 5.1 and 5.2. We suspect that the mixing process between the precursor (i.e., zinc nitrate), stabiliser (i.e., HMT), and dopant (i.e., aluminium nitrate) was improved by the stronger and more intense ultrasonic irradiation.

Figure 11 shows the photoresponse spectra of undoped and Al-doped ZnO nanorod array-based UV photoconductive sensors under 365 nm UV illumination with an optical power density of 750 $\mu W/cm^2$. The spectra reveal that the Al-doped ZnO nanorod array-based UV sensor almost doubled the magnitude of the photocurrent compared to the undoped ZnO nanorod. The responsivity of the Al-doped ZnO-based UV sensor was 3.24 A/W, while the responsivity of the undoped ZnO-based UV sensor was 1.60 A/W. From the calculation results, we found that the rise (decay) process time constant of the undoped and Al-doped ZnO nanorod array-based UV sensors were 16(16) and 3(10) s, respectively. This result suggests that the Al-doped ZnO nanorod array improved the photoresponse of the sensor by increasing the responsivity and decreasing the rise (decay) process time constant.

By adding Al, the carrier concentration of ZnO nanostructures were improved because the substitution of Al^{3+} at the Zn^{2+} site created an extra free carrier in the process (Mridha & Basak, 2007, Fournier et al., 2008). Because of the high electron concentration, this condition reduced the barrier height between the Al-doped ZnO nanorod and the seed layer and between the film and the Al metal contact interface. This reduction initially allowed the photogenerated electrons to move more easily from the Al-doped ZnO nanorods to the seed layer, then from the seed layer back to the Al-doped ZnO nanorods underneath the metal contact, and finally to the metal contact. The flow of the photogenerated electrons during UV illumination is depicted in Fig. 12. Additionally, Al ions doping led to a suppression of defects in the film, such as zinc interstitials and oxygen vacancies, which served to increase the stability and performance of the Al-doped ZnO film-based sensor (Mamat et al., 2011, Sharma & Khare, 2010).

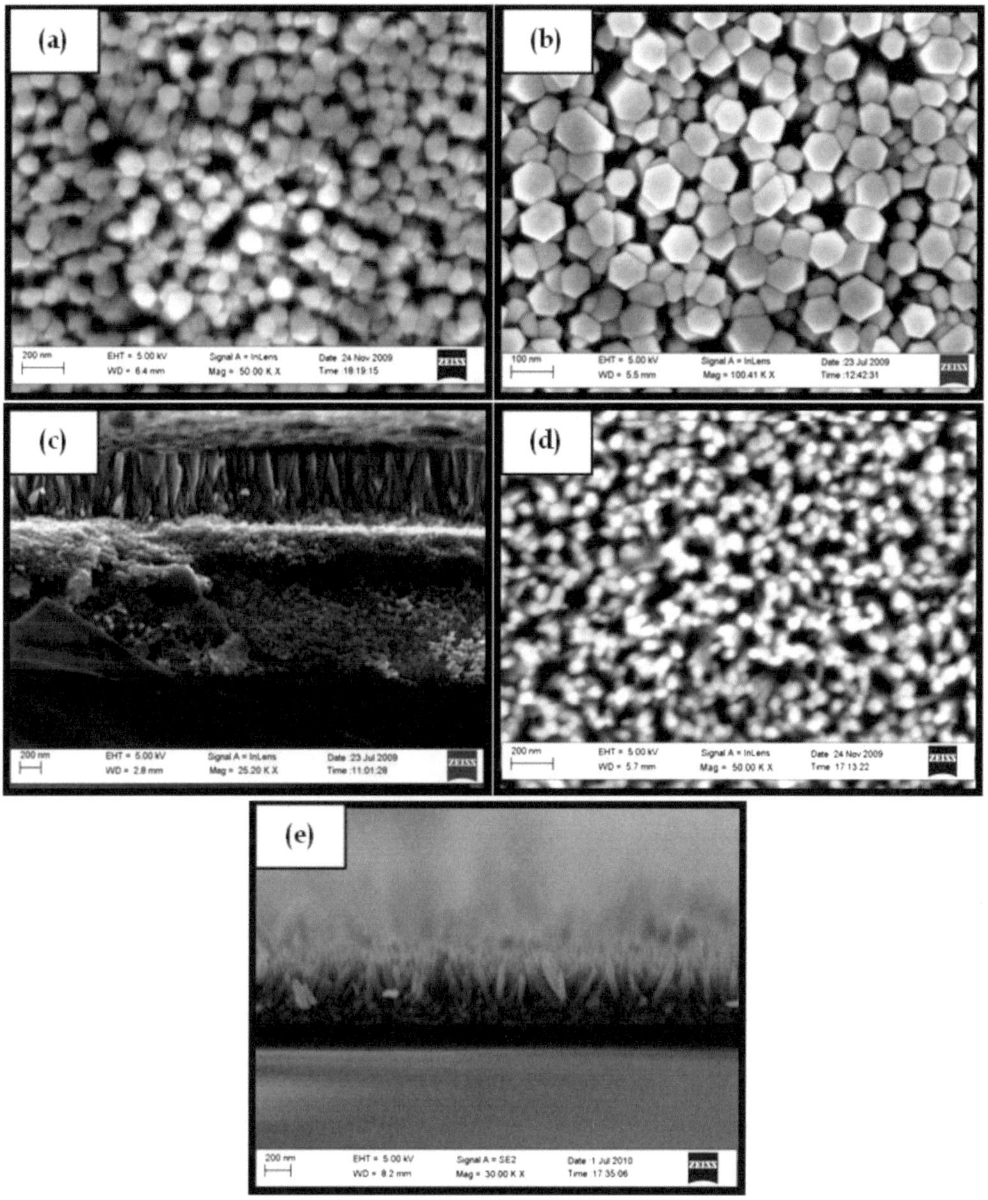

Fig. 10. FESEM images of undoped (a-c) and Al-doped ZnO (d-e) nanorod arrays prepared using 500 ml sonicated solution.

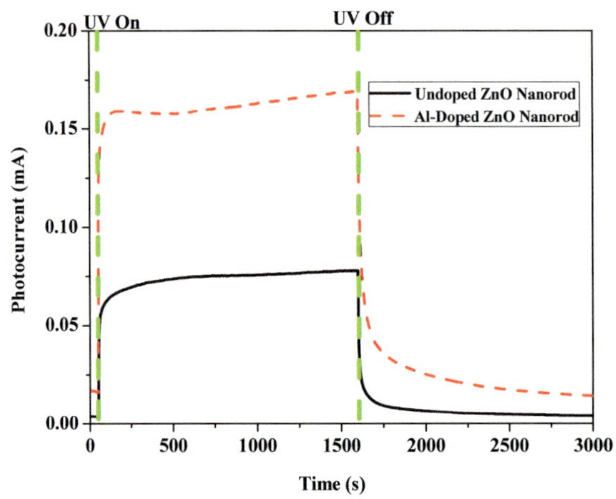

Fig. 11. The dependence of the photocurrent on operating time for undoped and Al-doped ZnO-nanorod-array-based UV photoconductive sensors under 365 nm UV light with a power density of 750 μW/cm2 and a bias of 10 V.

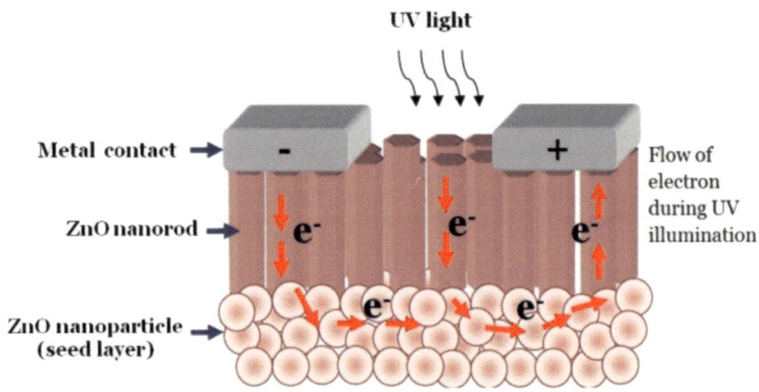

Fig. 12. Flow of photogenerated carrier during UV illumination in ZnO-nanorod-array-based UV photoconductive sensor

6. Conclusions

The performance of a UV photoconductive sensor using ZnO nanostructures, particularly ZnO nanorod arrays, was discussed. The problems related with grain boundaries and poor electron mobility have motivated researchers to develop ZnO nanostructure-based UV sensors. The ZnO nanostructures which could be prepared in many shapes such as nanorod and nanobelt might provide a way in improving electronic device performance through reducing grain boundaries concentration and increasing the electron mobility in the structure. Numerous techniques have been used to fabricate ZnO nanostructure-based UV sensors, such as radio frequency (RF) magnetron sputtering, chemical vapour deposition (CVD), sol-gel, molecular beam epitaxy (MBE), and pulsed laser deposition. Based on previous studies, the use of nanostructure materials for UV photoconductive sensor applications have many advantages over bulk structures, including high gain, low power consumption, high sensitivity, reduced dimensionality, and the use of an extremely small fraction of the device's active materials. There are a number of factors that contribute to the high photosensitivity of nanostructure-based devices, such as the surface-to-volume ratio, surface defects, light trapping, and porosity. Current research has mainly focused on the fabrication of UV photoconductive sensors using smaller ZnO nanostructures that contribute to the large surface area of the sensing element and prolonging the carrier lifetime of the device during UV illumination to lower the charge-carrier recombination. The nanostructures that have high carrier mobility also have been used to improve the transit time of carriers, which results in the improvement of the device performance. A review of the current status of the UV photoconductive sensor found that ZnO nanorods are very promising nanostructures for UV detection due to their large surface area, high mobility, and good surface photochemistry. However, there are still many challenges in the fabrication of ZnO nanorod-based UV sensors because of certain ZnO defects that are produced during the fabrication process. These defects eventually contribute to a lower photoresponsivity, a low photogenerated carrier lifetime, and less effective UV light absorption properties of the ZnO-nanorod-array-based UV sensor. In this chapter, we have discussed the fabrication of ZnO-nanorod-array-based UV photoconductive sensors via ultrasonic-assisted sol-gel and immersion methods. The fabrication of ZnO nanorod arrays using a solution-based method is very attractive because of its simplicity, versatility, and low-temperature processing. We found that several factors influence the UV sensor performance, including surface treatment, film thickness, and doping of the ZnO nanorod arrays.

7. Acknowledgements

The authors would like to thank Universiti Teknologi MARA (UiTM) Malaysia, Ministry of Higher Education (MOHE) Malaysia, Jabatan Perkhidmatan Awam (JPA) Malaysia and Research Management Institute (RMI) of UiTM for financial support. The authors would also like to thank the Faculty of Applied Sciences (Mr. Hayub) and Faculty of Mechanical Engineering, UiTM for their FESEM and XRD facilities, respectively. The authors thank Mr. Shuhaimi Ahmad (UiTM technician), Mrs. Nurul Wahida (UiTM Asst. Science Officer) and Mr. Mohd Azlan Jaafar (UiTM technician) for their kind support during this research.

8. References

Lee, J.M.; et al. (2009). ZnO Nanorod−Graphene Hybrid Architectures for Multifunctional Conductors. *The Journal of Physical Chemistry C*, Vol.113, No.44, pp.19134, ISSN 1932-7447.

Galoppini, E.; et al. (2006). Fast Electron Transport in Metal Organic Vapor Deposition Grown Dye-sensitized ZnO Nanorod Solar Cells. *The Journal of Physical Chemistry B*, Vol.110, No.33, pp.16159, ISSN 1520-6106.

Park, W.; et al. (2011). Enhancement in the photodetection of ZnO nanowires by introducing surface-roughness-induced traps. *Nanotechnology*, Vol.22, No.20, pp.205204, ISSN 0957-4484.

Hullavarad, S.; et al. (2007). Ultra violet sensors based on nanostructured ZnO spheres in network of nanowires: a novel approach. *Nanoscale Research Letters*, Vol.2, No.3, pp.161, ISSN 1931-7573.

Soci, C.; et al. (2010). Nanowire Photodetectors. *Journal of Nanoscience and Nanotechnology*, Vol.10, No.3, pp.1430, ISSN 1533-4880.

Pimentel, A.; et al. (2006). Role of the thickness on the electrical and optical performances of undoped polycrystalline zinc oxide films used as UV detectors. *Journal of Non-Crystalline Solids*, Vol.352, No.9-20, pp.1448, ISSN 0022-3093.

Xu, Z.-Q.; et al. (2006). Ultraviolet photoconductive detector based on Al doped ZnO films prepared by sol-gel method. *Applied Surface Science*, Vol.253, No.2, pp.476, ISSN 0169-4332.

Zheng, X.G.; et al. (2006). Photoconductive ultraviolet detectors based on ZnO films. *Applied Surface Science*, Vol.253, No.4, pp.2264, ISSN 0169-4332.

Jun, J.H.; et al. (2009). Ultraviolet photodetectors based on ZnO nanoparticles. *Ceramics International*, Vol.35, No.7, pp.2797, ISSN 0272-8842.

Liu, J.M.; et al. (2007). Effect of grain size on the electrical properties of ultraviolet photodetector with ZnO/diamond film structure. *Journal of Crystal Growth*, Vol.300, No.2, pp.353, ISSN 0022-0248.

Yang, W.; et al. (2003). Compositionally-tuned epitaxial cubic $Mg_xZn_{1-x}O$ on Si(100) for deep ultraviolet photodetectors *Applied Physics Letters*, Vol.82, pp.3424, ISSN 1077-3118

Yuan, B.; et al. (2011). High photosensitivity and low dark current of photoconductive semiconductor switch based on ZnO single nanobelt. *Solid-State Electronics*, Vol.55, No.1, pp.49, ISSN 0038-1101.

Jiang, D.; et al. (2011). Ultraviolet photodetectors with MgZnO nanowall networks grown by molecular beam epitaxy on Si(1 1 1) substrates. *Materials Science and Engineering: B*, Vol.176, No.9, pp.736, ISSN 0921-5107.

Soci, C.; et al. (2007). ZnO Nanowire UV Photodetectors with High Internal Gain. *Nano Letters*, Vol.7, No.4, pp.1003, ISSN 1530-6984.

Mamat, M.H.; et al. (2011). Fabrication of ultraviolet photoconductive sensor using a novel aluminium-doped zinc oxide nanorod-nanoflake network thin film prepared via ultrasonic-assisted sol-gel and immersion methods. *Sensors and Actuators A: Physical*, Vol.In Press, Corrected Proof, ISSN 0924-4247.

Su, Y.K.; et al. (2009). Ultraviolet ZnO Nanorod Photosensors. *Langmuir*, Vol.26, No.1, pp.603, ISSN 0743-7463.

Zheng, X.G.; et al. Photoconductive properties of ZnO thin films grown by pulsed laser deposition. *Journal of Luminescence*, Vol.122-123, pp.198, ISSN 0022-2313.

Lupan, O.; et al. (2010). Ultraviolet photoconductive sensor based on single ZnO nanowire. *physica status solidi (a)*, Vol.207, No.7, pp.1735, ISSN 1862-6319.

Chai, G.Y.; et al. (2011). Fabrication and characterization of an individual ZnO microwire-based UV photodetector. *Solid State Sciences*, Vol.13, No.5, pp.1205, ISSN 1293-2558.

Niarchos, G.; et al. (2010). Growth of ZnO nanorods on patterned templates for efficient, large-area energy scavengers. *Microsystem Technologies*, Vol.16, No.5, pp.669, ISSN 0946-7076.

Suslick, K.S.; et al. (1991). Sonochemical synthesis of amorphous iron. *Nature*, Vol.353, No.6343, pp.414, ISSN 0028-0836.

Mishra, P.; et al. (2010). Growth mechanism and photoluminescence property of flower-like ZnO nanostructures synthesized by starch-assisted sonochemical method. *Ultrasonics Sonochemistry*, Vol.17, No.3, pp.560, ISSN 1350-4177.

Jia, X.; et al. (2010). Using sonochemistry for the fabrication of hollow ZnO microspheres. *Ultrasonics Sonochemistry*, Vol.17, No.2, pp.284, ISSN 1350-4177.

Mazloumi, M.; et al. (2009). Ultrasonic induced photoluminescence decay in sonochemically obtained cauliflower-like ZnO nanostructures with surface 1D nanoarrays. *Ultrasonics Sonochemistry*, Vol.16, No.1, pp.11, ISSN 1350-4177.

Mamat, M.H.; et al. (2010). Novel synthesis of aligned Zinc oxide nanorods on a glass substrate by sonicated sol-gel immersion. *Materials Letters*, Vol.64, No.10, pp.1211, ISSN 0167-577X.

Mamat, M.H.; et al. (2010). Influence of doping concentrations on the aluminum doped zinc oxide thin films properties for ultraviolet photoconductive sensor applications. *Optical Materials*, Vol.32, No.6, pp.696, ISSN 0925-3467.

Lee, S.; et al. (2007). Fabrication of a solution-processed thin-film transistor using zinc oxide nanoparticles and zinc acetate. *Superlattices and Microstructures*, Vol.42, No.1-6, pp.361, ISSN 0749-6036.

Shinde, S.S. & Rajpure, K.Y. (2011). High-performance UV detector based on Ga-doped zinc oxide thin films. *Applied Surface Science*, Vol.257, No.22, pp.9595, ISSN 0169-4332.

Qiu, M.; et al. (2009). Growth and properties of ZnO nanorod and nanonails by thermal evaporation. *Applied Surface Science*, Vol.255, No.7, pp.3972, ISSN 0169-4332.

Mamat, M.H.; et al. (2011). Controllable Growth of Vertically Aligned Aluminum-Doped Zinc Oxide Nanorod Arrays by Sonicated Sol-Gel Immersion Method depending on Precursor Solution Volumes. *Japanese Journal of Applied Physics*, Vol.50, No.6, pp.06GH04, ISSN 1347-4065.

Khusaimi, Z.; et al. (2010). Controlled Growth of Zinc Oxide Nanorods by Aqueous-Solution Method. *Synthesis and Reactivity in Inorganic, Metal-Organic, and Nano-Metal Chemistry*, Vol.40, No.3, pp.190 ISSN 1553-3174.

Guo, M.; et al. (2005). Hydrothermal growth of well-aligned ZnO nanorod arrays: Dependence of morphology and alignment ordering upon preparing conditions. *Journal of Solid State Chemistry*, Vol.178, No.6, pp.1864, ISSN 0022-4596.

Chen, Y.W.; et al. (2009). Size-Controlled Synthesis and Optical Properties of Small-Sized ZnO Nanorods. *The Journal of Physical Chemistry C*, Vol.113, No.18, pp.7497, ISSN 1932-7447.

Giri, P.K.; et al. (2010). Effect of ZnO seed layer on the catalytic growth of vertically aligned ZnO nanorod arrays. *Materials Chemistry and Physics*, Vol.122, No.1, pp.18, ISSN 0254-0584.

Lupan, O.; et al. (2007). Nanofabrication and characterization of ZnO nanorod arrays and branched microrods by aqueous solution route and rapid thermal processing. *Materials Science and Engineering: B*, Vol.145, No.1-3, pp.57, ISSN 0921-5107.

Sugunan, A.; et al. (2006). Zinc oxide nanowires in chemical bath on seeded substrates: Role of hexamine. *Journal of Sol-Gel Science and Technology*, Vol.39, No.1, pp.49, ISSN 0928-0707.

Laudise, R.A. & Ballman, A.A. (1960). Hydrothermal synthesis of zinc oxide and zinc sulfide. *The Journal of Physical Chemistry*, Vol.64, No.5, pp.688, ISSN 0022-3654.

Laudise, R.A.; et al. (1965). Impurity content of synthetic quartz and its effect upon mechanical Q. *Journal of Physics and Chemistry of Solids*, Vol.26, No.8, pp.1305, ISSN 0022-3697.

De la Rosa, E.; et al. (2007). Controlling the Growth and Luminescence Properties of Well-Faceted ZnO Nanorods. *The Journal of Physical Chemistry C*, Vol.111, No.24, pp.8489, ISSN 1932-7447.

Lee, J.; et al. (2010). Improvement of optical properties of post-annealed ZnO nanorods. *Physica E: Low-dimensional Systems and Nanostructures*, Vol.42, No.8, pp.2143, ISSN 1386-9477.

Lin, B.; et al. (2001). Green luminescent center in undoped zinc oxide films deposited on silicon substrates. *Applied Physics Letters*, Vol.79, No.7, pp.943, ISSN 1077-3118

Li, Y.; et al. (2009). Fabrication of ZnO nanorod array-based photodetector with high sensitivity to ultraviolet. *Physica B: Condensed Matter*, Vol.404, No.21, pp.4282, ISSN 0921-4526.

Kim, K.-P.; et al. (2011). Thermal annealing effects on the dynamic photoresponse properties of Al-doped ZnO nanowires network. *Current Applied Physics*, Vol.11, No.6, pp.1311, ISSN 1567-1739.

Qin, L.; et al. (2011). Enhanced ultraviolet sensitivity of zinc oxide nanoparticle photoconductors by surface passivation. *Optical Materials*, Vol.33, No.3, pp.359, ISSN 0925-3467.

Fu, Y. & Cao, W. (2006). Preparation of transparent TiO_2 nanocrystalline film for UV sensor. *Chinese Science Bulletin*, Vol.51, No.14, pp.1657, ISSN 1001-6538.

Shinde, S.S. & Rajpure, K.Y. (2011). Fast response ultraviolet Ga-doped ZnO based photoconductive detector. *Materials Research Bulletin*, Vol.46, No.10, pp.1734, ISSN 0025-5408.

Yun, S. & Lim, S. (2011). Effect of Al-doping on the structure and optical properties of electrospun zinc oxide nanofiber films. *Journal of Colloid and Interface Science*, Vol.360, No.2, pp.430, ISSN 0021-9797.

Hsu, C.-H. & Chen, D.-H. (2010). Synthesis and conductivity enhancement of Al-doped ZnO nanorod array thin films. *Nanotechnology*, Vol.21, No.28, pp.285603, ISSN 0957-4484.

Mridha, S. & Basak, D. (2007). Aluminium doped ZnO films: electrical, optical and photoresponse studies. *Journal of Physics D: Applied Physics*, Vol.40, No.22, pp.6902, ISSN 0022-3727.

Fournier, C.; et al. (2008). Effects of substrate temperature on the optical and electrical properties of Al:ZnO films. *Semiconductor Science and Technology*, Vol.23, No.8, pp.085019, ISSN 0268-1242.

Sharma, B.K. & Khare, N. (2010). Stress-dependent band gap shift and quenching of defects in Al-doped ZnO films. *Journal of Physics D: Applied Physics*, Vol.43, No.46, pp.465402, ISSN 0022-3727.

6

Synthesis and Application of Nanorods

Babak Sadeghi
*Department of Chemistry, Tonekabon Branch,
Islamic Azad University, Tonekabon,
Iran*

1. Introduction

In nanotechnology, nanorods are morphology of nanoscale objects. Each of their dimension ranges from 1–100 nm. Nanorods may be synthesized from metals or semiconducting materials with ratios (length divided by width) are 3-5. One way for synthesis of nanorods is produced by direct chemical synthesis. The combinations of ligands act as shape control agents and bond to different facets of the nanorod with different strengths. This allows different faces of the nanorod to grow at different rates, producing an elongated object.

Gold nanorods are considered excellent candidates for biological sensing applications because the absorbance band changes with the refractive index of local material [1], allowing for extremely accurate sensing. In addition, nanorods with near-infrared absorption peaks can be excited by a laser at the absorbance band wavelength to produce heat, potentially allowing for the selective thermal destruction of cancerous tissues [2].

Nanoscale materials such as fullerenes, quantum dots and metallic nanoparticles have unique properties, because of their high surface area to volume ratio [3]. Gold nanospheres and nanorods also have unique optical properties, because of the quantum size effect [4]. Gold nanorods are cylindrical rods which range from less than ten to over forty nanometers in width and up to several hundred nanometers in length. These particles are typically characterized by their aspect ratio (length divided by width) [5-6].

In order to study and exploit the unique properties of nanorods, it is necessary to have a robust extinction coefficient which can predict the concentration of a solution at a particular absorbance. It is difficult to accurately obtain a measure of nanoparticle (as opposed to metal atom) concentration in moles per liter[2]. No spectroscopic device can provide concentration data, and only approximations are currently available.

Biomedical applications of nanoparticles require nanorods to be capped with biological molecules such as antibodies.

The El-Sayed method of nanorod concentration determination [7-8] is currently the standard way of measuring extinction coefficients (ε), and involves the coupling of bulk gold concentration, Transmission Electron Microscopy (TEM) size analysis and absorbance data. Recently, Liao and Hafner calculated ε values of nanorods by preparing films of immobilized nanorods [2]. Liao and Hafner note that spherical byproducts lower the extinction coefficient calculated by the El-Sayed method, but they do not discuss the sensitivity of ε to this kind of error.

Synthesis of gold nanorods has recently undergone dramatic improvements. It is possible to produce high yields of nearly monodispersed short gold nanorods [6-7,9]. The rods synthesized for this chapter were synthesized using Murphy's method [9]. First, a "seed" solution of spherical gold nanoparticles was prepared by adding the following:

Reagents	Quantity(ml)
0.01M $HAuCl_4.3H_2O$	0.250
0.1 M CTAB (cetyltrimethylammonium bromide)	7.5
0.01 M $NaBH_4$	0.6

Table. 1. Preparation of Gold seed

$NaBH_4$ reduces the gold salt to form nanoparticles, and cetyltrimethylammonium bromide (CTAB) (Fig.1) is a surfactant which stabilizes the seeds to prevent aggregation. To make nanorods, the reagents in Table.1 are added in order from top to bottom. Gold rods are synthesized with a small amount of silver to control rod size and make short rods [7,9]. CTAB is a directing surfactant; without it, only spheres would form. CTAB forms a rod shaped template that is filled with gold atoms as they are reduced by ascorbic acid. Ascorbic acid is a weaker reducing agent than $NaBH_4$, but in the presence of seeds and a CTAB template, it reduces gold ions at the seed surface [6].

2. Synthesis of nanorods

General idea is the same as the growth of nanorods (seed-mediated method)

To make surfactant-coated nanorods, the well-documented seed-mediated procedure developed by was employed. This method yields nanorods that are stabilized as verified from transmission electron microscopy (TEM) analysis [10].

Slightly change the conditions when growing nanorods (concentration of different reactants)

The nanorods were coated with a very thin (ca. 3-5 nm) silica film that

i. improved the colloidal stability of the nanorods by reducing aggregation,
ii. improved the shape stability of the nanorods, and
iii. allowed for further modification of the nanorod surface.

This silication method was first developed for citrate-stabilized (Fig.2.) gold NPs,[11-13] and has been applied successfully to gold nanorods[14-19].

Cubes, hexagon, triangle, tetropods, branched

A two-step growth method has been developed to grow nanorods by changing the oxygen content in gas mixture during nucleation and growth steps. This is based on our systematic studies of nucleation and growth under different conditions. Due to the large lattice mismatch (~;18%) between molecule and sapphire, the nucleation of molecule on sapphire follows the three-dimensional island growth; that is, the Volmer–Weber mode, as reported by Yamauchi et al. in their observation of plasma-assisted epitaxial growth of molecule on sapphire [20,21]. At high temperature, the nucleation of nanorods islands on the surface of substrates depends strongly on the amount of active oxygen. When grown entirely in 90% oxygen plasma, nanorods has a high nucleation density and forms.

Direct chemical synthesis and a combination of ligands are all that are required for production and shape control of the nanorods. Ligands also bond to different facets of the nanorod with varying strengths. This is how different faces of nanorods can be made to grow at different rates, thereby producing an elongated object of a certain desired shape.

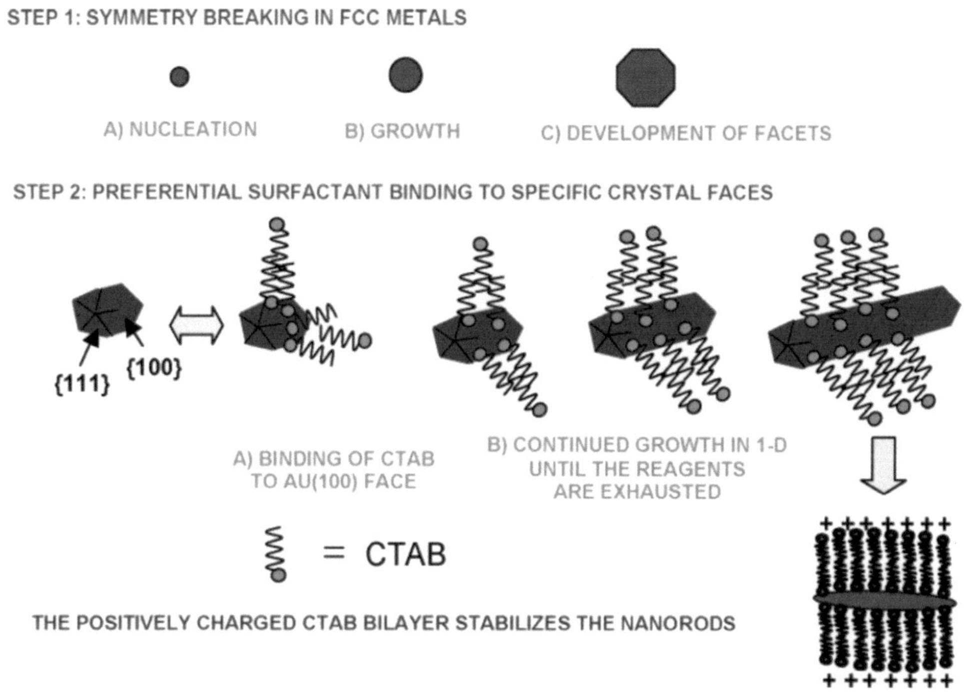

Fig. 1. Growth mechanism of nanorods

We synthesized silver nanorods with the average length of 280 nm and diameters of around 25 nm were synthesized by a simple reduction process of silver nitrate in the presence of polyvinyl alcohol (PVA) and investigated by means of scanning electron microscopy (SEM), X-ray diffraction (XRD), transmission-electron microscopy (TEM).

It was found out that both temperature and reaction time are important factors in determining the morphology and aspect ratios of nanorods (Fig.2.). TEM images showed the prepared silver nanorods have a face centered shape (fcc) with fivefold symmetry consisting of multiply twinned face centered cubes as revealed in the cross-section observations. The five fold axis, i.e. the growth direction, normally goes along the (111) zone axis direction of the basic fcc Ag-structure. Preferred crystallographic orientation along the (111), (200) or (220) crystallographic planes and the crystallite size of the silver nanorods are briefly analyzed [22-26].

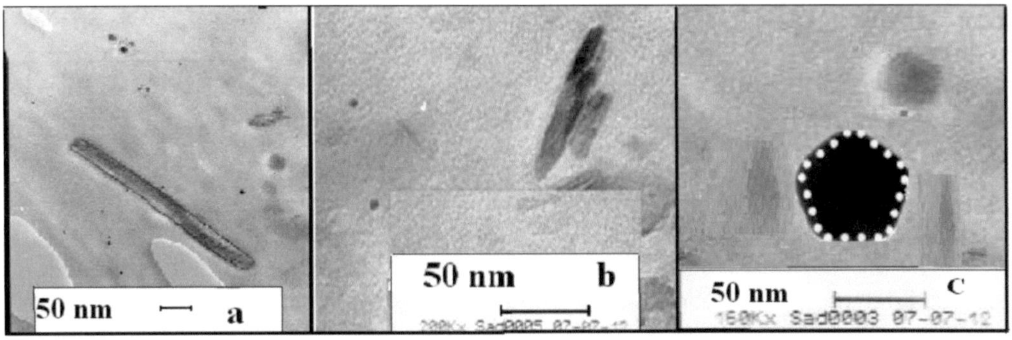

Fig. 2. Transmission electron microscopy (TEM) images of (a,b) individual and (c) cross section of Ag/PVA nanorods.

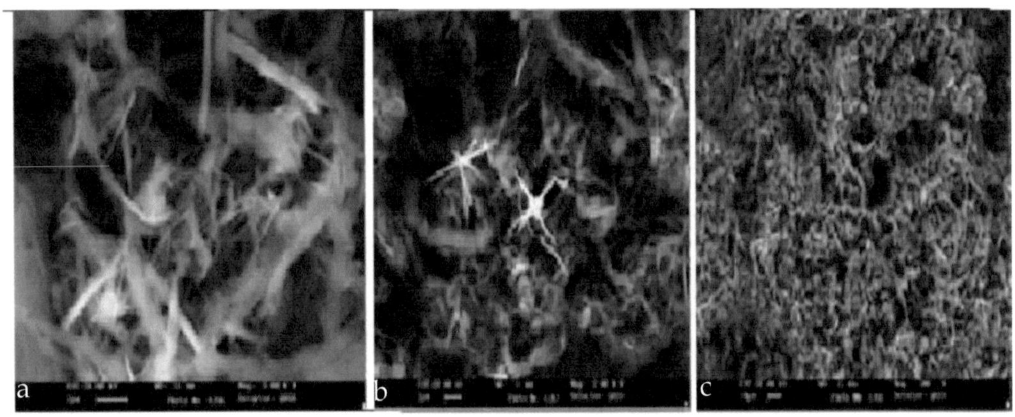

Fig. 3. (a,b,c) SEM image showing high concentrated distribution of Ag/PVA nanorods.

In our research, synthesis of silver nanorods with the controllable dimensions was described by using a reducing agent that involves the reduction of silver nitrate with N-N'-Dimethyl formamide (DMF) in the presence of PVA as a capping reagent. In this process the DMF is served as both reductant and solvent [27, 28]. SEM and TEM observations (fig.2,3) along a series of relevant directions show that the silver nanorods have an average length of 280 nm and diameters of around 25 nm. TEM observations from cross section of nanorods show that the transformation of silver nanospheres to silver nanorods is achieved by the oriented attachment of several spherical particles followed by their fusion. Resulted Ag-nanorods have a twinned fcc structure, appeared in a pentagonal shape with fivefold twinning. The fivefold axis, i.e. the growth direction, normally goes along the (110) zone axis direction of the fcc cubic structure. The XRD data (fig.4) confirmed that the silver nanorod is crystalline with fcc structure with a preferred crystallographic orientation along the (220) direction and straight, continuous, dense silver nanorod has been obtained with a diameter 25 nm. [22].

On the other hand, gold nanorods were prepared by adopting a photochemical method that employs UV-irradiation to facilitate slow growth of rods Tetraoctylammonium bromide as a co-surfactant used instead of tetradodecylammonium bromide. The growth solution was prepared by dissolving 440 mg of cetyltrimethylammonium bromide (CTAB) and 4.5 mg of tetraoctylammonium bromide (TOAB) in 15 mL of water and transferred to a cylindrical quartz tube (length 15 cm and diameter 2 cm). To this solution, 1.25 mL of 0.024 M $HAuCl_4$ solution was added along with 325 μL of acetone and 225 μL of cyclohexane. The formation of the gold nanorod and its aspect ratio was confirmed from Transmission Electron Microscopic analysis. A drop of a dilute solution of Au nanorods was allowed to dry on a carbon coated copper grid and then probed using a JEOL JEM-100sx electron microscope. The average length and diameter of rods employed in the present investigation are 50.0 nm and 20.0 nm, respectively and an average aspect ratio of 2.5 [29].

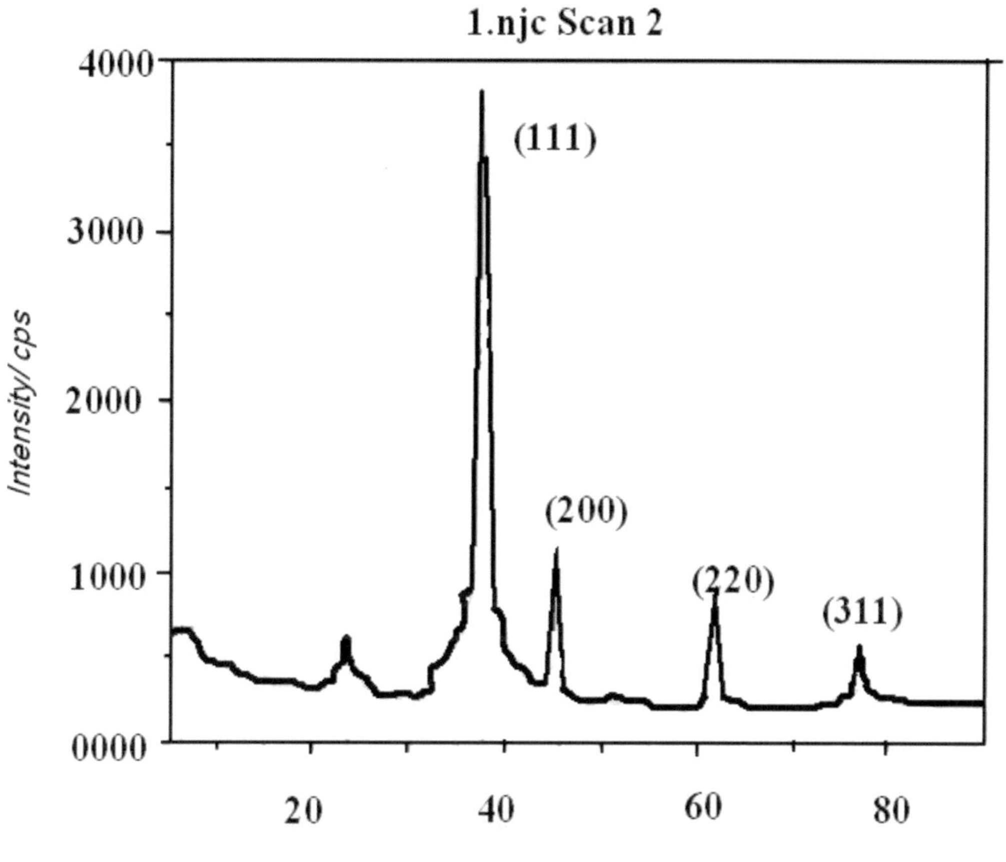

Fig. 4. XRD pattern of Ag/PVA nanocomposite indicative the face-centered cube structure.

Changes in nanorod dimensions have the greatest impact on extinction coefficients. Fig.5(a) shows the change in the Beer's law plot with increasing polydispersity. As variation in the rod dimensions increases, the extinction coefficient increases. However, these increases are not excessive - from 3.85 ± 0.07 to 4.90 ± 0.08 x 10^{-8} M is not significant, and errors in concentration resulting from this magnitude of change in ε would not be enough to cause aggregation or substantial miscalculations of antibodies. However, the effect of spherical byproducts is more pronounced, and the presence of spheres can alter the extinction coefficient enough to cause concern.

Fig. 5(b) shows that the extinction coefficient nearly doubles as percentage of spherical by products is increased from 5% to 40%. Spherical byproducts are an ongoing challenge in nanorods synthesis, and it is important to consider how they can change the extinction coefficient.

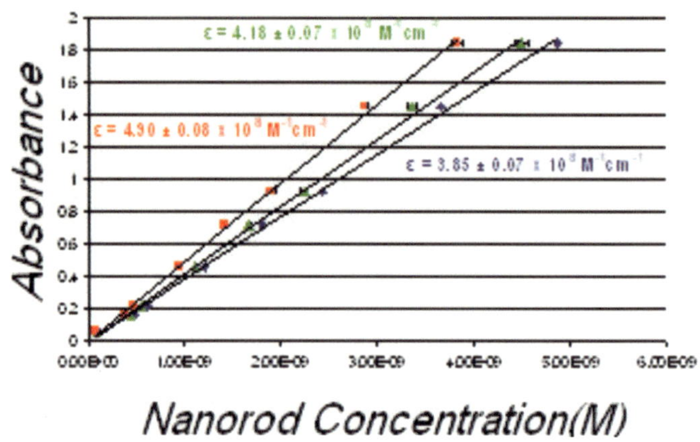

a

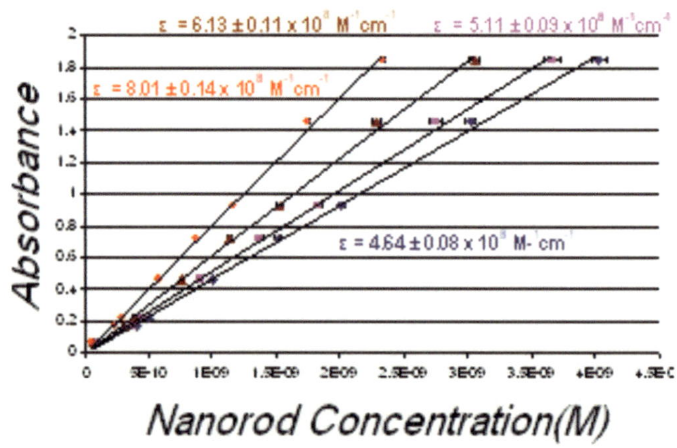

b

Fig. 5. Effect of error as (a) polydispersity of particles change and (b) number of spherical byproducts increases. In both cases, total gold concentration was held constant. Error bars are very narrow.

The ranges of extinction coefficients found in this model are quite different from the values calculated by [2] and [6], which were on the order of 109 M. However, the rods used in these calculations, while of approximately the same aspect ratio, had dimensions of 50 x 15 nm instead of 30 x 8 nm. The same aspect ratio implies the same peak wavelength, but can result in a different rod volume (and thus different concentration measurements). Small changes in aspect ratio can substantially change absorbance peak location [7]. Literature values must be used cautiously in nanorod studies.

4. Application of nanorod

Nanorods have wide application.

4.1 Nanorods for dye solar cells

The dye-solar cell (DSC or Grätzel) first presented in 1991 offers an interesting paradigm with regards the generation of electricity directly from sunlight via the photovoltaic effect. In essence the DSC is not a definite structure but more a design philosophy to mimicking nature in the conversion of solar energy [30]. The DCS is formed by an electrode, preferably with a large internal surface area onto which is attached a light absorbing dye. The dye upon absorption of a photon by photo-excitation of an electron (moving from the HOMO to the LUMO levels) will, if in favourable conditions, inject the photo-excited electron into the supporting structure. The dye is regenerated by a supporting Reduction & Oxidation (REDOX) electrolyte (or hole conducting semiconductor) which permeates the working electrode [31]. One of the many components of the DSC that can be altered is the working electrode. General requirements are that it be porous (i.e. large internal surface area) and be n-type semiconducting. Several metal-oxides fulfil these requirements (e.g., TiO_2, ZnO, WO_3) [32].

There are several reports describing the electrodeposition of ZnO with various conditions explored resulting in a variety of geometries as for e.g., nanorods, nanoneedles, nanotubes, nanoporous and compact layers [33]. In this work, a high-density vertically aligned ZnO nanorods arrays, Fig. 6, were prepared on fluorine doped SnO_2 (FTO) coated glass substrates, prepared at 70 °C from a neutral zinc nitrate solution, varying the deposition time.

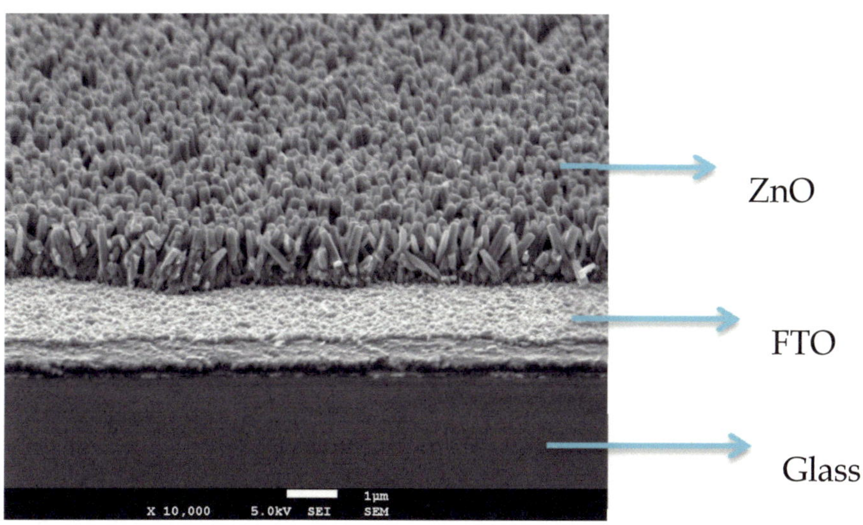

Fig. 6. Cross section view of ZnO nanorod arrays electrodeposited on FTO glass

4.2 The use of nanorods for oligonucleotide detection

Functionalization of nanoparticles with biomolecules (antibodies, nucleic acids, etc.) is of interest for many biomedical applications [34]. One of the most perspectives is application of gold nanorods (GNRs) for detecting target sequences of infecting agents of many dangerous diseases, for example, a HIV-1 [35]. The method is based on electrostatic interaction between GNRs and DNA molecules. As single-stranded DNA molecules are zwitterions, and double-stranded are polyanions, their affinity towards polycationic GNR stabilizer (cetiltrimetilammoniumbromide, CTAB) is various. The GNRs plasmon-resonant [7-8] labels are functionalized with probe single-stranded oligonucleotides by physical adsorption or by chemical attachment. By adding complement targets to the GNRs-probe conjugate solution a formation of aggregates is observed. It has been shown recently [35] that GNRs aggregation, induced by DNA-DNA hybridization on their surface, can be detected by extinction and scattering spectroscopy techniques. We have found that the characteristic parameter of biospecific interaction is change of amplitude and differential light scattering method for detection DNA-DNA interaction. This work is aimed at study of aggregation properties of single-stranded probe DNA-GNRs conjugates as applied to biospecific detection of target oligonucleotides (fig.7).

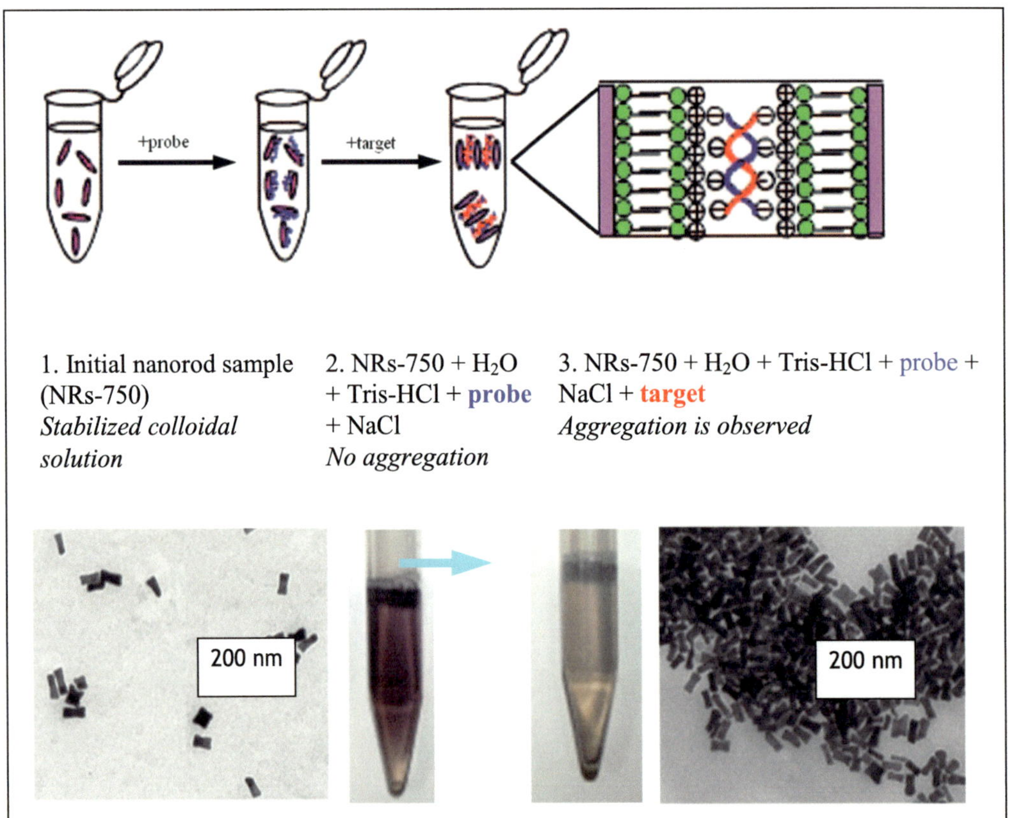

Fig. 7. Experimental diagram for oligonucleotide detection

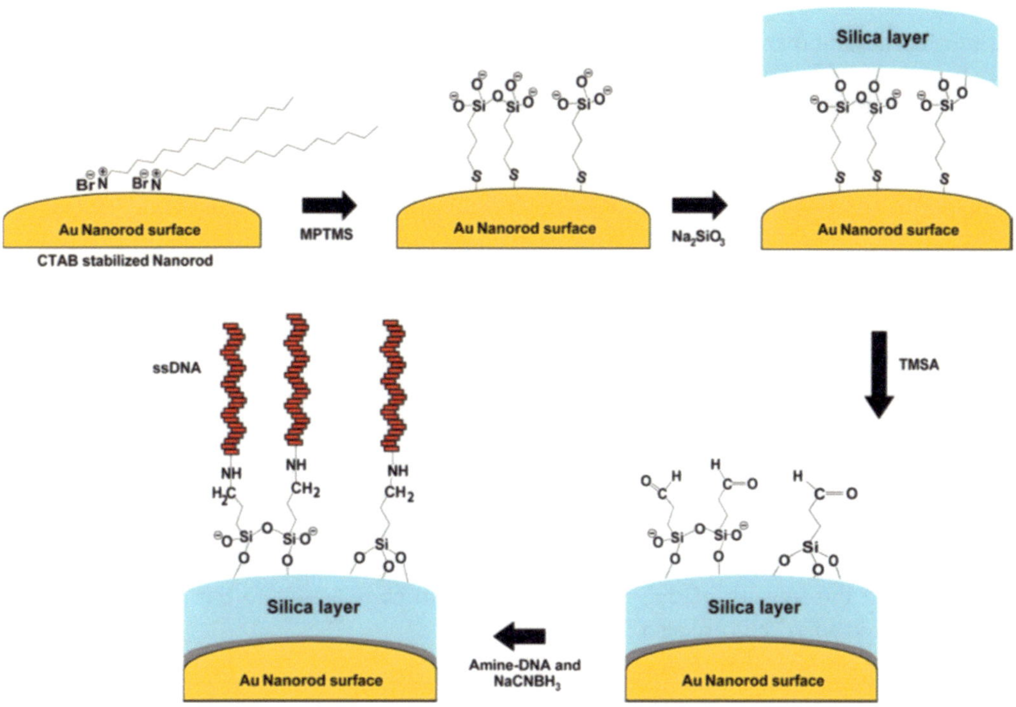

Fig. 8. Schematic diagram of the surface reactions employed to obtain DNA-functionalized gold nanorods.

In Fig.8, First the nanorods were coated with a thin silica layer using the silane coupling agent, MPTMS, followed by reaction with sodium silicate. Surface-bound aldehyde functional groups were attached to the silica film by using TMSA and then used to conjugate the amine-modified DNA in a reductive amination reaction.

GNRs were produced by seed-mediated growth method in presence of CTAB. Extinction and scattering spectra were measured in the wavelength range from 450 to 900 nanometers which includes the transversal and longitudinal plasmon resonance [7-8] bands of GNRs. The particle and aggregate average size were measured by the dynamic light scattering method (DLS), and the particle shape and cluster structure were examined by transmission electron microscopy (TEM). 21-mer oligonucleotide complement pair was taken as a biospecific pair, and the target sequence was related to human immunodeficiency virus-1 (HIV-1) genome. In this experiments demonstrated that reproducibility of aggregation test depends on the GNRs synthesis protocol. In particularly, the use of protocol [36] led to unsatisfactory results. According to TEM data, the method realization was successful while using «dog-bone» morphology particles.

The functionalized gold nanorod synthesis can be divided into three main steps:

- Gold Nanorod Fabrication.
- Silica Shell Formation.
- DNAFunctionalization.

4.3 The use of nanorods for applied electric field

Prominent among them is in the use in display technologies. By changing the orientation of the nanorods with respect to an applied electric field, the reflectivity of the rods can be altered, resulting in superior displays. Picture quality can be improved radically. Each picture element, known as pixel, is composed of a sharp-tipped device of the scale of a few nanometers. Such TVs, known as field emission TVs, are brighter as the pixels can glow better in every color they take up as they pass though a small potential gap at high currents, emitting electrons at the same time.

Nanorod-based flexible, thin-film computers can revolutionize the retail industry, enabling customers to checkout easily without the hassles of having to pay cash.

The semiconductor nanorod structure is based on a junction between nanorod structure and another window semiconductor layer for solar cell application. The possibility of band gap tuning by varying the diameter of the nanorods along the length, higher absorption coefficient at nanodimensions, the presence of a strong electrical field at the nanorod-window semiconductor nanojunctions and the carrier confinement in lateral direction are expected to result in enhanced absorption and collection efficiency in the proposed device. Process steps, feasibility, technological tasks needed for the realization of the proposed structure and the novelty of the present structure in comparison to the already reported nanostructured solar cells are also discussed [37].

4.4 The use of nanorods for applied humidity sensitive

ZnO nanorod and nanowire films were fabricated on the Si substrates with comb type Pt electrodes by the vapor-phase transport method, and their humidity sensitive characteristics have been investigated. These nanomaterial films show high-humidity sensitivity, good long-term stability and fast response time. It was found that the resistance of the films decreases with increasing relative humidity (RH). At room temperature (RT), resistance changes of more than four and two orders of magnitude were observed when ZnO nanowire and nanorod devices were exposed, respectively, to a moisture pulse of 97% relative humidity. It appears that the ZnO nanomaterial films can be used as efficient humidity sensors [38]. The gas sensor fabricated from ZnO nanorod arrays showed a high sensitivity to H_2 from room temperature to a maximum sensitivity at 250 °C and a detection limit of 20 ppm. In addition, the ZnO gas sensor also exhibited excellent responses to NH_3 and CO exposure. Our results demonstrate that the hydrothermally grown vertically aligned ZnO nanorod arrays are very promising for the fabrication of cost effective and high performance gas sensors [39].

To use nanorods in biomedical applications, it is advised that samples are produced in bulk and ε is calculated for each batch. In addition, Nanorod aggregation, induced by biospecific interaction, was shown by four methods (extinction and scattering spectroscopy, DLS, TEM).

5. Conclusion

The mechanism of the nanorod formation is not yet completely understood. However, it has now been demonstrated that the necessary requirements are at least (i) a mild reducing agent, (ii) a protecting agent (for example consisting of a solution of PVA in DMF), and (iii) the presence of metal ions in solutions. It is found that both temperature and reaction time are important factors in determining the morphology and aspect ratios of nanorods. Lower temperature and longer time are favorable to form polycrystalline metal nanorods with high uniformity and aspect ratios. In addition to the nanoparticles with controllable size, silver nanorods were successfully synthesized by converting nanoparticles at a relatively low temperature of 80°C. We believe that the detailed investigation of silver nanorods [22] will also help us to synthesize other metal nanorod more easily by using the similar technique.

6. Acknowledgment

The financial and encouragement support provided by the Research vice Presidency of Tonekabon Branch, Islamic Azad University and Executive Director of Iran-Nanotechnology Organization (Govt. of Iran).

7. References

[1] J. Pérez-Juste, I. Pastoriza-Santos, L. M. Liz-Marzàn, and P. Mulvaney. "Gold nanorods: Synthesis, characterization and applications", Coordination Chemistry Reviews, vol. 249, 2005, pp. 1870-1901.
[2] H. Liao and J. H. Hafner. "Gold Nanorod Bioconjugates". Chemistry of Materials, vol. 17, 2005, pp. 4636-4641.
[3] S. Link and M. A. El-Sayed. "Shape and size dependence of radiative, non-radiative and photothermal properties of gold nanocrystals" International Reviews in Physical Chemistry, vol. 19, 2000, pp. 409-453.
[4] M.-C. Daniel and D. Astruc. "Gold Nanoparticles: Assembly, Supramolecular Chemistry, Quantum-Size-Regulated Properties, and Applications toward Biology, Catalysis, and Nanotechnology" Chemical Reviews, vol. 104, 2004, pp. 293-346.
[5] C. J. Murphy and N. R. Jana. "Controlling the Aspect Ratio of Inorganic Nanorods and Nanowires" Advanced Materials, vol. 14, 2002, pp. 80-82.
[6] C. J. Murphy, T. K. Sau, A. M. Gole, C. J. Orendorff, J. Gao, L. Gou, S. E. Hunyadi, and T. Li. "Anisotropic Metal Nanoparticles: Synthesis, Assembly, and Optical Applications", Journal of Physical Chemistry B, vol. 109, 2005, pp. 13857-13870.
[7] S. Link and M. A. El-Sayed. "Spectral Properties and Relaxation Dynamics of Surface Plasmon Electronic Oscillations in Gold and Silver Nanodots and Nanorods", Journal of Physical Chemistry B., vol. 103, 1999, pp. 8410-8426.
[8] B. Nikoobakht, J. Wang and M. A. Al-Sayed. "Surface-enhanced Raman scattering of molecules adsorbed on gold nanorods: off-surface plasmon resonance condition", Chemical Physics Letters, vol. 366, 2002, pp. 17-23.
[9] T. K. Sau and C. J. Murphy. "Seeded High Yield Synthesis of Short Au Nanorods in Aqueous Solution", Langmuir, vol. 20, 2004, pp. 6414-6420.
[10] Nikoobakht, B.; El-Sayed, M. A. Chem. Mater. 2003, 15, 1957–1962.
[11] Liz-Marz_an, L. M.; Giersig, M.; Mulvaney, P. Chem. Commun. 1996, 6, 731- 732.
[12] Liz-Marz_an, L. M.; Giersig, M.; Mulvaney, P. Langmuir 1996, 12, 4329- 4335.
[13] Roca, M.; Haes, A. J. J. Am. Chem. Soc. 2008, 130, 14273–14279.
[14] Obare, S. O.; Jana, N. R.; Murphy, C. J. Nano Lett. 2001, 1, 601–603.
[15] Perez-Juste, J.; Correa-Duarte, M. A.; Liz-Marzan, L. M. Appl. Surf. Sci. 2004, 226, 137–143.
[16] Zhang, J. J.; Liu, Y. G.; Jiang, L. P.; Zhu, J. J. Electrochem. Commun. 2008, 10, 355–358.
[17] Wang, G. G.; Ma, Z. F.; Wang, T. T.; Su, Z. M. Adv. Funct. Mater. 2006, 16, 1673–1678.

[18] Omura, N.; Uechi, I.; Yamada, S. Anal. Sci. 2009, 25, 255–259.
[19] Niidome, Y.; Honda, K.; Higashimoto, K.; Kawazumi, H.; Yamada, S.; Nakashima, N.; Sasaki, Y.; Ishida, Y.; Kikuchi, J. Chem. Commun. 2007, 3777- 3779.
[20] S. Yamauchi, H. Handa, A. Nagayama, and T. Hariu, Thin Solid Films 345, 12 (1999).
[21] S. Yamauchi, T. Ashiga, A. Nagayama, and T. Hariu, J. Cryst. Growth 214/215, 63 (2000).
[22] M.A.S. Sadjadi, Babak Sadeghi, M. Meskinfam, K. Zare, J. Azizian. "Synthesis and characterization of Ag/PVA nanorods by chemical reduction method" Physica E: Low-dimensional Systems and Nanostructures, Vol.40, Issue 10,P 3183-3186, September 2008.
[23] Babak Sadeghi, M.A.S. Sadjadi, A.Pourahmad "Effects of protective agents (PVA & PVP) on the formation of silver nanoparticles" Internatinal journal of Nanoscience and Nanotechnology (IJNN), Vol.4.No.1,P 3-11, December 2008.
[24] Babak Sadeghi, M.A.S.Sadjadi, R.A.R.Vahdati, "Nanoplates controlled synthesis and catalytic activities of silver nanocrystals"Superlattices and Microstructures, Vol.46, Issue 6, P 858-863, December 2009.
[25] Babak Sadeghi, Afshin Pourahmad, "Synthesis of silver/poly (diallyldimethylammonium chloride) hybrid nanocomposite", Advanced Powder Technology, Vol.22, Issue.5, P 669-673, 2011.
[26] Babak Sadeghi, Sh.Ghammamy, Z.Gholipour, M.Ghorchibeigy, A.Amini Nia, " Gold/HPC hybrid nanocomposite constructed with more complete coverage of gold nano-shell " , Mic & Nano Letters, Vol.6, Issue 4 , P 209-213, 2011.
[27] P.K.Khanna, N.Singh, S.Charan, V.V.V.S.Subbarao, R.Gokhale, U.P.Mulik, Mater.Chem and Phys.2005; 93:117-121.
[28] I.Pastoriza-Santoz, M.Luiz, L.Marzan, Nano lett.2002; 2:903-905.
[29] Kim, F.; Song, J. H.; Yang, P. D. J. Am. Chem. Soc. 2002, 124, 14316.
[30] B. O'Regan, M. Grätzel, Nature 335 (1991) 737;
[31] Q. Zhang, G. Cao, Nano Today 6 (2011) 91;
[32] M. Grätzel, J. Photochem. Photobiol. C: Photochem. Rev. 4 (2003) 145;
[33] O. Lupan, V. M. Guérin, I. M. Tiginyanu, V. V. Ursaki, L. Chow, H. Heinrich, T. Pauporté, J. Photochem. Photobiol. A: Chem. 211 (2010) 65.
[34] Rosi N.L., Mirkin C.A. Nanostructures in biodiagnostics // Chem. Rev. 2005. V. 105, № 4. P. 1547–1562.
[35] He W., Huang C.Z., Li Y.F., Xie J.P., Yang R.G., Zhou P.F., Wang J. One-step label-free optical genosensing system for sequence-specific DNA related to the human immunodeficiency virus based on the measurements of light scattering signals of gold nanorods // Anal. Chem. 2008. V. 80, № 22. P. 8424–8430.
[36] He W., Huang C.Z., Li Y.F., Xie J.P., Yang R.G., Zhou P.F., Wang J. One-step label-free optical genosensing system for sequence-specific DNA related to the human immunodeficiency virus based on the measurements of light scattering signals of gold nanorods // Anal. Chem. 2008. V. 80, № 22. P. 8424–8430.
[37] B. R. Mehta, F. E. Kruis. A graded diameter and oriented nanorod–thin film structure for solar cell application: a device proposal Solar Energy Materials and Solar Cells, Volume 85, Issue 1, 1 January 2005, Pages 107-113.
[38] C.M. Chang, M.H. Hon, I.C. Leu .Preparation of ZnO nanorod arrays with tailored defect-related characterisitcs and their effect on the ethanol gas sensing performance Sensors and Actuators B: Chemical, Volume 151, Issue 1, 26 November 2010, Pages 15-20.[35] Yongsheng Zhang, Ke Yu, Desheng Jiang, Ziqiang Zhu, Haoran Geng, Laiqiang Luo . Zinc oxide nanorod and nanowire for humidity sensor. Applied Surface Science, Volume 242, Issues 1-2, 31 March 2005, Pages 212-217.

7

Charge Transfer Within Multilayered Films of Gold Nanorods

Mariana Chirea, Carlos M. Pereira and A. Fernando Silva
University of Porto, Faculty of Sciences, Chemistry and Biochemistry Department, Porto, Portugal

1. Introduction

Controlling the charge transport across hybrid nanostructures composed of metal nanoparticles, carbon nanonostructures or quantum dots is very important for developing functional sensing, optoelectronic, and photovoltaic devices (Kamat, 2008; Katz & Willner, 2004). The unique electronic, optical, and catalytic properties of metal nanoparticles (1–100 nm), together with the large variety of the synthetic methods available for their synthesis with precise control over the shape and size, provide exciting possibilities for the fabrication of nanoscale assemblies, structures, and devices (Feldheim & Foss Jr, 2004). The self-assembly of nanoparticles into hybrid nanostructures is achieved by using functional molecular linkers which exhibit specific interactions with both the nanoparticles and the substrate. The most stable hybrid nanostructures can be built based on covalent interactions. For example, self-assembled monolayers (SAMs) of aliphatic chain thiols form well defined films on gold surfaces based on spontaneous covalent bonding to the substrate (Love et al., 2005). The charge transfer through these types of molecular linkers is characterized by a decay of the tunnelling current with the length of the linker, the tunnelling coefficient being around $\beta \approx 0.8 Å^{-1}$ (Adams et al., 2003; Finklea, 1987, 1996). However, recent studies demonstrate that the tunnelling decay is not the dominant property of the films when the nanomaterials are tethered at the end of the molecular bridges determining a fast electron transfer process and moreover, this fast electron transfer process is dependent on the size and the shape of the nanomaterials (Bradbury et al., 2008; Chirea et al, 2009, 2010; Kissling et al., 2010). A more detailed research has to be developed in order to fully understand the mechanism behind this fast electron transfer processes taking place at electrodes modified with hydrid nanostructures. Various electrochemical techniques have been use for the investigation of the electron transfer processes taking place at hybrid nanostructured modified electrodes. Cyclic voltammetry, electrochemical impedance spectroscopy, square wave voltammetry, differential pulse voltammetry, potential step chronoamperometry and rotating disk electrodes are among the most used techniques for these electrochemical studies (Chen &Pei, 2001; Chirea et al, 2005, 2007; Hicks et al., 2002, Horswell et al.,2003). Electrochemical impedance spectroscopy (EIS) has been proven as one of the most powerful tools for the investigation of interfacial electrode reaction mechanisms (Katz & Willner, 2003, Yan & Sadik, 2001, Lasia, A. 1999). EIS measures the response (current and its phase) of an electrochemical system to an applied oscillating potential as a function of the frequency. The electrochemical behaviour of redox probes at hybrid nanostructures modified electrodes is rather complex and depends not only on the nature of the outermost layer but also on the composition of the whole nanostructured film. The EIS technique allows the accurate evaluation of electron transfer kinetics and mass transport despite the complexity of the systems studied (Katz & Willner, 2004 and references therein). In this chapter, it will be presented the fabrication procedure and the electrochemical properties of hybrid nanostructures composed of anisotropic gold nanomaterials, namely gold nanorods (AuNRs) and 1,6hexanedithiol (1,6HDT) which was used as a molecular bridge to covalently bond the rods to the electrodes. The gold nanorods were synthesized based on a seed mediated method using cetyltrimetylammonium bromide as the stabilizing and shape inducing agent (Nikoobakht &El-Sayed, 2003). Control over the size of the gold nanorods was achieved by varying the volume of silver nitrate used during the growth step of these nanomaterials. Multilayered films of gold nanorods and 1,6hexanedithiol were built-up on clean gold electrodes based on the layer–by-layer self-assembly method (Decher & Schlenoff, 2003). Quartz crystal microbalance (QCM), ellipsometry and atomic force microscopy (AFM) measurements were performed in order to determine the self-assembly time of AuNR and 1,6HDT layers, the films' thicknesses, the type of surface bonding of the AuNRs and the films' topographies. The charge transport mechanism through these films was studied as a function of the type of the outermost layer, number of layers within the multilayered films and size of the gold nanorods using $[Fe(CN)_6]^{3-/4-}$ as redox probes. Cyclic voltammetry, square wave voltammetry and electrochemical impedance spectroscopy were used for the electrochemical characterization of the film modified electrodes. Accurate fittings of the EIS spectra were obtained by using nonlinear square fit software. An excellent agreement between the experimental data and the fitted data has been obtained allowing an accurate correlation of the electrochemical parameters.

2. Experimental section

2.1 Chemicals

Hydrogen tetrachloroaurate (III) trihydrate ($HAuCl_4{\cdot}3H_2O$, 99,999 %, Sigma Aldrich), cetyltrimetyl ammonium bromide ($C_{19}H_{42}NBr$, Sigma, 99%), silver nitrate ($AgNO_3$ pa quality, Riedel-de Haën), sodium borohydride ($NaBH_4$, 96%, Sigma Aldrich), L-ascorbic acid ($C_6H_8O_6$, 99%, Aldrich), 1,6 hexanedithiol ($C_6H_{14}S_2$, Fluka, 97%) methanol (CH_3OH, 99.8%, Sigma), potassium hexacyanoferrate (III), potassium hexacyanoferrate(II) (pa quality, Merck, $NaClO_4$ (pa quality, Merck), $HClO_4$ (70%, redistilled, 99.999%, Aldrich), H_2O_2 30% (Fluka), H_2SO_4 (pure, Pronalab) were used as received. Millipore filtered water (resistivity > 18 MΩ·cm) was used to prepare all aqueous solutions and for rinsing. Prior use, all the glassware were cleaned with freshly prepared *aqua regia* (HNO_3: HCl =1:3, % v/v), rinsed abundantly with Millipore water and dried.

2.2 Synthesis of 2.44 and 3.48 aspect ratios gold nanorods

The gold nanorods were prepared following the procedure published in the literature (Nikoobakht, & El-Sayed, 2003). First, spherical gold nanoparticles were synthesized by reducing $HAuCl_4$ (aqueous solution of 5.0 mL, 0.00050M) with ice-cold $NaBH_4$ solution (0.60 mL, 0.01M) under strong stirring in the presence of cetyltrimetylammonium bromide (5 mL aqueous solution, 0.20 M). The resulted seeds were kept for 2 hours at 25°C. Consecutively, growth solutions were prepared by adding ascorbic acid (70 μL of 0.0788 M) into two solution mixtures composed of cetyltrimetylammonium bromide aqueous solution (5 mL, 0.20 M), 0.050 or 0.15 mL of $AgNO_3$ solution (0.0040 M) respectively, and $HAuCl_4$ (5.0 mL of 0.0010 M). To these mixtures 12 μL of the prepared seed solution were added at 30°C solution temperature, which determined the growth of the nanorods proved by the apparition of the specific colour for each sample within 10 minutes. The short rods featured a blue colour in aqueous solution whereas the large rods featured a pink-brownish colour (Figure 1a and 1d). The gold nanorods were purified by washing with Millipore water, centrifugation and decantation. This procedure of purification was performed twice.

2.3 Transmission electron microscopy measurements

A drop of each rod solution was cast on formvar copper carbon grids, let to dry for at least 24 hours and imaged. The TEM images were recorded using a transmission electron microscope Hitachi 8100 equipped with a Rontec Standard EDS detector and digital images acquisition, operating at 200 kV and having a point resolution of 1.6 nm. The length/width ratio of the rods was evaluated using Image J software.

2.4 UV-Vis spectroscopy measurements

The optical spectra of the AuNRs samples were recorded on a Hitachi U-3000 spectrophotometer in the range 200-900 nm, using quartz cuvettes with 1cm light path and freshly prepared Au NRs solutions.

2.5 Fabrication of 1,6hexanedithiol/gold nanorod multilayers

2.5.1 Quartz Crystal Microbalance measurements

The QCM measurements were performed with a QCM-Z500 apparatus (KSV, Finland) which can monitor ΔF changes for six overtone frequencies (5, 15, 25, 35, 45 and 55MHz). The QCM electrodes (gold coated quartz crystals, 0.785 cm^2 area) were washed with freshly prepared Piranha solution (3:1 mixture of concentrated sulphuric acid and hydrogen peroxide, 30%, *Caution! Piranha solution is corrosive and reacts violently with organic materials*) rinsed with large amounts of Millipore water, dried in a stream of nitrogen and placed in the QCM cell for surface modification. The base line was recorded in the presence of ethanol. Consecutively, the pure solvent was gently removed from the QCM cell and an ethanolic solution of 1,6 hexanedithiol (15 mM) was injected into the cell. The frequency change was recorded continuously during the adsorption process of the dithiol, at 25 °C solution temperature (results not shown). Consecutively, the SAM modified QCM electrode were washed with ethanol and Millipore water, dried in a stream of nitrogen and analyzed by tapping mode atomic force microscopy (AFM). After AFM imaging the 1,6HDT modified QCM electrode was reinserted in the QCM chamber and a second base line was recorded for 20 minute using Millipore water as the solvent. The Au-1,6HDT modified electrode was further modified with gold nanorods by gentle injection of freshly prepared $AuNR_1$ solution (0.50 mg/mL) at a solution temperature of 35°C. The resulted Au-1,6HDT-$AuNR_1$ modified electrode was removed from the QCM cell, washed persistently with Millipore water, dried in a stream of nitrogen and imagined by tapping mode AFM. The Au-1,6HDT-$AuNR_1$ modified electrode was reinserted in the QCM cell and a third layer, (a 1,6HDT layer) was self-assembled from ethanolic solution. These steps were repeated until a number of 8 layers were built-up on the QCM electrode. The same procedure was followed for the build-up of multilayers containing hexanedithiol and large gold nanorods ($AuNR_2$, 0.50mg/mL). The resulted QCM-multilayer modified electrodes were analyzed by atomic force microscopy. The QCM measurements were performed in order to estimate the necessary self-assembly time for each layer.

2.5.2 Atomic force microscopy measurements

The resulted 1,6HDT/AuNR multilayer modified QCM electrodes were imaged in air using a Molecular Imaging PicoLe AFM in tapping mode and Silicon cantilevers (Nanosensors) with a resonance frequency of 200-400 kHz. Topographic, amplitude and phase images were recorded for all the samples with a resolution of 512×512 pixels.

2.5.3 Ellipsometry measurements

The thickness of the multilayers was determined using a Multi-Wavelength Discrete Ellipsometer. First, the Brewster angle was determined for the bare gold QCM electrode previously washed with Piranha solution, large amounts of Millipore water and dried in a stream of nitrogen. The search of the Brewster angle was performed for several angles of the ellipsometer: 74°, 70°, 69°, 68° and the best results were obtained for the angle of 68°. The refraction index (n) and the extinction coefficient (K) determined for the bare gold QCM were $n = 0.496 \pm 0.011$ and $K = 2.323 \pm 0.006$. These parameters values were used for the determination of the multilayers' thicknesses.

2.5.4 Electrochemical measurements

Cyclic, square wave voltammetry and electrochemical impedance spectroscopy measurements were performed on a PGSTAT 302N potentiostat (EcoChemie B.V., The Netherlands) for each new layer self-assembled on the working electrodes. All electrochemical experiments were carried out in a conventional three-electrode cell equipped with a working electrode of gold (with an area of 0.0314 cm^2), a Pt wire as the counter electrode and an Ag/AgCl (3M KCl) electrode as the reference electrode. The electrochemical cell was placed in a Faraday cage in order to minimize the electrical noise. The gold electrodes were cleaned as previously explained (Chirea et al, 2005) and consecutively modified following the procedure used for the QCM experiments. The electrolyte solution consisted of 0.0005M $[Fe(CN)_6]^{3-/4-}$ and 0.1$NaClO_4$. Cyclic voltammograms were measured between -0.2 and 0.6 V at several scan rates (25, 50, 75, 100mV/s) for each layer self-assembled on the gold electrodes. The square-wave voltammograms were measured for the same potential window, at frequency varying between 10Hz to 75Hz, using 50mV amplitude and a step potential of 2mV. Electrochemical impedance spectroscopy (EIS) measurements were performed for a frequency range of 10000 to 0.1 Hz and amplitude of 20 mV at the formal potential of the $[Fe(CN)_6]^{3-/4-}$ redox couple (0.210V vs. Ag/AgCl (KCl, 3M). Before each measurement, N_2 was purged into the solution for 15 minutes for deaeration. The electrochemical measurements were performed twice on freshly modified electrodes and similar results were obtained.

3. Results and discussions

3.1 Characterization of gold nanorods

It has been demonstrated that the growth of gold nanorods in aqueous solution from spherical gold nanoparticles (4nm average diameters) in the presence of cetyltrimetyl ammonium bromide takes place simultaneously on all directions until the seeds are reaching a critical size allowing surfactant binding (Nikoobakht &El-Sayed, 2003).

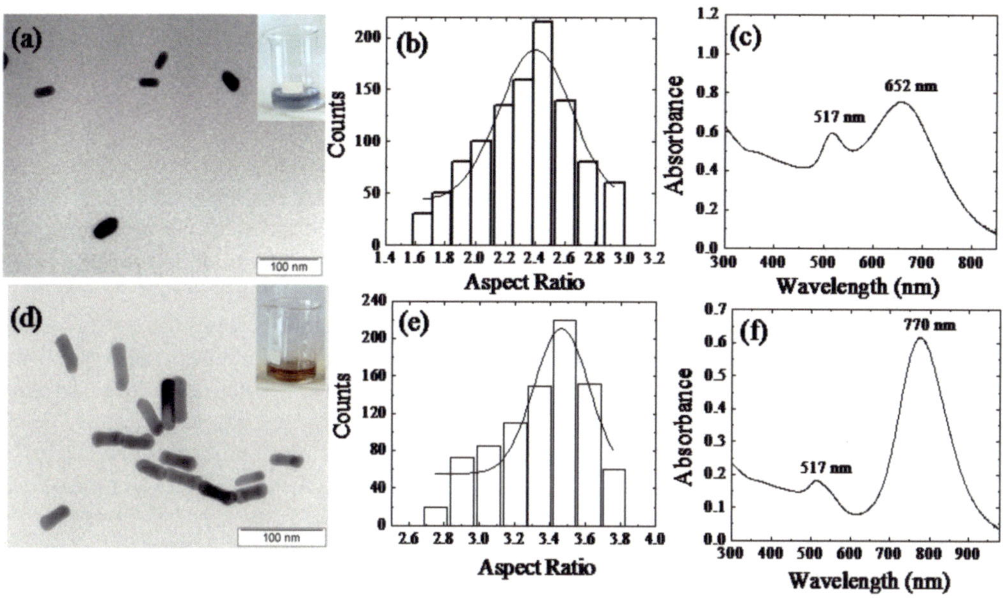

Fig. 1. Transmission electron microscopy images, size distribution histograms, and UV-Vis spectra of 2.44 aspect ratio gold nanorods (a, b, c) and 3.48 aspect ratio gold nanorods (d, e, f).Insets of a and d : digital images of the gold nanorods solutions.

Moreover, the growth in the longitudinal direction takes place parallel to the {001} planes, which is accompanied by formation of four relatively unstable {110} facets and four {111} facets. The surfactant has a stronger affinity to {110} facets determining a slower growth on the width of the rods than on the longitudinal direction characterized mainly by {111} facets. The growth rate on different facets determines the final shape of these nanomaterials. The presence of Ag ions in the growth solution (from $AgNO_3$ used as growth agent) generates Ag-Br pairs between the headgroups of the capping material. These Ag-Br pairs may decrease the charge density on the bromide ions determining less repulsion between the neighbouring headgroups on the rods surface and as a result controlling the size of the rods. This seed-mediated method allows control over the size of the rods by varying the $AgNO_3$ volume, at high concentration of surfactant. In accordance, two different sized gold nanorod samples were synthesized following this method. The morphology (size, shape) and the optical properties of the gold nanorods were analyzed by transmission electron microscopy and UV-Vis spectroscopy (Figure 1). An accurate size distribution was estimated based on TEM images. The gold nanorods were 2.44 aspect ratio ($AuNR_1$, 35.48 nm length per 14.52

nm width, Figure 1a and 1b) and 3.48 aspect ratio ($AuNR_2$, 51.56 nm length per 14.80 nm width, Figure 1d and 1e). Transverse plasmon bands at 517 nm and longitudinal plasmon bands at 652 nm (Figure 1c, $AuNR_1$, blue colour) or 770 nm (Figure 1f,$AuNR_2$, pink-brownish colour) were observed in the optical spectra of the rods demonstrating an evident increase of rods size with increased volumes of $AgNO_3$ added during the growth procedure (see section 2.2).

3.2 Characterization of 1,6hexanedithiol/gold nanorod multilayers by quartz crystal microbalance, ellipsometry and atomic force microscopy

In the following discussion n denotes the number of layers self-assembled on the QCM gold electrodes. The bare gold corresponds to n = 0, the 1,6hexanedithiol layers correspond to n odd (n = 1, 3, 5, 7) whereas the gold nanorods layers correspond to n even (n = 2, 4, 6, 8).In accordance, the first bilayer corresponds to n = 2, the second bilayer corresponds to n = 4, the third bilayer corresponds to n = 6 whereas the forth bilayer corresponds to n = 8. The 1,6HDT layers were self-assembled on the QCM gold electrodes from ethanolic solution at 25°C whereas the consecutive AuNRs layers were self-assembled from aqueous solution at 35°C. As determined previously, (Chirea et al., 2010) these are the adequate self-assembly condition in order to obtain dense and vertically aligned gold nanorods using 1,6hexanedithiol as a bridge molecule. The self-assembly time for each layer was determined by QCM measurements. The first hexanedithiol layer (n = 1) was self-assembled after 14 hours whereas the consecutive gold nanorod layer (n = 2), was self-assembled after 7 hour in the case of short rods or 15 hours in the case of large rods at a solution temperature of 35°C (Chirea, 2010). For the multilayer containing short rods ($AuNR_1$, 2.44 aspect ratio), the third layer self-assembled on the modified electrodes, a 1,6HDT layer (n = 3) was formed after 16 hours, the consecutive $AuNR_1$ layer (n = 4) was self-assembled after 10 hours, the 5th layer , a 1,6HDT layer, was self-assembled after 19 hours, the consecutive $AuNR_1$ layer was self-assembled after 12 hours (n = 6), whereas the last two layers were self-assembled after 19 hours (n = 7) and 15 hours (n = 8). The multilayer containing 3.48 aspect ratio gold nanorods was build-up for the self-assembly times of 14 hours (n = 1), 15 hours (n = 2), 11 hours (n = 3), 16 hours (n = 4), 18 hours (n = 5), 18 hours (n = 6), 15 hours (n = 7) and 6 hours (n = 8) respectively. The deposited mass has decreased with increasing number of layers self-assembled on the gold electrodes (results not shown). This decrease of deposited mass with each new layer self-assembled on the Au-QCM electrodes suggests a pyramidal growth of the multilayers. For example, the highest number of small gold nanorods was self-assembled on the first 1,6hexanedithiol monolayer formed on the Au-QCM electrodes which is a well organized film. The tapping mode AFM image of this bilayer illustrates a vertical alignment of the rods to the substrate surface (Figure 2a, n = 2, $AuNR_1$, 35.48 nm length per 14.52 nm width). The third layer self-assembled on the modified QCM electrode, a 1,6hexanedithiol layer (n = 3) determined no evident change in topography (results not shown).

Based on the strong affinity of thiols toward covalent bonding to gold surfaces (Love et al., 2005), it is expected that the 1,6HDT layer (n = 3) will cover the rods removing the cetyltrimetyl ammonium bromide from their surface through a place exchange reaction (Hostetler, M. J. et.al, 1999), and possibly will connect them on a horizontal plane if they are in very close vicinity. However, the chemical bonding of the rods on a horizontal plane may change or not the films topographies. The consecutive self-assembly of a fourth layer, namely an $AuNR_1$ layer, (Figure 2b, n = 4) has generated a similar film topography (vertically aligned nanorods) with a lower surface coverage of rods than the previous $AuNR_1$ layer (Figure 2a, n = 2). It seems that the only possibility to connect gold nanorods on an upper layer (n = 4) is at the ends of the gold nanorods from the previous layer (n = 2) through the 1,6hexanedithiol molecules acting as molecular bridges (n = 3). In consequence, the number of 1,6hexanedithiol molecules available for further chemical interactions are the molecules at the ends of each vertically aligned rod from a previous layer which implies already a lower number of 1,6HDT molecules than in the n = 1 layer. This chemical structure has determined a pyramidal growth of the multilayered film (Figure 2). Moreover, the AFM images have revealed the presence of neighbour rods within the same layer (the thicker "rods" features) together with a decreasing surface coverage of the gold nanorods in the upper layers of the multilayered film (Figure 2b, c, d). The sequential self-assembly of $AuNR_1$ and 1, 6 hexanedithiol layers up to a maximum 8 layers have generated similar topographies (Figure 2c, d). The surface roughness has increased with the numbers of bilayers self-assembled on the electrodes from 12.24 nm (Figure 2a, n = 2), to 15.36 nm (Figure 2b, n = 4), to 18.55 nm (Figure 2c, n = 6) and to 20.20 nm (Figure 2d, n = 8). The increase of surface roughness proves a progressive growth of the film upon self-assembly of each (1,6HDT/$AuNR_1$) bilayer. The AFM analysis of the film confirms its pyramidal growth with decreasing surface coverage of gold nanorods within consecutive layers (Figure 2). The tapping mode AFM analysis of the multilayer film containing layers of 1,6hexanedithiol and large rods ($AuNR_2$, 51.56 nm/14.80 nm average sizes) is illustrated in Figure 3.

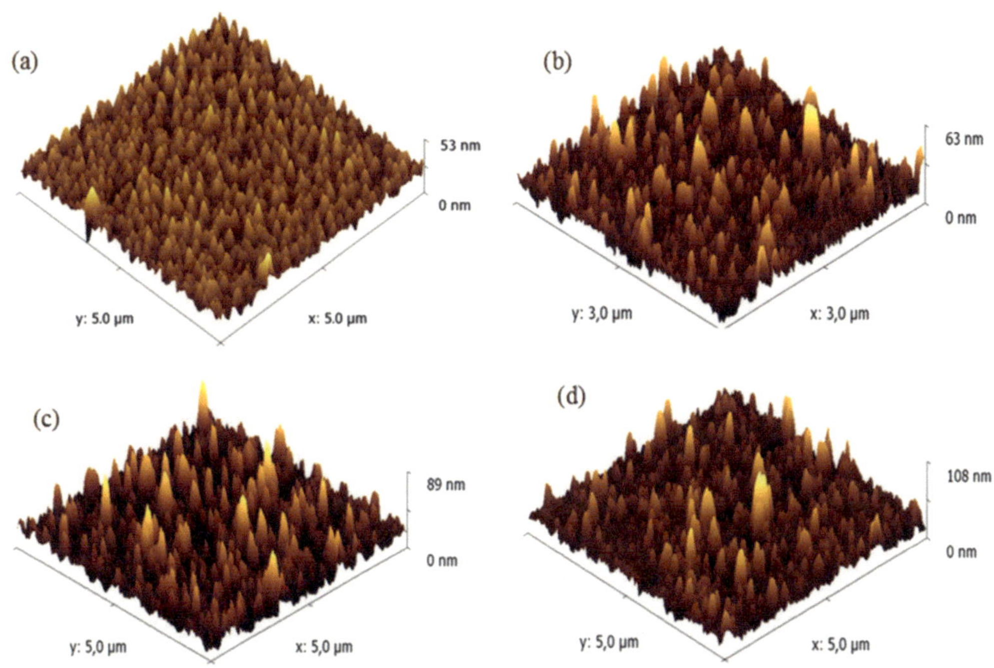

Fig. 2. Tapping mode atomic force microscopy images of Au-(1,6HDT-$AuNR_1$)$_4$ multilayers showing the progressive build-up of the film:(a) first bilayer, (n = 2), (b) second bilayer, (n = 4), (c) third bilayer (n = 6), and (d) fourth bilayer (n = 8). $AuNR_1$ were 35.48nm/14.52nm average sizes.

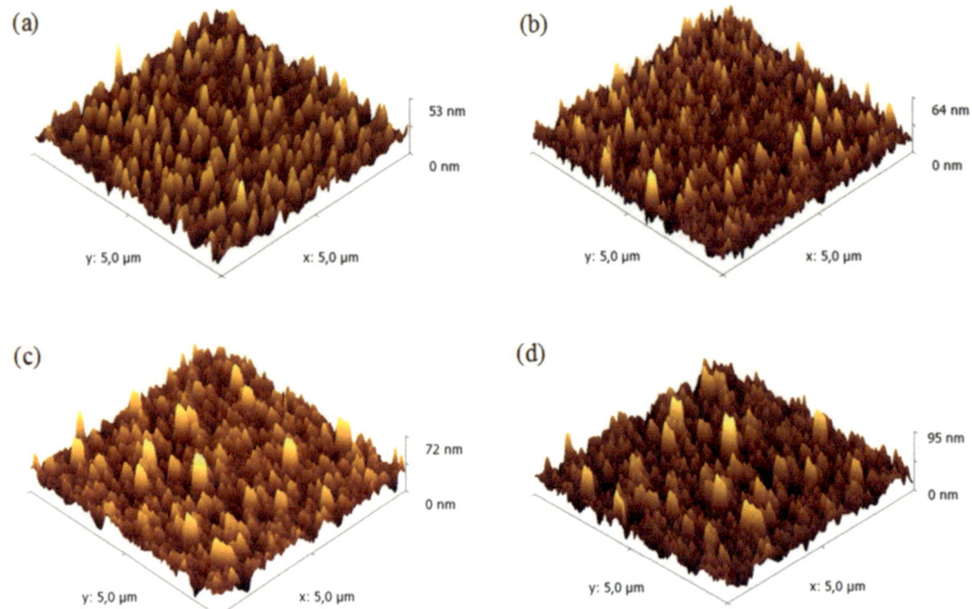

Fig. 3. Tapping mode atomic force microscopy images of Au-(1,6HDT-$AuNR_2$)$_4$ multilayers showing the progressive build-up of the film: (a) first bilayer, (n = 2), (b) second bilayer, (n = 4), (c) third bilayer (n = 6), and (d) fourth bilayer (n = 8). $AuNR_2$ were 51.56 nm/14.80nm average sizes.

The topographic features of this film suggest that the $AuNR_2$ were attached vertically at the ends of the rods from the previous layer with a decreased surface coverage (compare Figure 3b and 3a). The consecutive self-assembly of 1,6hexanedithiol and $AuNR_2$ layers (Figure 3c, n = 6) has generated a more dense structure implying a covalent bonding of the large rods both at the tips of the previous $AuNR_2$ layer (n = 4) and, in higher number, in-between the rods of the same n = 4 layer. A similar dense surface coverage was observed after the self-assembly of the next 1,6hexanedithiol/$AuNR_2$ bilayer (Figure 3d, n = 8). The surface roughness increased progressively with increasing number of bilayers from 14.26 nm (Figure 3a, n = 2), to 18.23 nm (Figure 3b, n = 4), to 24.20 nm (Figure 3c, n = 6) and to 28.0 nm (Figure 3d, n = 8), respectively. The progressive increase of surface roughness demonstrates also a progressive increase of film thickness. If the gold nanorods within a layer are in very close vicinity, then the 1,6hexanedithiol is connecting them on a horizontal plane generating quit ordered features with twin nanorods as illustrated in Figure 4. The topographic analysis of both types of films demonstrates that the film growth and film density depend on the size of the gold nanorods. The short rods tend to organize themselves within quit homogeneous films being covalently bonded at their ends through the 1,6hexanedithiol molecular bridges and generating a well organized network.

Contrarily, the large rods tend to fill-up first the free spaces in-between the rods in a previous layer and to covalently bond in an upper layer only with a low surface coverage. The different chemical structures of these two types of multilayers will determine a different electrochemical behaviour as it will demonstrate in the following paragraphs. The vertical alignment of the rods within these films is in accordance with previous observation related to the type of surface bonding (end-bonding) of the rods when self-assembled on gold electrodes at 35°C solution temperature and using the 1,6hexanedithiol as bridge molecules (Chirea et al., 2009, 2010). The thicknesses of the multilayers were determined by ellipsometry measurements. The average thickness of Au-(1, 6HDT-$AuNR_1)_4$ multilayered film was 188, 85 ± 1.05 nm yielding a refraction index of 2.78 ± 0.11 and an extinction coefficient of K = 3.42 ± 0.03 whereas the average thickness of the Au-(1, 6HDT-$AuNR_2)_4$ multilayered film was 227, 10 ±1.2 nm yielding a refraction index of n = 3.70 ± 0.15 and a extinction coefficient of K = 3.93 ± 0.05. The estimated film thicknesses are in accordance with a vertical alignment of the nanorods to the substrate surface as demonstrated also by AFM measurements (Figure 2, 3 and 4).

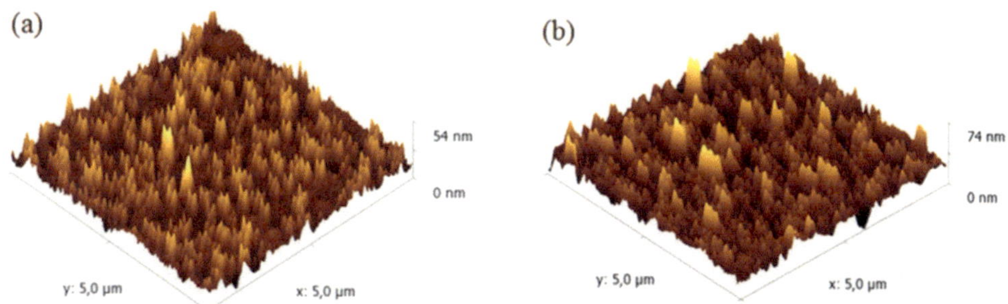

Fig. 4. Tapping mode atomic force microscopy images of the multilayered film illustrating the effect of the 1,6HDT monolayer as the outermost layer on the overall film topography: (a) Au-1,6HDT-$AuNR_2$-1,6HDT film (n =3) , (b) Au-1,6HDT-$AuNR_2$-1,6HDT-$AuNR_2$-1,6HDT-$AuNR_2$-1,6HDT film (n = 7). The surface roughness was 7.47nm (a) and 15.32 nm (b).

3.3 Electrochemical characterization of 1,6hexanedithiol/gold nanorod multilayers

As previously mentioned n denotes the number of layers self-assembled on the gold electrodes. The bare gold corresponds to n = 0, the 1,6hexanedithiol layers correspond to n odd (n = 1, 3, 5, 7) whereas the gold nanorods layers correspond to n even (n = 2, 4, 6, 8). The electrochemical properties of the 1,6hexanedithiol/gold nanorod multilayers were studied as a function of the number of layers self-assembled on the gold electrodes and as a function of the sizes of the gold nanorods. The electrolyte consisted of 0.0005M $K_3[Fe(CN)_6]^{3-}$ and 0.0005M $K_4[Fe(CN)_6]^{4-}$(0.0005M) dissolved in 0.1M $NaClO_4$ aqueous solution. Cyclic and square wave voltammetry measurements as well as electrochemical impedance spectroscopy measurements were performed for each layer self-assembled on the gold electrodes in order to elucidate the electron transfer process taking place at the 1,6hexanedithiol/gold nanorod modified electrodes.

3.3.1 Electrochemical characterization of multilayers containing short gold nanorods

Figure 5 shows the cyclic (Figure 5a and 5b) and square wave voltammograms (Figure 5c and 5d) recorded at the Au-(1,6HDT-$AuNR_1)_4$ multilayer modified gold electrodes in aqueous solution of 0.1M $NaClO_4$ and 0.0005M $[Fe(CN)_6]^{3-/4-}$. As mentioned previously, the average sizes of the rods incorporated into this film were 35.48 nm lengths per 14.52 nm width ($AuNR_1$, Figure 1a, b) with an aspect ratio of 2.44. The gold nanorods were self-assembled in end-bonding topography through 1,6hexanedithiol bridges generating a quit homogeneous film (Figure 2). The electrical properties of this film were highly dependent on the type of outermost layer. The Faradaic process was highly hindered by the 1,6hexanedithiol layers (Figure 5, *n*= 1, 3, 5, 7). This electrochemical behaviour is caused by the insulating properties of this dithiol when self-assembled on gold electrodes (Chirea et al., 2009, 2010; Chou et al., 2009). For example, the first layer self-assembled on the gold electrodes determined a disappearance of the Faradaic currents in the cyclic and square wave voltammograms (Figure 5, *n* = 1, black curve) as compared to the bare gold electrode (Figure 5, *n*=0, dashed black curve) illustrating a dense packing of the 1,6hexanedithiol layer on the gold electrodes. In other words, the 1,6hexanedithiol is blocking the electronic communication between the $[Fe(CN)_6]^{4-/3-}$ in solution and the underlying bare gold electrode. The consecutive self-assembly of an $AuNR_1$ layer determined the reappearance of the Faradic currents in the cyclic and square wave voltammograms (Figure 5a, c, *n* = 2, red curve) generating a quasi reversible behaviour of the $[Fe(CN)_6]^{3-/4-}$ redox couple. The third layer self-assembled on the modified electrodes (a 1,6HDT layer) has determined a drastic decrease of peak currents in the cyclic and square wave voltammograms (Figure 5b, d, *n* = 3, purple curve), again due to its insulating properties confirming a high coverage with 1,6HDT molecules on $AuNR_1$ surfaces from the previous layer. Nevertheless, the presence of the small gold nanorods in the inner layer (*n* = 2) seems to facilitate the electron transfer process despite the complete electronic blocking generated by the 1,6hexanedithiol layers (*n* = 1 and *n* = 3). At this stage of the multilayer build-up, the electron transfer process is based on electron tunnelling through the dithiol layers (very low peak currents (*n* = 3) or no peak currents (*n* = 1) in the CVs and SQWVs, Figure 5) and electron transport mediated by the gold nanorods from the inner layer (*n* = 2). A similar electrochemical behaviour has been observed at citrate stabilized gold nanoparticle/1,4 benzenedimethanethiol multilayers modified gold electrodes. Abdelrahman et al. have shown that these gold nanoparticles have a good catalytic activity toward $[Ru(NH_3)_6]^{3+}$ in solution, despite the electronic blocking induced by the 1,4 benzenedimethanethiol layers used to connect the gold nanoparticles into dense films. Moreover these gold nanoparticle/1,4 benzenedimethanethiol multilayers have proven also highly efficient for oxygen reduction (Abdelrahman et al., 2006).

In the present work, the self-assembly of a fourth layer, a $AuNR_1$ layer, re-established an efficient electron transfer process depicted through high peak currents in the CVs and SQWVs (Figure 5a, c , *n* = 4, green curve) demonstrating also a high catalytic activity of these rods. These electrical features demonstrate a stepped electron transfer process consisting of : (i) increase of the interfacial concentration of the negatively charged redox probes at the positively charge gold nanorods surfaces by electrostatic attraction (ii) electron transport mediated by the gold nanorods and (iii) and coherent electron tunnelling through the 1,6hexanedthiol layers toward the underlying bare gold electrode (Scheme 1). A remarkable feature is to be mentioned: the Faradaic currents are slightly lower at the four layers modified electrode than at the two layers modified electrode (in Figure 5 a, c compare *n* =2 and *n* = 4) but both films determined a faster electron transfer than at the bare gold electrode (higher Faradaic currents in SQWVs than at the bare gold, see Figure 5c). This electrochemical behaviour is caused by two features: (i) high catalytic activity of the short rods which is nullifying the electronic blocking induced by the 1,6hexanedithiol layers assuring an coherent electron tunnelling through these molecular bridges and (ii) the surface coverage with gold nanorods is decreasing as shown by AFM imaging (compare Figure 2a and Figure 2b).

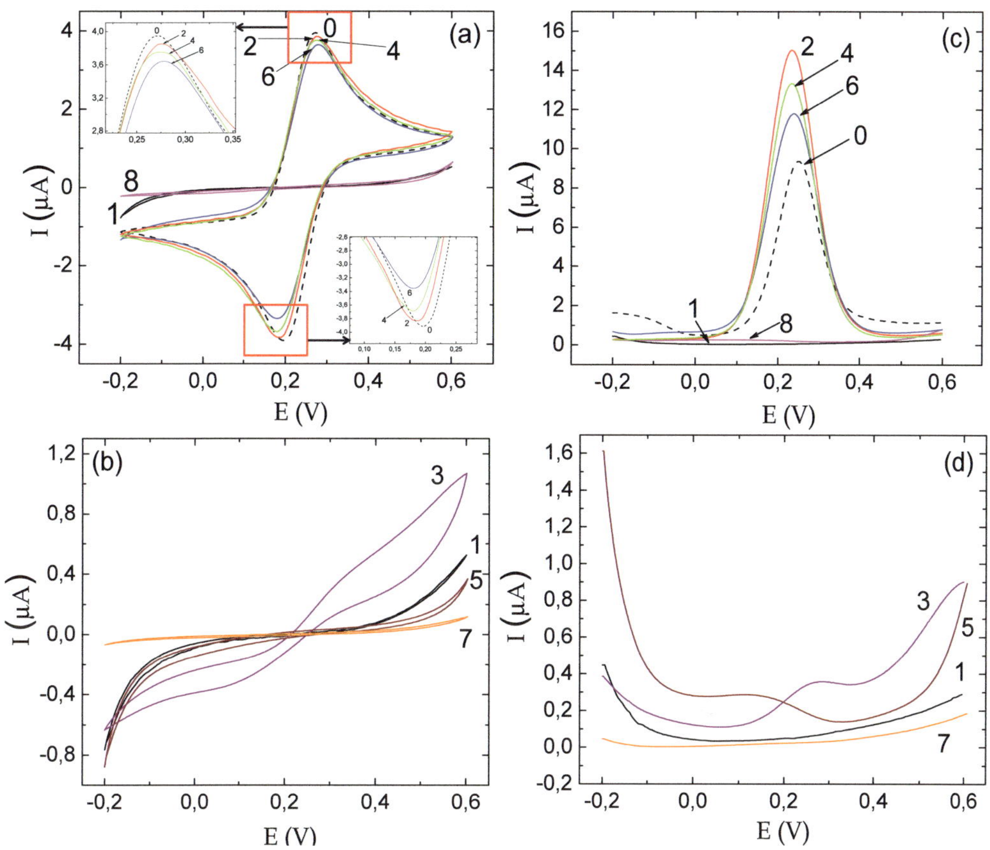

Fig. 5. Cyclic voltammograms (a, b) and square wave voltammograms (c, d) recorded at the Au-(1, 6HDT-AuNR$_1$)$_4$ multilayer modified gold electrodes in aqueous solution of 0.1M $NaClO_4$ and 0.0005M $[Fe(CN)_6]^{3-/4-}$: bare gold (n = 0), 1,6HDT outermost layers (n = 1,3,5,7) and AuNR$_1$ outermost layers (n = 2,4,6,8), respectively.

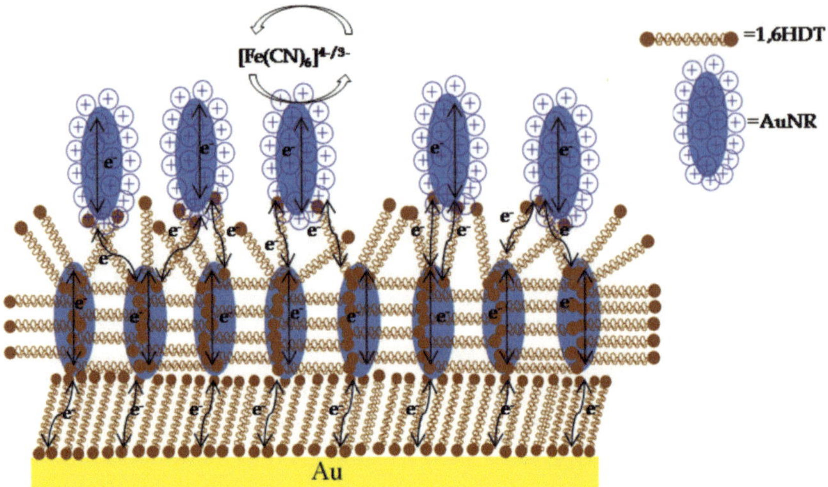

Scheme 1. Schematic representation of the electron transfer process at 1,6hexanedithiol/gold nanorods multilayers modified electrode.

The sequential self-assembly of the fifth layer, a 1,6hexanedithiol layer, determined a complete electronic blocking between the redox probes in solution and the underlying bare gold electrode depicted through the disappearance of the Faradaic current in the CV (Figure 5b, n = 5, wine curve) or a drastic decrease of Faradaic current in the SQWV (Figure 5d, n = 5, wine curve) and shifting of the peak current toward less positive potentials. This difference of the electrochemical features in the cyclic and square wave voltammograms for the same layer is due to the difference of electrochemical sensitivity between the two techniques. Square wave voltammetry has as main advantage the discrimination of the non-Faradaic (capacitative) currents which is achieved by measuring the current at the end of the imposed square-wave pulse. In this way the ratio between the Faradaic and non-Faradaic currents is very high because almost no capacitative currents are measured at the end of the measuring time. This cannot be achieved by cyclic voltammetry. The square wave signal consists of a symmetrical square wave pulse superimposed on a staircase waveform (Mirčeski et al.2007) which assures a higher sensitivity of this technique toward surface confined electron transfer processes than the cyclic voltammetry. In consequence, the square wave voltammograms illustrate more accurate the effect of each layer self-assembled on the gold electrodes. The sixth layer self-assembled on the gold electrodes, an $AuNR_1$ layer, restored the efficient electronic communication between the redox probes in solution and the underlying bare gold electrode generating cyclic and square wave voltammograms with well defined peak currents and diminished peak separations (Figure 5a, b n =6, blue curve). The 7th layer self-assembled on the gold electrodes (Figure 5, n = 7, orange curve) showed a similar electrochemical behaviour as the previous dithiol layers, namely an insulating effect, whereas the 8th layer determined no evident increase of Faradaic current (Figure 5a,c, n = 8, pink curve). At this stage of the multilayer build-up the low surface density of the rods does not allow a further improvement of the overall electron transfer process. The peak currents and the peak-to-peak separations in the cyclic voltammograms varied as follows: I_a= 3.95μA, ΔE_p= 0.071V, for n = 0 (bare gold electrode), I_a= 3.85μA, ΔE_p= 0.091V, for n = 2,

(Au-1,6HDT-$AuNR_1$), I_a= 0.58μA, ΔE_p= 0.369V, for n = 3, (Au-1,6HDT-$AuNR_1$-1,6HDT), I_a= 3.75μA, ΔE_p= 0.095V, for n = 4,(Au-1,6HDT-$AuNR_1$)$_2$, I_a= 3.64μA, ΔE_p= 0.097V, for n = 6 ,(Au-1,6HDT-$AuNR_1$)$_3$ whereas for the other layers (n = 1,5,7 and 8) there were no peak currents in the CVs (Figure 5a,b). As proven by these parameters value, the current densities in the CVs are decreasing with the number of $AuNR_1$ present in each layer. As seen by AFM analysis, the number of gold nanorods in direct contact with the electroactive species from solution is decreasing within each new layer self-assembled on the gold electrodes (Figures 2). Due to their high surface to volume ratio, the gold nanorods act as nanoelectrodes proving a strong catalytic effect in the overall electrochemical process. If their number decreases then the current generated at these nanoelectrodes is decreasing. In consequence the surface density of the rods has a strong influence on the overall electrochemical process. In the square wave voltammograms, the lowest peak current was recorded at the film containing all 8 layers (n = 8, Figure 5c), for which the number of $AuNR_1$ covalently bonded to the modified electrode is the smallest (Figure 3d) and the distance from the underlying bare gold surface is the longest. The corresponding peak currents and shifting of the peak potentials in the square wave voltammograms varied as follows: I_a= 9.40μA at 0.247V for n =0, I_a=15.04 μA at 0.233V for n =2, I_a= 0.35μA at 0.274V for n =3, I_a= 13.37μA at 0.232V for n =4, I_a= 0.28μA at 0.142V for n =5, I_a= 11.72μA at 0.236V for n =6, I_a= 0.32μA at 0.065V for n =8 (Figures 5c and 5d). For the 1,6hexanedithiol outermost layers having an insulating effect, the redox current at any potential is given by the expression (Xu et al., 1993):

$$I = I_0 e^{-\beta d} \tag{1}$$

where I_0 is the current measured at the bare electrode, β is the potential independent electron tunnelling coefficient, and d is the thickness of the monolayer. For an electrochemical reaction at equilibrium, using equation (1) it can be obtained:

$$k = k_0 e^{-\beta d} \tag{2}$$

where k_0 and k are the electron transfer rate constants at the bare and 1,6HDT-SAM modified electrode. The heterogeneous electron-transfer rate constant (k_{et}) at the modified electrodes can be calculated using equation (3) (Chirea et al, 2007and references therein):

$$k_{et} = (RT) / \left(n^2 F^2 A R_{CT} c^0\right) \tag{3}$$

where R is the gas constant, T is temperature (K), F is the Faraday constant, A is the electrode area (0.0314 cm^2), R_{CT} is the charge-transfer resistance, c^0 is the concentration of the redox couple in the bulk of solution (0.5×10^{-3} mol/cm^3) and n is the number of transferred electrons per molecule of the redox probe (n = 1 for the $[Fe(CN)_6]^{3-/4-}$probes). From the fittings of the impedance spectra (Figures 6, 7), it was obtained the apparent charge transfer resistance, R_{CT}, for each layer self-assembled on the gold electrodes (table 2). The heterogeneous electron-transfer rate constants, k_{et}, for each layer self-assembled on the gold electrode were estimated using R_{CT} parameters and based on equation (3), (table 1). The variation of the heterogeneous electron-transfer rate constant, k_{et}, shows a three order magnitude increase upon the self-assembly of the first $AuNR_1$ layer (k_{et}=1.63×10^{-5} cm×s^{-1}

n=2, table 1) as compared to the first 1,6HDT-SAM (k_{et} = 0.48×10^{-8} cm×s^{-1}, n = 1, table1). The consecutive self-assembly of a 1,6HDT layer (n = 3, table 1) determined two orders magnitude decrease of the k_{et}, followed by a two orders magnitude increase of the k_{et} for the next $AuNR_1$ self-assembled layer (n = 4, table1). This oscillation of the k_{et} is repeated up to the 7th layer. For the last layer self-assembled on the multilayer modified electrodes (n = 8, an $AuNR_1$ outermost layer) the k_{et} decrease was more pronounced (k_{et}=0.14×10^{-7}cm×s^{-1}) as compared to the previous $AuNR_1$ layer (k_{et}=0.35×10^{-5}cm×s^{-1}, n = 6, table1) confirming the presence of very few gold nanorods within this last layer. The variation of the heterogeneous electron-transfer rate constants at the Au-(1,6HDT-$AuNR_1$)$_4$ multilayer modified electrode confirms the alternative electrochemical behaviour observed in the cyclic and square wave voltammograms depicted through peak shaped and plateau shaped CVs and SQWVs.

n	k_{et} (cm×s^{-1})	self-assembly time (h)	β(Å^{-1})	Tunnelling distance (Å)
1	0.48×10^{-8}	14	0.92	17.04
2	1.63×10^{-5}	7		
3	0.62×10^{-7}	16	0.77	17.04
4	0.94×10^{-5}	10		
5	0.22×10^{-7}	19	0.83	17.04
6	0.35×10^{-5}	12		
7	0.20×10^{-7}	19	0.86	17.04
8	0.14×10^{-7}	15		

Table 1. Heterogeneous electron-transfer rate constants, k_{et} and tunnelling parameter, β, calculated for each self-assembled layer within the Au-(1,6HDT-$AuNR_1$)$_4$ multilayer, using the R_{CT} parameters obtained from the fittings of the EIS spectra (table 2) and based on equations (1)-(3).

The electron transfer process is switched on when the outermost layer is an $AuNR_1$ layer (high k_{et}, table 1, n = 2, 4, 6) and switched off when the outermost layer is a 1,6HDT layer (low k_{et}, table1, n = 1, 3, 5, 7). As the distance from the underlying bare gold electrode is increasing and the surface coverage with rods is decreases, the electron transfer process becomes slower (low k_{et}, table 1, n = 8, total film thickness: 188,85±1.05 nm). The tunnelling distance was estimated taking into consideration the length of the C-C, C-S and S-Au bonds which are 1.54Å, 1.82 Å and respectively 2.85 Å (Jiang, J.; Lu, W. & Luo, Y., 2004). The variation of the tunnelling parameter, β, suggest that the electron tunnelling process is more efficient at the thiol layers self-assembled on the gold nanorods than at the 1,6HDT self-assembled on the bare gold electrode (lower values of β). Overall, the electron transfer process is faster at Au-(1,6 HDT-$AuNR_1$)$_4$ multilayer modified electrode than at bare gold electrode (up to n = 6) demonstrating a catalytic effect of the small rods as outermost layers. Electrochemical impedance spectroscopy (EIS) measurements were performed for each layer self-assembled on the gold electrodes. The Nyquist plots of the bare gold (n = 0) and Au-1,6HDT-SAM modified electrode (n =1) are presented in Figure 6 whereas the Nyquist plots recorded at the Au-(1, 6HDT-$AuNR_1$)$_4$ multilayer modified electrode (n = 2-8) are presented in Figure 7. The fittings of the EIS spectra were performed using the non-linear least square-fit procedure and the equivalent electrical circuits presented in Figure 8. Randles circuit was used for fittings of the bare gold EIS spectra (n =0, Figure 6), whereas the other two equivalent electrical circuits were used for fittings of the EIS spectra of the film modified electrodes. The impedance responses varied with the type of layer self-assembled on the gold electrodes. The Warburg impedance for semi-infinite planar diffusion is the sum of concentration impedances which are linked to the mass transport (ionic diffusion) and to the kinetics of the charge transfer being given by the equation (4):

$$Z_d = \frac{\sigma}{\sqrt{w}} - \frac{i\sigma}{\sqrt{w}} \tag{4}$$

where

$$\sigma = \frac{RT}{n^2F^2A\sqrt{2}}\left\{\frac{1}{c_R\sqrt{D_R}} - \frac{1}{c_o\sqrt{D_o}}\right\} \tag{5}$$

The parameters D_R and D_O represent the diffusion coefficients of the reduced and oxidized species (1.76×10^{-6} cm^2s^{-1}), c_R and c_O are the solution concentrations of the reduced and oxidized species ($c_R = c_O = 0.5\times10^{-3}$ mol/cm^3), σ is the mass transfer coefficient and ω is the angular frequency ($\omega=2\pi f$, f is the alternative voltage signal frequency, Hz). The other parameters have the same meaning as previously mentioned. The total impedance at the bare gold electrode is described by the equation (6) and the Randles electrical circuit pictured in Figure 8a:

$$Z_a = R_s + \frac{R_{CT}+\sigma\omega^{-\frac{1}{2}}-i(2\sigma^2C_{dl}+\omega C_{dl}R_{CT}^2+2\omega^{\frac{1}{2}}R_{CT}\sigma C_{dl}-\sigma\omega^{-\frac{1}{2}})}{(1+\omega^{\frac{1}{2}}\sigma\ \ C_{dl})^2+(\omega\ \ C_{dl}R_{CT}+\omega^{1/2}\sigma C_{dl})^2} \tag{6}$$

where R_s (Ω) is the solution resistance, R_{CT} (Ω) is the charge transfer resistance and C_{dl} (F) is the double layer capacitance. The Nyquist plots (Figure 6, n =1) shows that the self-assembly of the 1,6hexanedithiol on bare gold electrode generates a very high resistance to the charge transfer (table 2), depicted through depressed semicircle for all range of frequencies with no evident diffusion profile at low frequencies. The 1,6hexanedithiol monolayer is changing the dielectric properties of the electrode surface generating a new film resistance R_f and film capacitance C_f. The total impedance at the Au-1,6HDT-SAM and Au-(1,6HDT-AuNR$_1$)$_3$ modified electrodes is given by the equation (7). The detailed expression of the real (Z_b') and imaginary (Z_b'') complex impedances are presented in Appendix A, at the end of this chapter. The complex impedances where written by taking R_{CT}, R_f, as resistances, 1/iωCdl, 1/iωCf, as capacitances and applying Ohm's and Kirchhoff's laws to the connection of these elements in the electrical circuit. The parameter "i" is $\sqrt{-1}$.

$$Z_b = Z_b' + Z_b'' \tag{7}$$

For a higher number of layers (n = 7, 8) the fittings of the EIS spectra were performed using the electrical circuit depicted in Figure 8c. The equation describing the total impedance Z_c, corresponding to this electrical circuit is presented in Appendix A. The electronic blocking effect induced by the 1,6hexanedithiol monolayer determined an increase of 3 orders magnitude in the charge transfer resistance, R_{CT} (n = 1, table 2, Figure 6) as compared to the bare gold electrode (n = 0, table 2, Figure 6) and ~16 times decrease of C_{dl} demonstrating that the distance between the redox probes in solution and the underlying bare gold surface has increased.

This effect of the aliphatic chain thiol has been frequently observed at SAM modified electrodes (Chou et al., 2009; Su et al.,2006). The sequential self-assembly of the $AuNR_1$ layer has determined a decrease of the charge transfer resistance by 3 orders of magnitude (n = 2, table 2). The corresponding EIS spectrum featured a small semicircle at high and intermediate frequencies, which is due to the coupling between R_{CT} and C_{dl} parameters and a straight line at low frequencies (Figure 7a, n = 2, red circles) implying semi finite planar diffusion to the film (slope unity). The R_{CT} parameter relates to surface modifications that enhance/hinder the electron transfer process at the electrode/solution interface. The strong decrease of R_{CT} upon self-assembly of the $AuNR_1$ layer proves that these small rods (35.48 nm length per 14.52 nm width, Figure 1a and 1b) have a high catalytic activity for the $[Fe(CN)_6]^{3-/4-}$ redox probes facilitating the electron transfer at the electrode/solution interface. Additionally, the strong decrease in film resistance, Rf, which is a tunnelling resistance in this case, proves that the gold nanorods assures a highly efficient electron tunnelling through the underlying 1,6HDT-SAM.

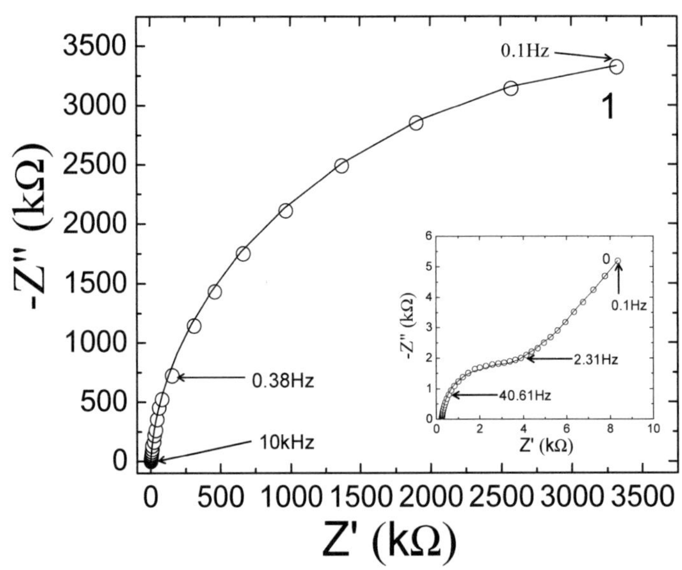

Fig. 6. Nyquist diagrams recorded at bare gold (n =0) and Au-1,6HDT-SAM modified electrode (n = 1) in aqueous solution of 0.1M $NaClO_4$ and 0.0005M $[Fe(CN)_6]^{3-/4-}$.The experimental data are represented by circles whereas the theoretical curves calculated using the parameters values in Table (2) are represented by lines.

The film capacitance, C_f, has slightly increased upon $AuNR_1$ self-assembly implying an increase of electrode area. Further build-up of the film on the gold electrodes, by the self-assembly of a third layer (1,6HDT) has determined a two orders magnitude increase in the charge transfer resistance (n = 3, table 2, Figure 7b, wine circles plot) as compared to the previous $AuNR_1$ layer (n = 2, table 2, Figure 7a) demonstrating a hindered electron transfer process. It is interestingly to remark the fact that the 1,6hexanedithiol seems to have a weaker insulating effect when self-assembled on gold nanorods surface than on bulk gold. The decrease of C_{dl} for n = 3, (table 2) as compared to the C_{dl} of the previous layer (n =2, table 2) demonstrates an increase of distance between the redox probes in solution and the underlying gold nanorod layer. This decrease of C_{dl} for the dithiol layers confirms the supposition that the cetyltrimetyl ammonium bromide stabilizing the rods will be removed during the self-assembly of the dithiol layers. Moreover, the consecutive increase of film resistance, R_f, and decrease of film capacitance, C_f, (n = 3, table 2) demonstrates that the electron tunnelling process is slow and the increase of area of the electrode is smaller at the 1,6HDT outermost layer. These alternating insulating/conductive features of the multilayer modified electrode are repeated through similar variation of the R_{CT}, C_{dl}, R_f, C_f parameters up to the 6 layers (table 2).

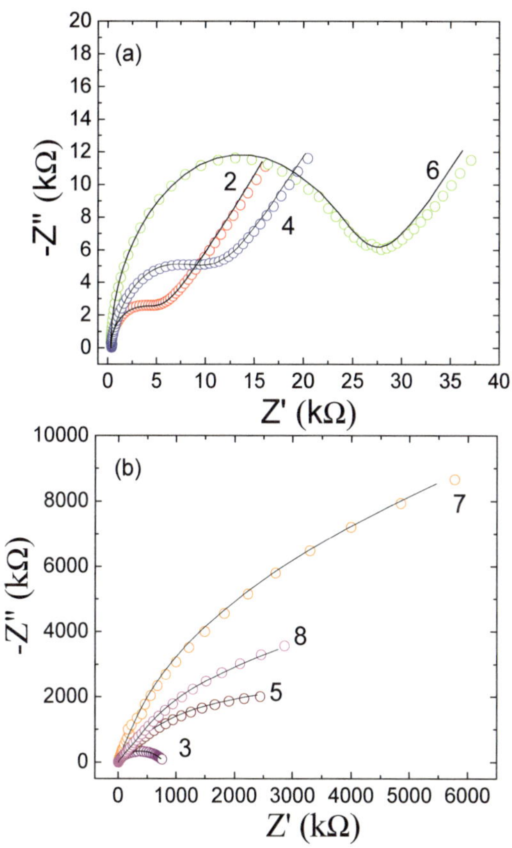

Fig. 7. Nyquist diagrams recorded at the Au-(1, 6HDT-$AuNR_1$)$_4$ multilayered modified gold electrodes in aqueous solution of 0.1M $NaClO_4$ and 0.0005M $[Fe(CN)_6]^{3-/4-}$: 1,6HDT outermost layers (n = 3,5,7) and $AuNR_1$ outermost layers (n = 2,4,6,8), respectively.

The charge transfer resistance has drastically increased from 7th to the 8th layer self-assembled on the gold electrodes demonstrating that the last layer of gold nanorods has a low surface coverage with rods (n = 7,8, table 2) and in consequence the electron transfer process is slow. The Cdl has practically no variation passing from the 7th to the 8th layer proving that the main effect in the electrochemical process is caused by the 1,6HDT layer (*n* = 7, hindered electron transfer) than by the few rods from the 8th layer. Similarly, the capacitance of the film, C_f, is practically constant for the two last layers suggesting that area of the electrode has no variation upon self-assembly of the 8th layer ($AuNR_1$ layer).The film resistance has increased (compare Rf , $n = 7$ and Rf n=8) implying that some material is still attached to the substrates. The variation of these parameters confirms the pyramidal growth of the multilayered film on gold electrodes, as shown by AFM imaging (Figures 2, 3). The additional R'_f and C'_f parameters may suggest that the electron transfer process becomes very slow when the film thickness is too high. Overall, the variation of R_{CT}, C_{dl}, R_f, C_f parameters demonstrate that a maximum of 6 layers self-assembled on the gold electrodes is the optimal number of layers which assures a very efficient electron transfer through the small gold nanorods layers and efficient electron tunnelling through the dithiol layers in the electrochemistry of $[Fe(CN)_6]^{3-/4-}$ redox probes. Moreover, the overall electrochemical process is strongly dependent of the nature of the outermost layer. This dependence of the electrochemical process on the nature of the outermost layer has been observed for the first time at mercaptosuccinic stabilized gold nanoparticle (Au-MSA)/poly-Lysine multilayers self-assembled on gold electrodes (Chirea et al., 2005). The difference between these two types of multilayers is in the nature of the interaction between the redox probes in solution and the outermost layers. The depressed semicircles observed in the EIS spectra of the 1,6HDT outermost layers are due to the insulating properties of the aliphatic chain of the dithiol which is blocking the electronic communication between the redox probes in solution and the underlying bare gold electrodes (Chou et al., 2009; Su et al.,2006).

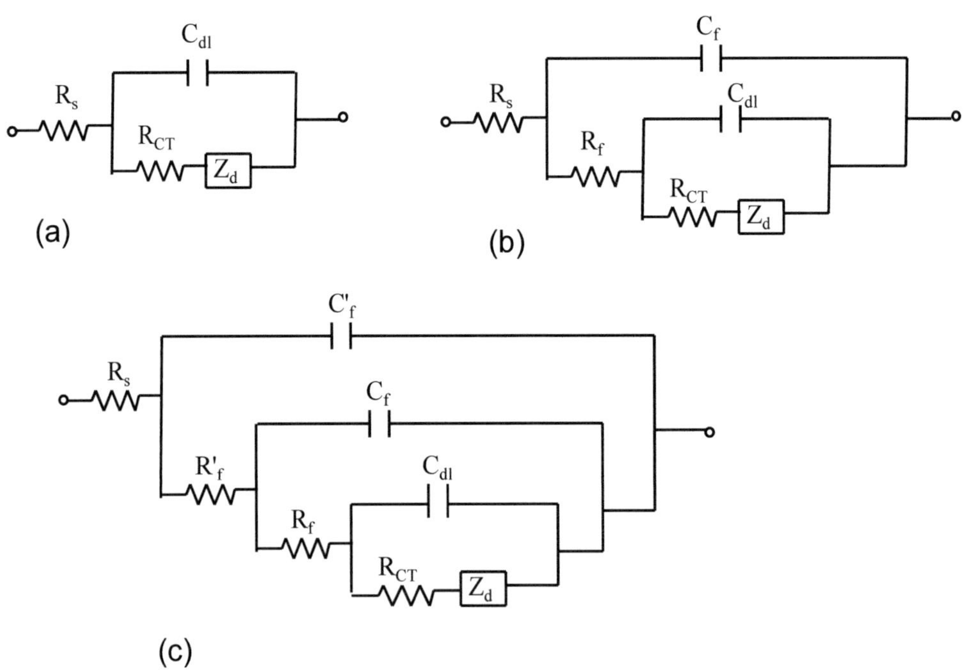

Fig. 8. Equivalent electrical circuits used for the fitting of impedance spectra represented in Figures 6, 7 and 10.

n	R_s (kΩ)	C'_f (μF)	R'_f(kΩ)	C_f (μF)	R_f(kΩ)	C_{dl} (μF)	R_{CT} (kΩ)
0	0.292 (1.05)					0.83 (0.72)	2.61 (0.58)
1	1.20 (0.98)			0.14 (1.25)	2674 (0.28)	0.05 (0.35)	3510 (0.68)
2	0.297 (1.17)			0.97 (0.70)	3.58 (0.43)	2.48 (1.06)	1.03 (0.88)
3	1.24 (1.3)			0.13 (0.60)	430 (0.56)	0.18 (1.58)	269.3
4	0.303 (1.59)			0.43 (0.45)	19.87 (0.74)	1.56 (1.20)	1.78 (0.32)
5	1.73 (1.64)			0.07 (0.37)	2.60 (1.08)	0.03 (1.06)	772 (0.91)
6	0.34 (1.50)			1.57 (0.94)	4.38 (0.60)	1.77 (0.52)	4.82 (0.84)
7	2.14 (1.04)	0.16 (0.55)	2012 (1.2)	0.08 (1.16)	207.6 (0.54)	0.06 (0.38)	818.3 (0.55)
8	7.67 (1.20)	0.11 (0.51)	3306 (1.69)	0.07 (0.76)	324.5 (0.53)	0.08 (0.61)	1234 (0.73)

Table 2. Parameter values obtained from the fittings of the impedance spectra presented in Figures 6 and 7 using the nonlinear least-square fit. The percent errors from the fit for each element are given in parentheses.

In our previous work (Chirea et al., 2005), the depressed semicircles observed in the EIS spectra were due to a repelling effect (electrostatic repulsion) between the outermost layers of negatively charged gold nanoparticles (Au-MSA NPs) and negatively charge redox probes ($[Fe(CN)_6]^{3-/4-}$) in solution. It has to be mentioned also that the electron transfer was mediated by the Au-MSA NPs when they were covered by a positively charged poly-Lysine layer which facilitated the diffusion of the electroactive species through the film based on electrostatic attraction. In the case of the gold nanorod outermost layers, the diffusion of the electroactive species is favoured by the positively charged cetyltrimetyl ammonium bromide stabilizing the rods (proven by the diffusion profiles with slope unity at low frequencies in Figure 7a, n=2, 4, 6) whereas the electron transfer process is mediated by all the gold nanorods from a layer (proven by high Faradaic currents in the CVs and SQWV and decreased peak-to-peak separations, Figure 5a, c). The surface density of the rods is also decisive for the overall electrochemical process, an optimal surface coverage assuring a fast electron transfer (Chirea et al., 2010).

3.3.2 Electrochemical characterization of multilayers containing large gold nanorods

A different electrochemical behaviour was observed at the Au-(1,6HDT-AuNR$_2$)$_4$ modified electrodes. In this case, the size of the gold nanorods seems to have a high influence on the film growth and implicitly on the overall electron transfer process. Figure 9 shows the cyclic voltammograms (Figure 9a,b) and the square wave voltammograms (Figure 9c,d) recorded at the Au-(1,6HDT-AuNR$_2$)$_4$ multilayer modified gold electrodes in aqueous solution of 0.1M $NaClO_4$ and 0.0005M $[Fe(CN)_6]^{3-/4-}$. The average sizes of the rods incorporated into this film were 51.56 nm lengths per 14.80 nm width (AuNR$_2$, Figure 1d, e) with an aspect ratio of 3.48. Although these large rods were self-assembled in an end topography (vertically aligned to the substrate surface) as the small rods, the film growth was different starting with the 4th layer self-assembled on the electrodes when the film density has increased upon the self-assembly of the $AuNR_2$ layer (compare Figure 3a and Figure 3b). The electrical properties of Au-1,6HDT-$AuNR_2$ multilayered modified electrodes were highly dependent on the type of outermost layer and the surface coverage with rods. As seen previously, the Faradic process was highly hindered by the 1,6hexanedithiol layers and this is proven by either a disappearance or a drastic decrease of peak currents in the CVs and SQWVs (Figure 9b, d, *n* =1, 3, 5, 7). The electrochemistry of $[Fe(CN)_6]^{3-/4-}$ redox probes was recovered upon the self-assembly of a $AuNR_2$ layers. For example, cyclic and square wave voltammograms with high peak currents and decreased peak separation were recorded at the Au-1,6HDT-$AuNR_2$ bilayer modified electrode (*n* = 2, Figure 9a,c red curve). The consecutive self-assembly of a 1,6HDT layer has diminished the Faradaic currents in the CVs and SQWVs (*n* = 3, Figure 9b, d, purple curve) which is equivalent to a hindered electron transfer process. The 4th layer self-assembled on the modified gold electrodes (an $AuNR_2$ layer) has restored the Faradaic currents in the CVs and SQWVs, but the electrochemical process seems to be slower than at the previous $AuNR_2$ outermost layer due to lower peak currents and higher peak to peak separation in the voltammograms (compare *n* = 4 and *n* = 2 in Figure 9a, c). The 5th layer self-assembled on the electrodes determined a complete blocking of the electronic communication between the $[Fe(CN)_6]^{3-/4-}$ couple in solution and the underlying bare gold electrode (*n* = 5, Figure 9b,d wine curve). At the 6th layer (an $AuNR_2$ outermost layer) the electrochemical process is improved although it is even slower than at the previous $AuNR_2$ outermost layers (lowest Faradaic peaks and highest increase of peak-to-peak separations in the CVs and SQWVs, *n* = 6, Figure 9a, c). The last two layers bring no evident improvement of the electron transfer process at the 1,6HDT-$AuNR_2$ multilayer modified electrodes (*n* = 7, 8, Figure 9, orange and pink curves). At this stage of the multilayer build up, the surface coverage with rods is the lowest as compared to the previous $AuNR_2$ layers and the film density is the highest (Figure 3d). The Faradaic currents and the peak-to-peak separations in the CVs were: I_a= 3.95μA, ΔE_p= 0.071V for *n* = 0 (bare gold electrode), I_a= 3.41μA, ΔE_p= 0.100V for *n* = 2, (Au-1,6HDT-$AuNR_2$), I_a= 0.33μA and ΔE_p= 0.368V for *n* = 3, (Au-1,6HDT-$AuNR_2$-1,6HDT), I_a= 2.97μA and ΔE_p= 0.172V for *n* = 4,(Au-1,6HDT-$AuNR_2)_2$, I_a= 2.61μA and ΔE_p= 0.244V for *n* = 6 ,(Au-1,6HDT-$AuNR_2)_3$ whereas for the other layers (n = 1,5,7 and 8) there were no peak currents in the CVs (Figure 9a,b). A more evident effect of each outermost layer self-assembled on the gold electrodes is proven by square wave voltammetry measurements which are showing a progressive decrease of Faradaic currents with increasing number of layers (Figure 9c,d) for both types of outermost layers. The peak currents and peak potential shifts were: I_a= 9.40μA at 0.247V for *n* =0, I_a= 11.8μA at 0.232V for *n* =2, I_a= 0.180μA at 0.283V for *n* =3, I_a= 5.12μA at 0.246V for *n* =4, I_a= 0.107μA at 0.283V for *n* =5, I_a= 2.79μA at 0.286V for *n* =6, I_a= 0.084μA at 0.267V for *n* =7 (Figure 9c and 9d).

The highest peak currents in the SQWVs were recorded at the Au-1,6HDT-$AuNR_2$ modified electrode (*n* =2, Figure 9c) for which the electron transfer process is faster than at the bare gold electrode. A lower surface coverage of gold nanorods in the upper layers of the Au-(1,6HDT-$AuNR_2)_4$ film (Figure 3) determined a slightly slower electron transfer (lower peak currents in the CVs and SQWVs, *n* = 4,6,8, Figure 9c) than at the Au-1,6HDT-$AuNR_2$ bilayer modified electrode (*n* =2, Figure 9c).The variation of the heterogeneous electron-transfer rate constants (table 3) estimated based on equation (1) -(3) and using the R_{CT} parameters obtained from the fittings of the EIS spectra (Figure 10, table 4) demonstrate that the electron transfer process at the Au-(1,6HDT-AuNR$_2$)$_4$ modified electrodes is slower than at the at the Au-(1,6HDT-AuNR$_1$)$_4$ modified electrodes in the presence of the $[Fe(CN)_6]^{3-/4-}$ redox couple. The heterogeneous electron-transfer rate constants estimated for the AuNR$_2$ outermost layers (*n* =2, 4, 6, 8, table 3) are slightly lower than the heterogeneous electron-transfer rate constants estimated for the AuNR$_1$ outermost layers (*n* = 2,4,6,8, table 1). This difference demonstrates a size effect of the gold nanorods, the short rods having a stronger catalytic activity toward the electrochemical process than the large rods. The k_{et} variations are consistent with previous published work in which was demonstrated a size effect of the rods for low and high surface coverages with rods at Au-1,6HDT-AuNR bilayers modified electrodes (Chirea et al. 2009, 2010). In those two reports was demonstrated for the first time the dependence of the electron transfer process on the size of these anisotropic gold nanorods and on the type of surface bonding (end or side surface bonding of rods). Other authors are starting to pursue similar studies. For example, it has been demonstrated a size effect toward electrochemistry of the same $[Fe(CN)_6]^{3-/4-}$ couple at CdTe QD self-assembled on gold electrodes by means of mercaptoundecanoic acid and poly(diallyldimethylammonium chloride) (Kissling et al, 2010). The CdTe dots being very small (2.2nm, 2.5nm 2.9 nm) have slightly enhance mainly the reduction process. A size of 3nm or higher average diameters of the nanomaterials induces a higher enhancement of the electron transfer process as it was demonstrated in another report (Chirea et al., 2007).

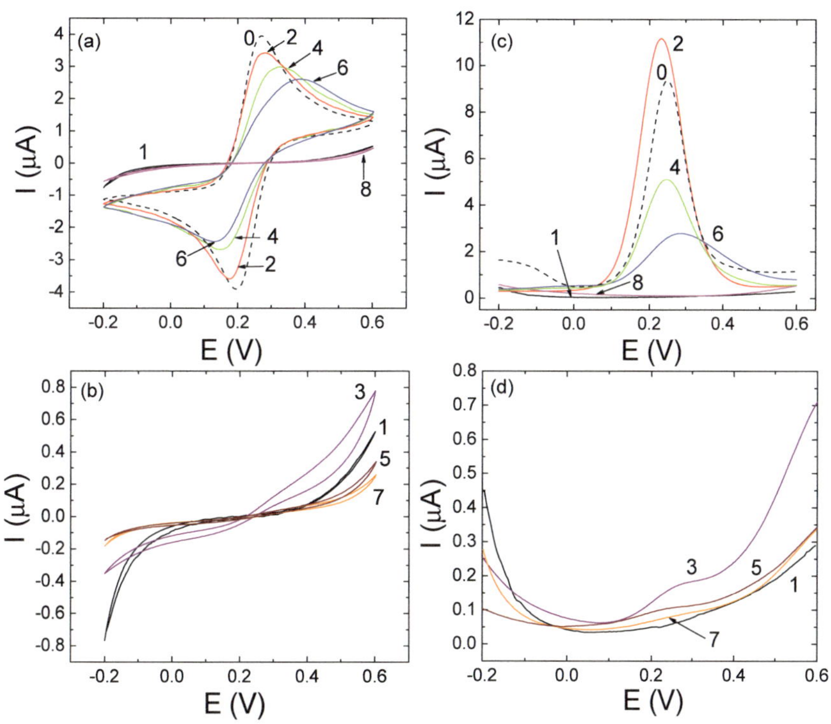

Fig. 9. Cyclic voltammograms (a, b) and square wave voltammograms (c, d) recorded at the Au-(1, 6HDT-AuNR$_2$)$_4$ multilayered modified gold electrodes in aqueous solution of 0.1M $NaClO_4$ and 0.0005M $[Fe(CN)_6]^{3-/4-}$: bare gold (*n* = 0), 1,6HDT outermost layers (*n* = 1,3,5,7) and AuNR$_2$ outermost layers (*n* = 2,4,6,8), respectively.

n	k_{et} (cm×s^{-1})	self-assembly time (h)	β(Å^{-1})	Tunnelling distance (Å)
1	0.50×10^{-8}	14	0.91	17.04
2	1.02×10^{-5}	15		
3	0.25×10^{-7}	11	0.82	17.04
4	0.45×10^{-5}	16		
5	0.42×10^{-8}	18	0.92	17.04
6	0.15×10^{-5}	18		
7	0.30×10^{-8}	15	0.96	17.04
8	0.89×10^{-8}	6		

Table 3. Heterogeneous electron-transfer rate constants, k_{et} and tunnelling parameter, β, calculated for each self-assembled layer within the Au-(1,6HDT-AuNR$_2$)$_4$ multilayer, using the R_{CT} parameters obtained from the fittings of the EIS spectra (table 4) and based on equations (1)-(3).

The impedance responses recorded at the Au-(1,6HDT-AuNR$_2$)$_4$ multilayered modified electrodes in aqueous solution of $[Fe(CN)_6]^{3-/4-}$ redox probes were dependent on the type of outermost layer (Figure 10). The EIS spectra featured depressed semicircles at 1,6HDT outermost layers (Figure 10b, *n*=3, 5, 7) for all range of frequencies. The charge transfer resistance R_{CT} has increased 3 or 2 orders of magnitude at the 1,6HDT outermost layers (*n*=1, 5, 7 and *n*=3, table 4) than at the bare gold electrode or AuNR$_2$ outermost modified electrodes (*n* =2, 4, 6, table 4). The C_{dl} varied with the insulating (decrease) or conductive properties (increase) of the outermost layers (table 4). Although, the charge transfer resistance R_{CT} has decreased at the AuNR$_2$ outermost layers, its tendency is to increase with increasing numbers of layers self-assembled on the gold electrodes implying that the electron tunnelling through the aliphatic chains of the dithiol and the electron transport mediated by the large gold nanorods is less efficient when the number of gold nanorods is decreasing due to the pyramidal growth of the film (Figure 3). The R_f has also a tendency to increase with increasing number of layers self-assembled on the electrodes being either 1,6HDT layers (higher R_f, *n* =1, 3, 5, 7, table 4) or AuNR$_2$ layers (lower R_f, n =2, 4, 6, 8, table 4).

The variation of these parameters demonstrates that the electron transfer is faster at the AuNR$_2$ outermost layers than at the 1,6HDT outermost layers and overall the increasing film thickness is affecting the overall electron transfer process. Interestingly, the additional parameters R'_f and C'_f have an evident effect on the overall electron transfer process starting the 3rd layer (table 4). These parameters are related to the electron tunnelling on a horizontal plane through the 1,6HDT layer connecting neighbours AuNR$_2$ (Scheme 1). As shown in the AFM images (Figure 4) these large rods can be covalently bonded on the horizontal plane, not only on a vertical plane. The increase of both film resistances R_f and R'_f and progressive decrease of the capacitances C_f and C'_f demonstrates that the efficiency of the electron tunnelling process is progressively diminished at low surface coverage with rods (in the upper layers) and the area of the electrodes is increasing. These electrochemical features do not imply that the Au-1,6HDT-AuNR$_2$ multilayers modified electrodes can be less efficient for other electrochemical processes. For example, Chen-Zhong et al. have demonstrated that gold nanoparticles of 20-40 nm average diameters electrochemically deposited on gold electrodes and chemically modified by the self-assembly of thiolated DNA layers specific for anticancer drugs detection have determined a 20-40 fold increase of the detection limit (Chen-Zhong et al.,2005).

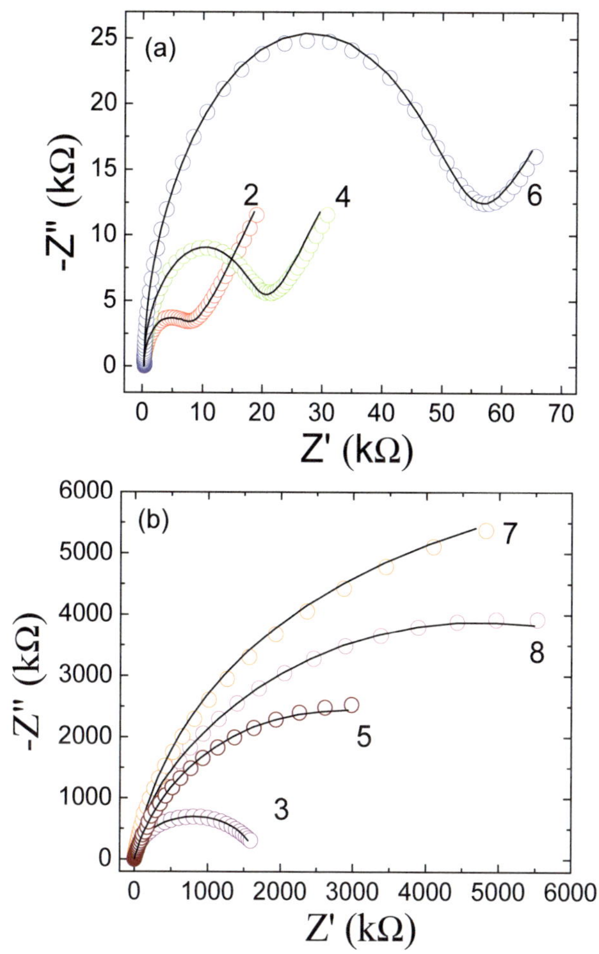

Fig. 10. Nyquist diagrams recorded at the Au-(1, 6HDT-AuNR$_2$)$_4$ multilayered modified gold electrodes in aqueous solution of 0.1M $NaClO_4$ and 0.0005M $[Fe(CN)_6]^{3-/4-}$: 1,6HDT outermost layers (n = 3,5,7) and AuNR$_2$ outermost layers (n = 2,4,6,8), respectively.

Furthermore, recent studies have demonstrated that citrate stabilized gold nanoparticles embedded into poly-Lysine matrix through electropolymerization on glassy carbon electrodes from aqueous solution have improved the electron transfer kinetics of the $[Fe(CN)_6]^{3-/4-}$ redox probes (Chirea et al., 2011a). Additionally, the citrate gold nanoparticles/poly-Lysine films where used for the electrostatic self-assembly of dsDNA. The resulted hybrid nanostructure has proven an improved sensitivity and a good detection limit for an anticancer drug, namely actinomycin D (Chirea et al., 2011b).

n	R_s (kΩ)	C'_f (μF)	R'_f(kΩ)	C_f (μF)	R_f(kΩ)	C_{dl} (μF)	R_{CT} (kΩ)
0	0.290 (0.95)					0.78 (0.93)	2.53 (0.89)
1	1.05 (0.87)			0.18 (0.73)	2523 (0.62)	0.04 (0.67)	3427 (0.98)
2	0.290 (1.06)			0.90 (0.53)	5.53 (0.55)	1.79 (1.34)	1.64 (0.60)
3	1.44 (0.93)	0.81 (1.51)	274.7 (1.63)	0.10 (0.50)	625.6 (0.52)	0.05 (0.97)	682.4 (0.80)
4	0.296 (1.30)			0.58 (0.42)	14.65 (1.26)	1.43 (1.05)	3.75 (1.16)
5	4.77 (1.68)	0.17 (0.83)	2650 (0.83)	0.08 (0.83)	1330 (1.74)	0.03 (0.57)	3973.0 (0.91)
6	0.33 (1.18)			0.56 (0.32)	39.64 (1.16)	1.02 (1.36)	11.29 (0.65)
7	10.23 (0.95)			0.011 (0.69)	293 (0.63)	0.05 (0.65)	5558.3 (1.20)
8	7.07 (1.31)	0.014 (1.38)	2254 (1.32)	0.013 (0.82)	459 (0.98)	0.072 (0.61)	1899 (1.05)

Table 4. Parameter values obtained from the fittings of the impedance spectra represented in Figures 10 using the nonlinear least-square fit.

4. Conclusion

The dynamics of charge transfer at multilayered films composed of 1,6hexanedithiol/gold nanorods sequentially self-assembled on gold electrodes have been investigated as a function of the nature of the outermost layer and the size of the gold nanorods. The increase of R_{CT} upon covalent self-assembly of the 1,6hexanedithiol layers on bulk gold and gold nanorods layers is rationalized in terms of electronic blocking between the redox probes in solution and underlying gold nanorod layers or bare gold surface. The 3 orders magnitude decrease of R_{CT} upon consecutive covalent self-assembly of gold nanorods as outermost layers was rationalized in terms of increase in the interfacial concentration of the electroactive species at the gold nanorods surface (due to electrostatic attraction) , electron transport mediated by the rods and coherent electron tunnelling through the underlying 1,6hexanedithiol layers. The electron transfer process was faster (higher k_{et} values) at the multilayers containing short rods (35.48nm/14.52nm average sizes) than at the multilayers containing large rods (slightly lower k_{et} values, 51.56nm/14.80nm average sizes). An evident size effect, due to higher catalytic activity of the short rods, was observed in the electrochemical process. An optimal number of maximum 6 layers within the multilayered films have determined a fast electron transfer and coherent electron tunnelling at the hybrid nanostructured modified electrodes. The chemical structures of these films with an end bonding topography of the rods and their enhanced electrochemical properties make them very appealing for the fabrication of electronic devices or biosensors applications (Katz, E.&Willner , I., 2003).

5. Appendix A

The total impedance corresponding to the electrical circuit from Figure 8b, Z_b is written below:

$$Z_b = Z_b' + Z_b''$$

$$Z_b' = R_s + \frac{\frac{R_f+\sigma+R_{ct}}{C_{dl}^4\omega^4} + \frac{C_fR_{ct}^3 + C_f\sigma R_{ct}^2 + C_fR_{ct}^2R_f - R_{ct}^2R_fC_f - R_{ct}^3C_f - R_{ct}^4}{C_{dl}^3\omega^2} + \frac{2\sigma^2 + 2\sigma R_{ct} + 4\sigma R_f}{C_{dl}^3\omega^{7/2}} + \frac{2R_fR_{ct}^2 + 2R_fC_f\sigma^3 + R_{ct}^3 + R_{ct}^2\sigma}{C_{dl}^2\omega^2} + \frac{8\sigma^2 + 2\sigma^3 + 2R_{ct}\sigma^2}{C_{dl}^2\omega^3} + \frac{4\sigma R_{ct}R_f + 2\sigma R_{ct}^2 + 2\sigma^2R_{ct}}{C_{dl}^2\omega^{5/2}} - \frac{2R_fC_f\sigma^3}{C_{dl}^2\omega}}{\left(C_f\omega R_fR_{ct}^2 + 2R_fC_f\omega\sigma^2 + 2\sigma R_{ct}R_fC_f\omega^{1/2} + \frac{2\sigma R_fC_f}{C_{dl}\omega^{1/2}} + \frac{R_{ct}C_f + R_fC_f + \sigma C_f}{C_{dl}^2\omega}\right)^2 + \left(\frac{C_f\sigma}{C_{dl}^2\omega} + \frac{C_fR_{ct}^2}{C_{dl}} + \frac{C_f2\sigma R_{ct}}{C_{dl}\omega^{1/2}} + \frac{C_f2\sigma^2}{C\omega} + R_{ct}^2 + 2\left(\frac{\sigma}{\sqrt{\omega}}\right)^2 + 2\frac{\sigma R_{ct}}{\sqrt{\omega}} + \left(\frac{1}{C_{dl}\omega}\right)^2 + \frac{2\sigma}{C_{dl}\omega^{3/2}}\right)^2}$$

$$+ \frac{-\frac{2R_fC_f\sigma^2R_{ct}^2}{C_{dl}} + \frac{4C_f\sigma R_fR_{ct}^3 - 4\sigma R_fC_fR_{ct}^3 - 4\sigma^3R_{ct}R_fC_f}{C_{dl}\omega^{1/2}} + R_fR_{ct}^4 + \frac{8\sigma^2R_fR_{ct}^2}{\omega} + \frac{2\sigma R_fR_{ct}^3 + 2\sigma R_{ct}^2R_f}{\omega^{1/2}} + \frac{4\sigma R_fR_{ct}^2 + 4R_fC_f\sigma^3R_{ct} + 4\sigma^2C_fR_{ct}R_f - 4\sigma^3R_{ct}R_fC_f}{C_{dl}\omega^{3/2}}}{\left(C_f\omega R_fR_{ct}^2 + 2R_fC_f\omega\sigma^2 + 2\sigma R_{ct}R_fC_f\omega^{1/2} + \frac{2\sigma R_fC_f}{C_{dl}\omega^{1/2}} + \frac{R_{ct}C_f + R_fC_f + \sigma C_f}{C_{dl}^2\omega}\right)^2 + \left(\frac{C_f\sigma}{C_{dl}^2\omega} + \frac{C_fR_{ct}^2}{C_{dl}} + \frac{C_f2\sigma R_{ct}}{C_{dl}\omega^{1/2}} + \frac{C_f2\sigma^2}{C\omega} + R_{ct}^2 + 2\left(\frac{\sigma}{\sqrt{\omega}}\right)^2 + 2\frac{\sigma R_{ct}}{\sqrt{\omega}} + \left(\frac{1}{C_{dl}\omega}\right)^2 + \frac{2\sigma}{C_{dl}\omega^{3/2}}\right)^2}$$

$$+ \frac{\frac{2R_fC_f\sigma^2R_{ct}^2 - 4\sigma^4R_fC_f}{C_{dl}\omega} + \frac{4R_fC_f\sigma^4 + 4\sigma^2R_{ct}R_f + 4\sigma^3R_{ct}R_f}{C_{dl}\omega^2} + \frac{4R_f\sigma^4}{\omega^2} + \frac{8\sigma^3R_{ct}R_f}{\omega^{3/2}} + \frac{8R_f\sigma^3}{C_{dl}\omega^{5/2}}}{\left(C_f\omega R_fR_{ct}^2 + 2R_fC_f\omega\sigma^2 + 2\sigma R_{ct}R_fC_f\omega^{1/2} + \frac{2\sigma R_fC_f}{C_{dl}\omega^{1/2}} + \frac{R_{ct}C_f + R_fC_f + \sigma C_f}{C_{dl}^2\omega}\right)^2 + \left(\frac{C_f\sigma}{C_{dl}^2\omega} + \frac{C_fR_{ct}^2}{C_{dl}} + \frac{C_f2\sigma R_{ct}}{C_{dl}\omega^{1/2}} + \frac{C_f2\sigma^2}{C\omega} + R_{ct}^2 + 2\left(\frac{\sigma}{\sqrt{\omega}}\right)^2 + 2\frac{\sigma R_{ct}}{\sqrt{\omega}} + \left(\frac{1}{C_{dl}\omega}\right)^2 + \frac{2\sigma}{C_{dl}\omega^{3/2}}\right)^2}$$

$$Z_b'' = -i\frac{R_f^2C_f\omega R_{ct}^4 + 2R_f^2R_{ct}^2C_f\omega\sigma^2 + 4\sigma C_f\omega^{1/2}R_f^2R_{ct}^3 + 6\sigma^2C_fR_f^2R_{ct}^2 + 4R_f^2C_f\sigma^4 + 4\sigma^3R_f^2C_f\omega^{1/2}R_{ct} + \frac{4\sigma^3R_{ct}R_f^2C_f}{\omega^{1/2}}}{\left(C_f\omega R_fR_{ct}^2 + 2R_fC_f\omega\sigma^2 + 2\sigma R_{ct}R_fC_f\omega^{1/2} + \frac{2\sigma R_fC_f}{C_{dl}\omega^{1/2}} + \frac{R_{ct}C_f + R_fC_f + \sigma C_f}{C_{dl}^2\omega}\right)^2 + \left(\frac{C_f\sigma}{C_{dl}^2\omega} + \frac{C_fR_{ct}^2}{C_{dl}} + \frac{C_f2\sigma R_{ct}}{C_{dl}\omega^{1/2}} + \frac{C_f2\sigma^2}{C\omega} + R_{ct}^2 + 2\left(\frac{\sigma}{\sqrt{\omega}}\right)^2 + 2\frac{\sigma R_{ct}}{\sqrt{\omega}} + \left(\frac{1}{C_{dl}\omega}\right)^2 + \frac{2\sigma}{C_{dl}\omega^{3/2}}\right)^2}$$

$$-i\frac{\frac{4\sigma R_f^2C_fR_{ct}^2 + 4\sigma^3R_f^2C_f}{C_{dl}\omega^{1/2}} + \frac{8\sigma^2R_f^2C_fR_{ct} + R_{ct}^4}{C_{dl}\omega} + \frac{4\sigma^3R_f^2C_f + 4\sigma R_{ct}^3}{C_{dl}\omega^{3/2}} + \frac{8\sigma^2R_{ct}^2}{C_{dl}\omega^2} + \frac{8\sigma^3R_{ct}}{C_{dl}\omega^{5/2}} + \frac{4\sigma^4}{C\omega^3} + \frac{R_fR_{ct}^3C_f + R_f^2R_{ct}^2C_f + \sigma R_fR_{ct}^2C_f + C_fR_{ct}^4}{C_{dl}^2\omega} + \frac{2\sigma^2R_{ct} + 8C_f\sigma^3R_{ct}}{C_{dl}^2\omega^{5/2}}}{\left(C_f\omega R_fR_{ct}^2 + 2R_fC_f\omega\sigma^2 + 2\sigma R_{ct}R_fC_f\omega^{1/2} + \frac{2\sigma R_fC_f}{C_{dl}\omega^{1/2}} + \frac{R_{ct}C_f + R_fC_f + \sigma C_f}{C_{dl}^2\omega}\right)^2 + \left(\frac{C_f\sigma}{C_{dl}^2\omega} + \frac{C_fR_{ct}^2}{C_{dl}} + \frac{C_f2\sigma R_{ct}}{C_{dl}\omega^{1/2}} + \frac{C_f2\sigma^2}{C\omega} + R_{ct}^2 + 2\left(\frac{\sigma}{\sqrt{\omega}}\right)^2 + 2\frac{\sigma R_{ct}}{\sqrt{\omega}} + \left(\frac{1}{C_{dl}\omega}\right)^2 + \frac{2\sigma}{C_{dl}\omega^{3/2}}\right)^2}$$

$$-i\frac{\dfrac{2\sigma^2R_fR_{ct}C_f + 6\sigma^2{R_f}^2C_f + 2\sigma^3R_fC_f + C_f\omega R_f{R_{ct}}^3 + 2\sigma^2R_{ct}R_fC_f\omega + 2\sigma{R_{ct}}^2R_fC_f\omega^{\frac{1}{2}} + \sigma C_f\omega R_f{R_{ct}}^2 + 2\sigma^3R_fC_f\omega + 2\sigma^2R_{ct}R_fC_f\omega^{\frac{1}{2}} + C_f\omega{R_f}^2{R_{ct}}^2 + 2{R_f}^2C_f\omega\sigma^2 + 2\sigma R_{ct}{R_f}^2C_f\omega^{\frac{1}{2}} + \sigma R_{ct}^2 + 8C_f\sigma^2{R_{ct}}^2}{C_{dl}^2\omega^2}}{\left(C_f\omega R_f{R_{ct}}^2 + 2R_fC_f\omega\sigma^2 + 2\sigma R_{ct}R_fC_f\omega^{1/2} + \dfrac{2\sigma R_fC_f}{C_{dl}\omega^{1/2}} + \dfrac{R_{ct}C_f + R_fC_f + \sigma C_f}{C_{dl}^2\omega}\right)^2 + \left(\dfrac{C_f\sigma}{C_{dl}^2\omega} + \dfrac{C_f{R_{ct}}^2}{C_{dl}} + \dfrac{C_f2\sigma R_{ct}}{C_{dl}\omega^{1/2}} + \dfrac{C_f2\sigma^2}{C\omega} + {R_{ct}}^2 + 2\left(\dfrac{\sigma}{\sqrt{\omega}}\right)^2 + 2\dfrac{\sigma R_{ct}}{\sqrt{\omega}} + \left(\dfrac{1}{C_{dl}\omega}\right)^2 + \dfrac{2\sigma}{C_{dl}\omega^{3/2}}\right)^2}$$

$$-i\frac{\dfrac{2\sigma R_{ct}{R_f}^2C_f + 2\sigma{R_{ct}}^2R_fC_f + 2\sigma^2C_fR_{ct}R_f + 4\sigma C_f{R_{ct}}^3}{C_{dl}^2\omega^{3/2}} + \dfrac{2\sigma^3 + 4\sigma^2R_{ct} + 4C_f\sigma^4}{C_{dl}^2\omega^3} + \dfrac{2\sigma R_{ct} + 2\sigma{R_{ct}}^2 + 4\sigma^3}{C_{dl}^2\omega^{7/2}} + \dfrac{4\sigma{R_f}^2C_f + 4\sigma R_fR_{ct}C_f + 4\sigma^2R_fC_f + 4C_f\sigma^2R_{ct}}{C_{dl}^3\omega^{5/2}}}{\left(C_f\omega R_f{R_{ct}}^2 + 2R_fC_f\omega\sigma^2 + 2\sigma R_{ct}R_fC_f\omega^{1/2} + \dfrac{2\sigma R_fC_f}{C_{dl}\omega^{1/2}} + \dfrac{R_{ct}C_f + R_fC_f + \sigma C_f}{C_{dl}^2\omega}\right)^2 + \left(\dfrac{C_f\sigma}{C_{dl}^2\omega} + \dfrac{C_f{R_{ct}}^2}{C_{dl}} + \dfrac{C_f2\sigma R_{ct}}{C_{dl}\omega^{1/2}} + \dfrac{C_f2\sigma^2}{C\omega} + {R_{ct}}^2 + 2\left(\dfrac{\sigma}{\sqrt{\omega}}\right)^2 + 2\dfrac{\sigma R_{ct}}{\sqrt{\omega}} + \left(\dfrac{1}{C_{dl}\omega}\right)^2 + \dfrac{2\sigma}{C_{dl}\omega^{3/2}}\right)^2}$$

$$-i\frac{\dfrac{2\sigma C_f{R_{ct}}^2}{C_{dl}^3\omega^2} + \dfrac{{R_{ct}}^2 + 4\sigma^3C_f}{C_{dl}^3\omega^3} + \dfrac{2\sigma^2}{C_{dl}^3\omega^{7/2}} + \dfrac{2\sigma^2}{C_{dl}^3\omega^4} + \dfrac{2R_{ct}R_fC_f + {R_f}^2C_f + {R_{ct}}^2C_f + 2\sigma R_{ct}C_f + 2\sigma R_fC_f + 2\sigma^2C_f + \sigma}{C_{dl}^4\omega^3}}{\left(C_f\omega R_f{R_{ct}}^2 + 2R_fC_f\omega\sigma^2 + 2\sigma R_{ct}R_fC_f\omega^{1/2} + \dfrac{2\sigma R_fC_f}{C_{dl}\omega^{1/2}} + \dfrac{R_{ct}C_f + R_fC_f + \sigma C_f}{C_{dl}^2\omega}\right)^2 + \left(\dfrac{C_f\sigma}{C_{dl}^2\omega} + \dfrac{C_f{R_{ct}}^2}{C_{dl}} + \dfrac{C_f2\sigma R_{ct}}{C_{dl}\omega^{1/2}} + \dfrac{C_f2\sigma^2}{C\omega} + {R_{ct}}^2 + 2\left(\dfrac{\sigma}{\sqrt{\omega}}\right)^2 + 2\dfrac{\sigma R_{ct}}{\sqrt{\omega}} + \left(\dfrac{1}{C_{dl}\omega}\right)^2 + \dfrac{2\sigma}{C_{dl}\omega^{3/2}}\right)^2}$$

The total impedance corresponding to the electrical circuit from Figure 8c, Z_c is given below:

$$Z_c = R_s + \frac{-\dfrac{iA}{C_f'\omega}}{-\dfrac{i}{C_f'\omega}\left(C_f\omega R_f{R_{ct}}^2 + 2R_fC_f\omega\sigma^2 + 2\sigma R_{ct}R_fC_f\omega^{1/2} + \dfrac{2\sigma R_fC_f}{C_{dl}\omega^{1/2}} + \dfrac{R_{ct}C_f + R_fC_f + \sigma C_f}{C_{dl}^2\omega}\right)^2 + -\dfrac{i}{C_f'\omega}\left(\dfrac{C_f\sigma}{C_{dl}^2\omega} + \dfrac{C_f{R_{ct}}^2}{C_{dl}} + \dfrac{C_f2\sigma R_{ct}}{C_{dl}\omega^{1/2}} + \dfrac{C_f2\sigma^2}{C\omega} + {R_{ct}}^2 + 2\left(\dfrac{\sigma}{\sqrt{\omega}}\right)^2 + 2\dfrac{\sigma R_{ct}}{\sqrt{\omega}} + \left(\dfrac{1}{C_{dl}\omega}\right)^2 + \dfrac{2\sigma}{C_{dl}\omega^{3/2}}\right)^2 + A}$$

where

$$
\begin{aligned}
A = {} & \frac{R_f + \sigma + R_{ct}}{C_{dl}^4 \omega^4} + \frac{C_f R_{ct}{}^3 + C_f \sigma R_{ct}{}^2 + C_f R_{ct}{}^2 R_f - R_{ct}{}^2 R_f C_f - R_{ct}{}^3 C_f - R_{ct}{}^4}{C_{dl}^3 \omega^2} + \frac{2\sigma^2 + 2\sigma R_{ct} + 4\sigma R_f}{C_{dl}^3 \omega^{7/2}} + \frac{2R_f R_{ct}{}^2 + 2R_f C_f \sigma^3 + R_{ct}{}^3 + R_{ct}{}^2 \sigma}{C_{dl}^2 \omega^2} + \frac{8\sigma^2 + 2\sigma^3 + 2R_{ct}\sigma^2}{C_{dl}^2 \omega^3} + \frac{4\sigma R_{ct} R_f + 2\sigma R_{ct}{}^2 + 2\sigma^2 R_{ct}}{C_{dl}^2 \omega^{5/2}} - \frac{2R_f C_f \sigma^3}{C_{dl}^2 \omega} \\
& + R_f' \left(C_f \omega R_f R_{ct}{}^2 + 2R_f C_f \omega \sigma^2 + 2\sigma R_{ct} R_f C_f \omega^{1/2} + \frac{2\sigma R_f C_f}{C_{dl} \omega^{1/2}} + \frac{R_{ct} C_f + R_f C_f + \sigma C_f}{C_{dl}^2 \omega} \right)^2 + R_f' \left(\frac{C_f \sigma}{C_{dl}^2 \omega} + \frac{C_f R_{ct}{}^2}{C_{dl}} + \frac{C_f 2\sigma R_{ct}}{C_{dl} \omega^{1/2}} + \frac{C_f 2\sigma^2}{C\omega} + R_{ct}{}^2 + 2\left(\frac{\sigma}{\sqrt{\omega}}\right)^2 + 2\frac{\sigma R_{ct}}{\sqrt{\omega}} + \left(\frac{1}{C_{dl}\omega}\right)^2 + \frac{2\sigma}{C_{dl}\omega^{3/2}} \right)^2 - \frac{2R_f C_f \sigma^2 R_{ct}{}^2}{C_{dl}} \\
& + \frac{4C_f \sigma R_f R_{ct}{}^3 - 4\sigma R_f C_f R_{ct}{}^3 - 4\sigma^3 R_{ct} R_f C_f}{C_{dl} \omega^{1/2}} + \frac{2R_f C_f \sigma^2 R_{ct}{}^2 - 4\sigma^4 R_f C_f}{C_{dl} \omega} + \frac{4R_f C_f \sigma^4 + 4\sigma^2 R_{ct} R_f + 4\sigma^3 R_{ct} R_f}{C_{dl} \omega^2} + \frac{4R_f \sigma^4}{\omega^2} + \frac{8\sigma^3 R_{ct} R_f}{\omega^{3/2}} + \frac{8R_f \sigma^3}{C_{dl} \omega^{5/2}} \\
& - i \Bigg(R_f{}^2 C_f \omega R_{ct}{}^4 + 2R_f{}^2 R_{ct}{}^2 C_f \omega \sigma^2 + 4\sigma C_f \omega^{1/2} R_f{}^2 R_{ct}{}^3 + 6\sigma^2 C_f R_f{}^2 R_{ct}{}^2 + 4R_f{}^2 C_f \sigma^4 + 4\sigma^3 R_f{}^2 C_f \omega^{1/2} R_{ct} + \frac{4\sigma^3 R_{ct} R_f{}^2 C_f}{\omega^{1/2}} + \frac{4\sigma R_f{}^2 C_f R_{ct}{}^2 + 4\sigma^3 R_f{}^2 C_f}{C_{dl} \omega^{1/2}} + \frac{8\sigma^2 R_f{}^2 C_f R_{ct} + R_{ct}{}^4}{C_{dl} \omega} + \frac{4\sigma^3 R_f{}^2 C_f + 4\sigma R_{ct}{}^3}{C_{dl} \omega^{3/2}} + \frac{8\sigma^2 R_{ct}{}^2}{C_{dl} \omega^2} \\
& + \frac{8\sigma^3 R_{ct}}{C_{dl} \omega^{5/2}} + \frac{4\sigma^4}{C\omega^3} + \frac{R_f R_{ct}{}^3 C_f + R_f{}^2 R_{ct}{}^2 C_f + \sigma R_f R_{ct}{}^2 C_f + C_f R_{ct}{}^4}{C_{dl}^2 \omega} + \frac{2\sigma^2 R_{ct} + 8C_f \sigma^3 R_{ct}}{C_{dl}^2 \omega^{5/2}} \\
& + \frac{2\sigma^2 R_f R_{ct} C_f + 6\sigma^2 R_f{}^2 C_f + 2\sigma^3 R_f C_f + C_f \omega R_f R_{ct}{}^3 + 2\sigma^2 R_{ct} R_f C_f \omega + 2\sigma R_{ct}{}^2 R_f C_f \omega^{\frac{1}{2}} + \sigma C_f \omega R_f R_{ct}{}^2 + 2\sigma^3 R_f C_f \omega + 2\sigma^2 R_{ct} R_f C_f \omega^{\frac{1}{2}} + C_f \omega R_f{}^2 R_{ct}{}^2 + 2R_f{}^2 C_f \omega \sigma^2 + 2\sigma R_{ct} R_f{}^2 C_f \omega^{\frac{1}{2}} + \sigma R_{ct}^2 + 8C_f \sigma^2 R_{ct}{}^2}{C_{dl}^2 \omega^2} \\
& + \frac{2\sigma R_{ct} R_f{}^2 C_f + 2\sigma R_{ct}{}^2 R_f C_f + 2\sigma^2 C_f R_{ct} R_f + 4\sigma C_f R_{ct}{}^3}{C_{dl}^2 \omega^{3/2}} + \frac{2\sigma^3 + 4\sigma^2 R_{ct} + 4C_f \sigma^4}{C_{dl}^2 \omega^3} + \frac{2\sigma R_{ct} + 2\sigma R_{ct}{}^2 + 4\sigma^3}{C_{dl}^2 \omega^{7/2}} + \frac{4\sigma R_f{}^2 C_f + 4\sigma R_f R_{ct} C_f + 4\sigma^2 R_f C_f + 4C_f \sigma^2 R_{ct}}{C_{dl}^3 \omega^{5/2}} + \frac{2\sigma C_f R_{ct}{}^2}{C_{dl}^3 \omega^2} + \frac{R_{ct}{}^2 + 4\sigma^3 C_f}{C_{dl}^3 \omega^3} + \frac{2\sigma^2}{C_{dl}^3 \omega^{7/2}} + \frac{2\sigma^2}{C_{dl}^3 \omega^4} \\
& + \frac{2R_{ct} R_f C_f + R_f{}^2 C_f + R_{ct}{}^2 C_f + 2\sigma R_{ct} C_f + 2\sigma R_f C_f + 2\sigma^2 C_f + \sigma}{C_{dl}^4 \omega^3} \Bigg)
\end{aligned}
$$

6. Acknowledgment

Financial support from Fundação para a Ciência e a Tecnologia (FCT) of Portugal through the fellowship number SFRH/BPD/39294/2007 is gratefully acknowledged.

7. References

Abdelrahman, A.I.; A.I.; Mohammad,A.M; Okajima, T.; Ohsaka,T.(2006).Fabrication and electrochemical application of three-dimensional gold nanoparticles: self-assembly., *J.Phys.Chem.B*, 2006, 110, pp.2798-2803, ISSN 1520-5207

Bradbury, C.R.; Zhao,J.; ,Fermin, D.J.(2008). Distance-independent charge-transfer resistance at gold electrodes modified by thiol monolayers and metal nanoparticles. *J.Phys.Chem.C*, 2008, 112, pp.10153-10160, ISSN 1932-7455

Chen, S.W.; Pei, R.J., (2001). Ion-induced rectification of the nanoparticle quantized capacitance charging in aqueous solution. *J.Am.Chem.Soc.*2001, 123, pp.10607-10615, ISSN 1520-5126

Chen-Zhong, L., Liu, Y., Luong, J.H.T. (2005). Impedance sensing of DNA binding drugs using gold substrates modified with gold nanoparticles. *Anal.Chem.*, 2005, 77, pp.478-485, ISSN 0003-2700

Chirea, M.; Garcia-Morales, V.; Manzanares, J.A.; Pereira, C.; Gulaboski, R.; Silva, F.(2005) Electrochemical characterization of polyelectrolyte/gold nanoparticle multilayers self-assembled on gold electrodes. *J.Phys.Chem.B*,2005,109,pp.21808-21817, ISSN 1520-5207

Chirea, M.; Pereira, C.M.; Silva, F.(2007). Catalytic effect of gold nanoparticles self-assembled into multilayered polyelectrolyte films, *J.Phys.Chem.C*, 2007, 111, pp.9255-9266, ISSN 1932-7455

Chirea , M.; Cruz, A., Pereira, M.C., Silva, A.F., (2009). Size dependent electrochemical properties of gold nanorods. *J.Phys.Chem.C*, 2009, 113, pp.13077-13087, ISSN 1932-7455

Chirea, M.; Borges, J Pereira, M.C., Silva, A.F., (2010). Density dependent electrochemical properties of vertically aligned gold nanorods. *J.Phys.Chem.C*, 2010, 114, pp.9478-9488, ISSN 1932-7455

Chirea, M., Pereira, E.M., Pereira, C.M., Silva, F.(2011a). Synthesis of poly-Lysine/gold nanoparticles films and their electrocatalytic properties. *Biointerface Research in Applied Chemistry*, 2011, 1,(4) pp.119-126, ISSN 2069-5837

Chirea, M., Pereira, E.M., Pereira, C.M., Silva, F.(2011b). DNA Biosensor for the Detection of Actinomycin D. *Biointerface Research in Applied Chemistry*, 2011, 1,(4), pp.151-159, ISSN 2069-5837

Chou, A.; Eggers, P.K.; Paddon-Row, M.N.; Gooding, J.J.(2009).Self-assembled carbon nanotube electrode arrays: effect of length of the linker between nanotubes and electrode.J.*Phys.Chem.*C, 2009, 113, pp.3203-3211, ISSN 1932-7455

Decher, G., Schlenoff, J. B., Eds.(2002).*Multilayer thin films: Sequentially assembly of nanocomposite materials*. Wiley-VCH:Weinheim, ISBN 9783527600571, Germany

Feldheim, D.L. & Foss Jr, C. A.(2002). *Metal Nanoparticles: Synthesis, Characterization and Applications*, Marcel Dekker, Inc., ISBN 0-8247-0604-8.New York,2002

Finklea, H.O.,Avery, S., Lynch, M., Furtsch, T. (1987).Blocking oriented monolayers of alkyl mercaptans on gold electrodes. *Langmuir* 1987,3,pp.409-413, ISSN: 1520-5827

Finklea, H.O.(1996). In *Electroanalytical Chemistry: A Series of Advances*, A.J.Bard, I.Rubinstein, Eds. Marcel Dekker, Inc. New York, Basel, 1996, Vol.19, pp.110-318, ISBN 0-8247-4679-1

Hicks, J.F.; Zamborini, F.P.;Murray, R.W.(2002).Dynamics of electron transfer between electrodes and monolayers of nanoparticles.*J.Phys.Chem*.B.,2002,106,pp.7751-7757,ISSN 1520-5207

Horswell, S.L.; O'Neil, I.A.; Schiffrin, D.J.(2003).Kinetics of electron transfer at Pt nanostructured film electrode. *J.Phys.Chem.B*.,2003,107,pp.4844-4854, ISSN 1520-5207

Jiang, J. Lu,W, Luo, Y .(2004). Length dependence of coherent electron transportation in metal-alkanedithiol-metal and metal-alkanemonothiol-metal junctions. *Chemical Physics Letters*, 2004, 400, pp 336-340, ISNN 0009-2614

Kamat,P.V.(2008). Quantum Dot Solar Cells. Semiconductor nanocrystals as light harvesters, *J. Phys. Chem. C.*, 2008, *112*,pp 18737–18753, ISNN1932-7455

Katz,E.&Willner, I. (2003). Probing biomolecular interactions at conductive and semiconductive surfaces by impedance spectroscopy: routes to impedimetric immunosensors, DNA sensors and enzyme biosensors, *Electroanalysis*, 2003, 15, pp. 913-947, ISSN 1521-4109

Katz,E. &Willner, I. (2004). Integrated Nanoparticle–Biomolecule Hybrid Systems: Synthesis, Properties and Applications, *Angew. Chem. Int. Ed.* (2004), 43, pp.6042 - 6108,ISSN 1521-3773

Kissling, G.P.; Bünzly, C.; Fermin, D.J. (2010). Tuning electrochemical rectification via quantum dot assemblies.*J.Am.Chem.Soc.*,2010, 132, pp.16855-16861, ISSN 1520-5126

Lasia, A.(1999).Electrochemical Impedance Spectroscopy and Its Applications. In Modern Aspects of Electrochemistry, Conway.B.E.; Bockris, J.; White, R.E., Edts., Kluwer Academic/Plenum Publishers, New York, 1999, 32, pp.143-248. URL: http://www.chem.uw.edu.pl/studokt/wyklady/lasia/p1.pdf

Love, C.; Estroff, L.A; Kriebel,J.K.; Nuzzo,R.G.; Whitesides G. M. (2005).Self-assembled Monolayers of Thiolates on Metals as a Form of Nanotechnology, *Chem. Rev.*,2005,105,pp.1103-1169, ISNN 1520-6890

Mirčeski, V.; Komorsky-Lovrić, S.; Lovrić, M.(2007). *Square wave voltammetry*. In :*Monographs in Electrochemistry*.F.Scholz (Ed.), Springer. 1st edition, 2007,Berlin, Germany, ISBN 978-3-642-09292-3.

Nikoobakht, B. .&El-Sayed ; M, A., (2003). Preparation and growth mechanism of gold nanorods using seed-mediated growth method, *Chem.Mater.*, 2003, 15,pp.1957-1962, ISSN 1520-5002

Su, L.; Gao,F.; Mao, L.; (2006). Electrochemical properties of carbon nanotubes (CNT) film electrodes prepared by controllable adsorption of CNTs onto an alkanethiol monolayer self-assembled on gold electrodes, *Anal. Chem.*, 2006, 78,pp. 2651-2657, ISSN 0003-2700

Xu., J.; Li, H. L.; Zhang, Y.(1993). Relationship between electronic tunnelling coefficient and electrode potential investigated using self-assembled alkanethiol monolayers on gold electrodes, *J. Phys. Chem.* 1993, *97*, pp.11497–11500

Yan, F., Sadik, O.A., (2001).Enzyme-modulated cleavage of dsDNA for studying interfacial biomolecular interactions. *J.Am.Chem.Soc.*,2001, 123, pp.11335-11340, ISSN 1520-5126

8

Preparation and Characterization of Gold Nanorods

Qiaoling Li and Yahong Cao
Hebei University of Science and Technology, China

1. Introduction

Numerous characteristics of nanomaterials depend on size and shape, including their catalytic, optical, electronic, chemical and physical properties. The shape and crystallographic facets are the major factors in determining the catalytic and surface activity of nanoparticles. The size can influence the optical properties of metal nanoparticles. This is especially important when the particles have aspect ratios (length/diameter, L/D) larger than 1. So, in the synthesis of metal nanoparticles, control over the shape and size has been one of the important and challenging tasks.

A number of chemical approaches have been actively explored to process metal into one-dimensional (1 D) nanostructures. Among these objects of study, rodlike gold nanoparticles are especially attractive, due to their unique optical properties and potential applications in future nanoelectronics and functional nanodevices. Gold nanorods show different color depending on the aspect ratio, which is due to the two intense surface plasmon resonance peaks (longitudinal surface plasmon peak and transverse surface plasmon peak corresponding to the oscillation of the free electrons along and perpendicular to the long axis of the rods) (Kelly et al., 2003). The color change provides the opportunity to use gold Nanorods as novel optical applications. Gold nanorods are used in molecular biosensor for the diagnosis of diseases such as cancer, due to this intense color and its tunablity. Nanorods also show enhanced fluorescence over bulk metal and nanospheres, which will prove to be beneficial in sensory applications. The increase in the intensity of the surface plasmon resonance absorption results in an enhancement of the electric field and surface enhanced Raman scattering of molecules adsorbed on gold nanorods. All theses properties make gold nanorod a good candidate for future nanoelectronics (Park, 2006). In this chapter, we will describe the preparation and characterization of gold nanorods.

2. Preparation of gold nanorods

Although quite a few approaches have been developed for the creation of gold nanorods, wet chemistry promises to become the preferred choice, because of its relative simplicity and use of inexpensive materials. There are three main methods used to produce gold rods through wet chemistry. Chronological order is followed, which in turn implies successive improvement in material quality. Each new method is also accompanied by a decrease in difficulty of the preparation (Pérez-Juste et al., 2005).

2.1 Template method

The template method for the preparation of gold nanorods was first introduced by Martin and co-workers (Foss, 1992; Martin, 1994, 1996). The method is based on the electrochemical deposition of Au within the pores of nanoporous polycarbonate or alumina template membranes. The rods could be dispersed into organic solvents through the dissolution of the appropriate membrane followed by polymer stabilization (Cepak & Martin, 1998). The method can be explained as follows: initially a small amount of Ag or Cu is sputtered onto the alumina template membrane to provide a conductive film for electrodeposition. This is used as a foundation onto which the Au nanoparticles can be electrochemically grown (stage I in Fig. 1). Subsequently, Au is electrodeposited within the nanopores of alumina (stage II). The next stage involves the selective dissolution of both the alumina membrane and the copper or silver film, in the presence of a polymeric stabilizer such as poly(vinyl pyrrolidone) (III and IV in the Fig. 1). In the last stage, the rods are dispersed either in water or in organic solvents by means of sonication or agitation (Pérez-Juste et al., 2005).

The length of the nanorods can be controlled through the amount of gold deposited within the pores of the membrane (van der Zande et al., 2000). The diameter of the gold nanoparticles thus synthesized coincides with the pore diameter of the alumina membrane. So, Au nanorods with different diameters can be prepared by controlling the pore diameter of the template (Hulteen & Martin, 1997; Jirage et al., 1997). The fundamental limitation of the template method is the yield. Since only monolayers of rods are prepared, even milligram amounts of rods are arduous to prepare. Nevertheless, many basic optical effects could be confirmed through these initial pioneering studies.

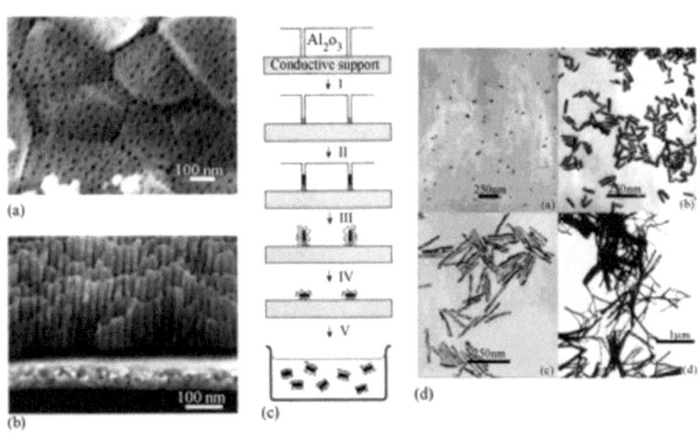

Fig. 1. (a and b) Field emission gun-scanning electron microscopes images of an alumina membrane. (c) Schematic representation of the successive stages during formation of gold nanorods via the template method. (d) TEM micrographs of gold nanorods obtained by the template method (van der Zande et al., 2000).

2.2 Electrochemical method

An electrochemical route to gold nanorod formation was first demonstrated by Wang and co-workers (Chang et al., 1997, 1999). The method provides a synthetic route for preparing high yields of Au nanorods. The synthesis is conducted within a simple two-electrode type electrochemical cell, as shown in the schematic diagram in Fig. 2A.

In the representative electrochemical process, the following conditions are necessary and important:

1. A gold metal plate (3 cm×1 cm×0.05 cm) as a sacrificial anode
2. A platinum plate similar as a cathode (3 cm×1 cm×0.05 cm)
3. A typical current of 3 mA and a typical electrolysis time of 30 min
4. Electrolytic solutions to immerse the both electrodes at 36 °C, it contained:

A cationic surfactant, for example: hexadecyltrimethylammonium bromide ($C_{16}TAB$) to support the electrolyte and to behave as the stablilizer for the nanoparticles to prevent aggregation.

A small amount of a tetradodecylammonium bromide ($TC_{12}AB$), which acts as a rod-inducing cosurfactant.

Appropriate amount of acetone added to the electrolytic solution for loosening the micellar framework to assist the incorporation of the cylindrical-shape-inducing cosurfactant into the $C_{16}TAB$ micelles.

Suitable amount of cyclohexane to enhance the formation of elongated rod-like $C_{16}TAB$ micelles.

A silver plate is gradually immersed close to the Pt electrode to control the aspect ratio of Au nanorods.

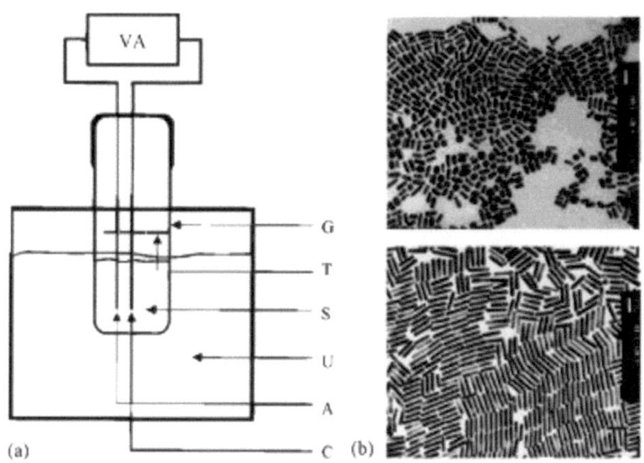

Fig. 2. (a) Schematic diagram of the set-up for preparation of gold nanorods via the electrochemical method containing; VA, power supply; G, glassware electrochemical cell; T, teflon spacer; S, electrode holder; U, ultrasonic cleaner; A, anode; C, cathode. (b) TEM micrographs of Au nanorods with different aspect ratios 2.7 (top) and 6.1 (bottom). Scale bars represent 50 nm (Chang, 1999).

During the synthesis, the bulk gold metal anode is initially consumed, forming $AuBr_4^-$. These anions are complexed to the cationic surfactants and migrate to the cathode where reduction occurs. It is unclear at present whether nucleation occurs on the cathode surface or within the micelles. Sonication is needed to shear the resultant rods as they form away from the surface or possibly to break the rod off the cathode surface. Another important factor controlling the aspect ratio of the Au nanorods is the presence of a silver plate inside the electrolytic solution, which is gradually immersed behind the Pt electrode. The redox reaction between gold ions generated from the anode and silver metal leads to the formation of silver ions (Pérez-Juste et al., 2005). Wang and co-workers found that the concentration of silver ions and their release rate determined the length of the nanorods. The complete mechanism, as well as the role of the silver ions, is still unknown.

2.3 Seeded growth method

Seeded growth of monodisperse colloid particles dates back to the 1920s. Recent studies have successfully led to control of the size distribution in the range 5-40 nm, whereas the sizes can be manipulated by varying the ratio of seed to metal salt (Jana et al., 2001). In the presence of seeds can make additional nucleation takes place. Nucleation can be avoided by controlling critical parameters such as the rate of addition of reducing agent to the metal seed, metal salt solution and the chemical reduction potential of the reducing agent. The step-by-step particle enlargement is more effective than a one step seeding method to avoid secondary nucleation. Gold nanorods have been conveniently fabricated using the seeding-growth method (Carrot et al., 1998).

The preparation of 3.5 nm seed solution can be explained as follows: $C_{16}TAB$ solution (5.0 mL, 0.20 M) was mixed with 2.0 mL of 5.0×10^{-4} M $HAuCl_4$. To the stirred solution, 0.60mL of ice-cold 0.010 M $NaBH_4$ was added, which resulted in the formation of a brownish yellow solution. After vigorous stirring of the seed solution for 2 min, it was kept at 25 °C without further stirring. The seed solution was used between 2 and 48 h after its preparation (Jana et al., 2001).

By controlling the growth conditions in aqueous surfactant media it was possible to inhibit secondary nucleation and synthesize gold nanorods with tunable aspect ratio. Some reserches showed addition of $AgNO_3$ influences not only the yield and aspect ratio control of the gold nanorods but also the mechanism for gold nanorod formation, correspondingly its crystal structure and optical properties (Pérez-Juste et al., 2005). At this point, it is thus convenient to differentiate seed-mediated approaches performed in the absence or in the presence of silver nitrate.

2.3.1 Preparation of gold nanorods without $AgNO_3$

Murphy and co-workers were able to synthesize high aspect ratio cylindrical nanorods using 3.5 nm gold seed particles prepared by sodium borohydride reduction in the presence of citrate, through careful control of the growth conditions, i.e., through optimization of the concentration of $C_{16}TAB$ and ascorbic acid, and by applying a two- or three- step seeding process (see Fig.3).

(1) Preparation of 4.6±1 Aspect Ratio Rod.

In a clean test tube, 10 mL of growth solution, containing 2.5×10^{-4} M $HAuCl_4$ and 0.1 M $C_{16}TAB$, was mixed with 0.05 mL of 0.1 M freshly prepared ascorbic acid solution. Next, 0.025 mL of the 3.5 nm seed solution was added without further stirring or agitation. Within 5-10 min, the solution color changed to reddish brown. The solution contained 4.6 aspect ratio rods, spheres, and some plates. The solution was stable for more than one month (Jana et al., 2001).

(2) Preparation of 13±2 Aspect Ratio Rod.

A three-step seeding method was used for this nanorod preparation. Three test tubes (labeled A, B, and C), each containing 9 mL growth solution, consisting of 2.5×10^{-4} M $HAuCl_4$ and 0.1 M $C_{16}TAB$, were mixed with 0.05 mL of 0.1 M ascorbic acid. Next, 1.0 mL of the 3.5 nm seed solution was mixed with sample A. The color of A turned red within 2-3 min. After 4-5 h, 1.0 mL was drawn from solution A and added to solution B, followed by thorough mixing. The color of solution B turned red within 4-5 min. After 4-5 h, 1 mL of B was mixed with C. Solution C turned red in color within 10 min. Solution C contained gold nanorods with aspect ratio 13. All of the solutions were stable for more than a month (Jana et al., 2001).

(3) Preparation of 18 ±2.5 Aspect Ratio Rod.

This procedure was similar to the method for preparing 13 aspect ratio rods. The only difference was the timing of seed addition in successive steps. For 13 aspect ratio rods, the seed or solutions A and B were added to the growth solution after the growth occurring in the previous reaction was complete. But to make 18 aspect ratio rods, particles from A and B were transferred to the growth solution while the particles in these solutions were still growing. Typically, solution A was transferred to B after 15 s of adding 3.5 nm seed to A, and solution B was transferred to C after 30 s of adding solution A to B (Jana et al., 2001).

In the above method, the yield of the nanorods thus synthesized is ca. 4 % (Jana et al., 2001). The long rods can be concentrated and separated from the spheres and excess surfactant by centrifugation. Later, the same group reported an improved methodology to produce monodisperse gold nanorods of high aspect ratio in 90 % yield (Busbee et al., 2003), just through pH control. In the new proposed protocol, addtion of sodium hydroxide, equimolar in concentration to the ascorbic acid, to the growth solution raised the pH. The pH of the growth solution was changed from 2.8 to 3.5 and 5.6, which led to the formation of gold nanorods of aspect ratio 18.8±1.3 and 25.1±5.1, respectively. The newer procedure also, resulted in a dramatic increase in the relative proportion of nanorods and reduced the separation steps necessary to remove smaller particles.

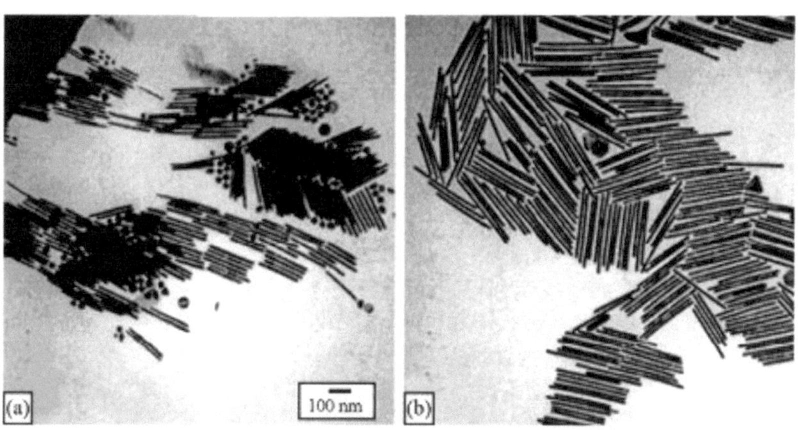

Fig. 3. TEM images of shape-separated 13 (a) and 18 (b) aspect ratio gold nanorods prepared by the seed-mediated method (Jana et al., 2001).

The mechanism of formation of rod-shaped nanoparticles in aqueous surfactant media remains unclear. Based on the idea that $C_{16}TAB$ absorbs onto gold nanorods in a bilayer fashion, with the trimethylammonium headgroups of the first monolayer facing the gold surface (Nikoobakht et al., 2001), Murphy and co-workers (Johnson et al., 2002) proposed that the $C_{16}TAB$ headgroup preferentially binds to the crystallographic faces of gold existing along the sides of pentahedrally twinned rods, as compared to the faces at the tips. The growth of gold nanorods would thus be governed by preferential adsorption of $C_{16}TAB$ to different crystal faces during the growth, rather than acting as a soft micellar template (Johnson et al., 2002). The influence of C_nTAB analogues in which the length of the hydrocarbon tails was varied, keeping the headgroup and the counterion constant was also studied (Gao et al., 2003). It was found that the length of the surfactant tail is critical for controlling not only the length of the nanorods but also the yield, with shorter chain lengths producing shorter nanorods and longer chain lengths leading to longer nanorods in higher yields (Pérez-Juste et al., 2005).

Considering the preferential adsorption of $C_{16}TAB$ to the different crystal faces in a bilayer fashion (Nikoobakht et al., 2001; Johnson et al., 2002; Gao et al., 2003), a "zipping" mechanism was proposed taking into account the van der Waals interactions between surfactant tails within the surfactant bilayer, on the gold surface, that may promote the formation of longer nanorods from more stable bilayers (see Fig. 4) (Gao et al., 2003).

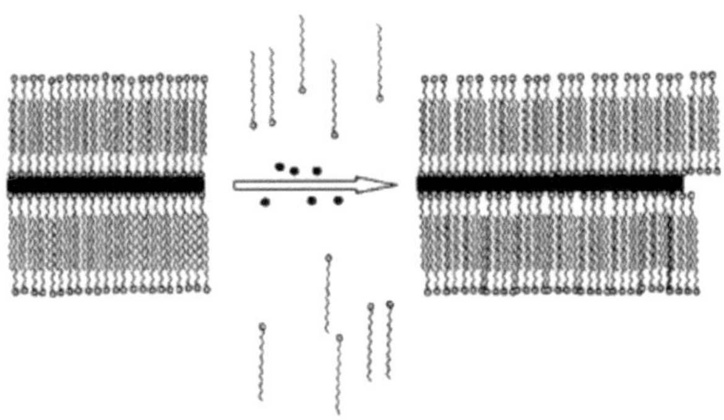

Fig. 4. Schematic representation of "zipping": the formation of the bilayer of CnTAB (squiggles) on the nanorod (black rectangle) surface may assist nanorod formation as more gold ions (black dots) are introduced (Gao et al., 2003).

Recently, Pérez-Juste et al. investigated the factors affecting the nucleation and growth of gold nanorods under similar conditions (Pérez-Juste et al., 2004). They showed that the aspect ratio, the monodispersity and the yield could be influenced by the stability of the seed, temperature, the nature and concentration of surfactant. The yield of nanorods prepared from $C_{16}TAB$ capped seeds is much higher than that from naked (or citrate stabilized) seeds. This indicates that the more colloidally stable the gold seed nanoparticles are, the higher the yield of rods.

2.3.2 Preparation of Gold Nanorods with $AgNO_3$

The presence of silver nitrate allows better control of the shape of gold nanorods synthesized by the electrochemical method, and Murphy and co-workers proposed a variation of their initial procedure for long nanorods (Jana et al., 2001), in order to increase the yield of rod-shaped nanoparticles (up to 50 %) and to control the aspect ratio of shorter nanorods and spheroids (Jana et al., 2001). Under identical experimental conditions, a small amount of silver nitrate is added (5×10^{-6} M) prior to the growth step. The aspect ratio of the spheroids and nanorods can be controlled by varying the ratio of seed to metal salt, as indicated in the spectra of Fig.5. The presence of the seed particles is still crucial in the growth process, and there is an increase in aspect ratio when the concentration of seed particles is decreased.

The mechanism by which Ag^+ ions modify the metal nanoparticle shape is not really understood. It has been hypothesized that Ag^+ adsorbs at the particle surface in the form of AgBr (Br^- coming from $C_{16}TAB$) and restricts the growth of the AgBr passivated crystal facets (Jana et al., 2001). The possibility that the silver ions themselves are reduced under these experimental conditions (pH 2.8) can be neglected since the reducing power of ascorbate is too positive at low pH (Pal et al., 1998). This shape effect depends not only on the presence of $AgNO_3$, but also on the nature of the seed solution. By simply adjusting the amount of silver ions in the growth solution, a fine-tuning of the aspect ratio of the nanorods can be achieved, so that an increase in silver concentration (keeping the amount of seed solution constant) leads to a redshift in the longitudinal plasmon band. Interestingly, the aspect ratio can also be controlled by adjusting the amount of seed solution added to the growth solution in the presence of constant Ag^+ concentration (Pérez-Juste et al., 2004). Contrary to expectations, an increase in the amount of seed produces a red-shift in the longitudinal plasmon band position, as shown in Fig. 6, pointing toward an increase in aspect ratio.

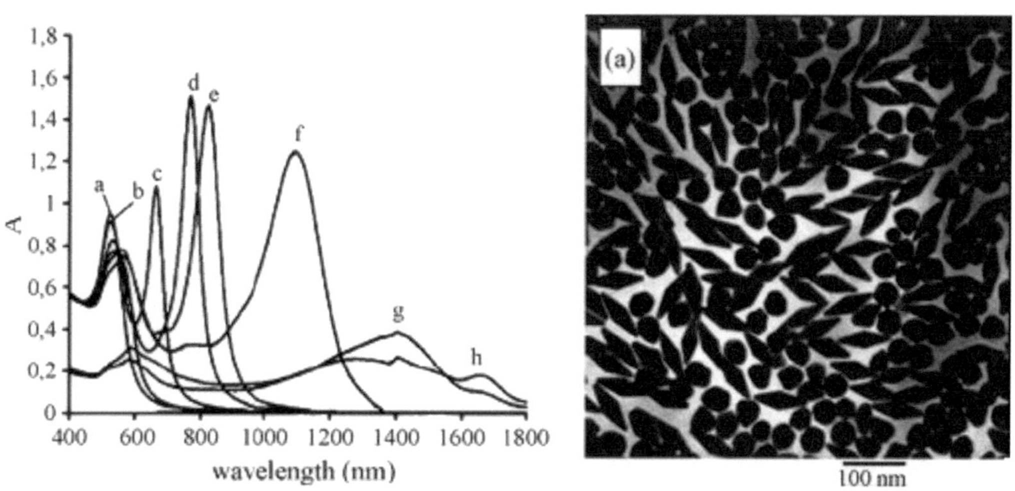

Fig. 5. UV-vis spectra of Au nanorods with increasing aspect ratios (a-h) formed by decreasing the amount of added seed (left). TEM image of Au nanorods synthesized in the presence of silver nitrate (right) (Jana et al., 2001; Jana et al., 2002).

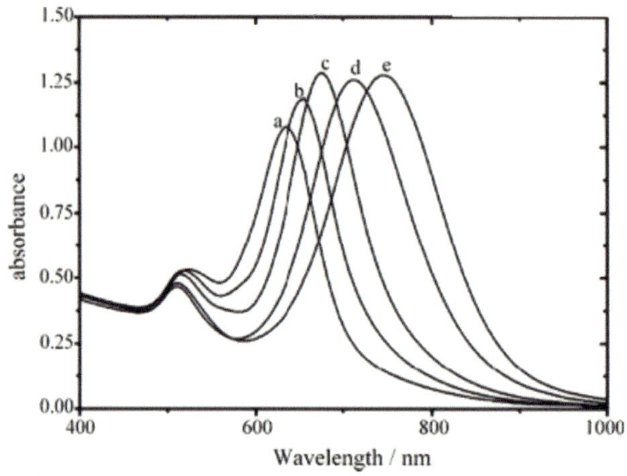

Fig. 6. UV–vis spectra of Au nanorods prepared in the presence of silver nitrate by the El-Sayed's protocol (Pérez-Juste et al., 2004).

2.4 Photochemical method

Yang and co-workers developed a photochemical method for the synthesis of gold nanorods (Kim et al., 2002), which is performed in a growth solution similar to that described for the electrochemical method (Chang et al., 1997), in the presence of different amounts of silver nitrate and with no chemical reducing agent.

The growth solution containing gold salts and others such as surfactants and reducing agents, was irradiated with a 254 nm UV light (420 µW/cm^2) for about 30 h. The resulting solution was centrifuged at 3000 rpm for 10 minutes, and the supernatant was collected, and then centrifuged again at 10,000 rpm for 10 minutes. The precipitate was collected and redispersed in deionized water. The colour of the resulting solution varies with the amount of silver ions added, which is indicative of gold nanorods with different aspect ratios (Boyes & Gai, 1997) as shown in Fig. 7.

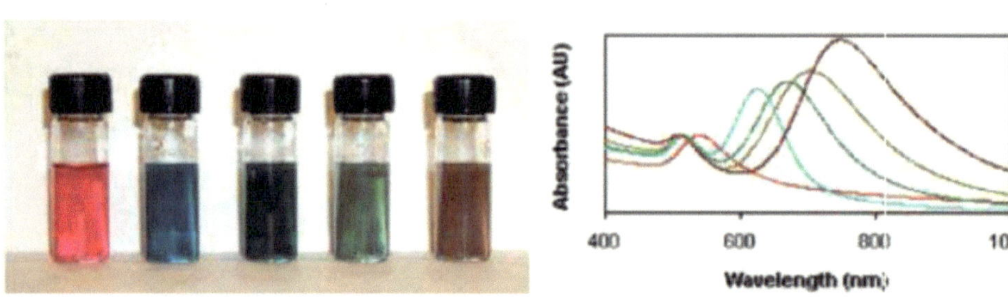

Fig. 7. (a) Image of photochemically prepared gold nanorods solution, and (b) corresponding UV-vis spectrum. The left most solution was prepared with no silver ion addition. The other solutions were prepared with addition of 15.8, 31.5, 23.7, 31.5 µL of silver nitrate solution, respectively. The middle solution was prepared with longer irradiation time (54 h) compared to that for all other solutions (30 h), and the transformation into shorter rods can be seen (Gai, 1998).

Seen from Fig. 7 two absorption peaks were obtained, which resulted from the longitudinal and transverse surface plasmon (in the UV-vis spectrum) that indicates gold nanorods are formed when silver ions are added (Gai, 1998). The aspect ratio increases when more silver ions are added, and this is accompanied by a decrease in rod width, while in the absence of silver ions, spherical particles are obtained. Therefore, the possibility of a rod-like micellar template mechanism can be discarded and these experiments indicate the critical role played by silver ions in determining the particle morphology.

2.5 Other methods

Markovich and co-workers adapted the seed-mediated method in the absence of silver nitrate proposed by Murphy and co-workers (Jana et al., 2001) for the growth of gold nanorods directly on mica surfaces (Taub et al., 2003). The method involves the attachment of the spherical seed nanoparticles to a mica surface, which is then dipped in a $C_{16}TAB$ surfactant growth solution. About 15 % of the surface-bound seeds are found to grow as nanorods. This yield enhancement of nanorods, compared to that obtained for the solution growth technique (ca. 4 %) (Jana et al., 2001), was attributed to a change in the probability of the growing seed to develop twinning defects. Subsequently, Wei et al. adapted the method to grow nanorods directly on glass surfaces (Wei et al., 2004). They studied the influence of the linker used to attach the seed particles and the gold salt concentration in the growth solution on the formed gold nanostructures.

3. Optical properties of gold nanorods

Gold nanorods show unique optical properties depending on the size and the aspect ratio (the ratio of longitudinal-to-transverse length). Although the spherical gold nanoparticle (nanosphere) has only one surface plasmon (SP) band in the visible region, the nanorod has a couple of SP bands. One SP band corresponding to the transverse oscillation mode locates in the visible region at around 520 nm, while the other corresponding to the longitudinal oscillation mode between far-red and near-infrared (near-IR) region. This is the distinctive optical characteristic of the nanorod as compared with the nanosphere. So, nanosphere may have electronic, crystallographic, mechanical or catalytic properties that are different to the nanorods. Such differences may be probed through optical measurements. Spectroscopic measurements are often the easiest methods for monitoring surface processes such as dissolution and precipitation, adsorption and electron transfer. If nanocrystals of any specific geometry could be grown then it is conceivable that optical materials could be designed from scratch. Photonic devices could be created from molecular growth reactors. In the section, we will only describe the optical properties of gold nanorods.

3.1 Plasmon resonance for ellipsoidal nanoparticles

For gold nanorods, the plasmon absorption splits into two bands (Fig. 8) corresponding to the oscillation of the free electrons along and perpendicular to the long axis of the rods (Link and EL-Sayed, 1999). The transverse mode (transverse surface plasmon peak: TSP) shows a resonance at around 520 nm, while the resonance of the longitudinal mode (longitudinal surface plasmon peak: LSP) occurs at higher wavelength and strongly depends on the aspect ratio of nanorods. As aspect ratio is increased, the longitudinal peak is redshifted. To account for the optical properties of Nanorods, it has been common to treat them as ellipsoids, which allows the Gans formula (extension of Mie theory) to be applied. Gans'formula (Gans, 1912) for randomly oriented elongated ellipsoids in the dipole approximation can be written as

$$\frac{\gamma}{N_p V} = \frac{2\pi \varepsilon_m^{3/2}}{3\lambda} \sum_{j=A}^{C} \frac{\left(1/P_j^2\right)\varepsilon^2}{\left[\varepsilon_1 + \left(\left(1-P_j\right)/P_j\right)\varepsilon_m\right]^2 + \varepsilon_2^2} \tag{1}$$

where N_p represents the number concentration of particles, V the single particle volume, λ the wavelength of light in vacuum, and ε_m the dielectric constant of the surrounding medium and ε_1 and ε_2 are the real ($n^2 - k^2$) and imaginary ($2nk$) parts of the complex dielectric function of the particles. The geometrical factors Pj for elongated ellipsoids along the A and B/C axes are respectively given by

$$P_A = \frac{1-e^2}{e^2}\left[\frac{1}{2e}\ln\left(\frac{1+e}{1-e}\right) - 1\right]$$
$$P_B = P_C = \frac{1-P_A}{2} \text{ and } e = \left(\frac{L^2 - d^2}{L^2}\right)^{1/2} \tag{2}$$

Fig. 9 shows the absorbance spectra for gold nanorods with varied aspect ratio calculated using the Gans expressions. The dielectric constants used for bulk gold are taken from the measurements done Johnson and Christy (Johnson, 1972), while the refractive index of the medium was assumed to be constant and same as for H_2O (1.333). The maximum of the longitudinal absorbance band shifts to longer wavelengths with increasing aspect ratio. There is the small shift of the transverse resonance maximum to shorter wavelengths with increasing aspect ratio. Electron microscopy reveals that most nanorods are more like cylinders or sphero-capped cylinders than ellipsoids. However, an analytical solution for such shapes is not derived yet, and so while the results are compared to the formula given by ellipsoids, such comparisons are somewhat approximate (Sharma et al., 2009).

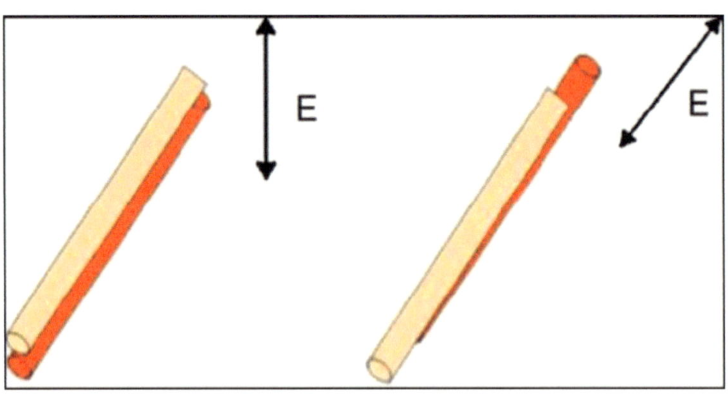

Fig. 8. Transverse and longitudinal modes of plasmon resonance in rod-like particles (Sharma et al., 2009).

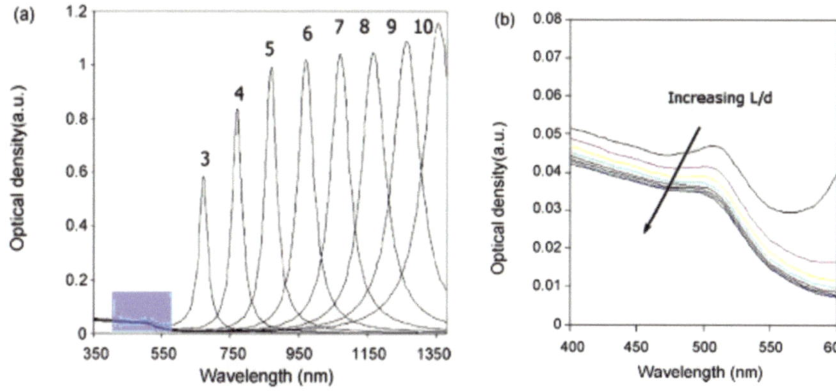

Fig. 9. Absorbance spectra calculated with the expressions of Gans for elongated ellipsoids using the bulk optical data for gold. (a) The numbers on the spectral curves indicate the aspect ratio (L/D). (b) Enlargement of the shaded area of (a) showing slight blue shift of transverse plasmon resonance peak on increasing aspect ratio (Park, 2006).

3.2 Absorption spectrum of colloidal dispersions of gold nanorods

The longitudinal and transverse plasmon resonance can be computed as a function of aspect ratio either by using analytical expression put forth by Gans in 1912 (Gans, 1912) or by using one of numerical techniques (Bohren, 1983; Kelly, 2001). Sharma et al. describe the how the absorption spectrum measured experimentally compares to the results from Gans theory (Gans, 1912; Sharma et al., 2009) and DDA simulations (Kelly, 2001). The gold nanorods cited from their research were synthesized using a seed-mediated method based on use of binary surfactant and all UV-vis-NIR spectra were acquired with a Cary 5G UV-visible-near-IR spectrophotometer. Even though optical properties of pure water were used for calculating the spectrum, the peak resonance measured experimentally show a remarkable agreement with theoretical and simulation results (Fig. 10). Several groups have observed similar trends (Murphy, 2005; Link, 1999).

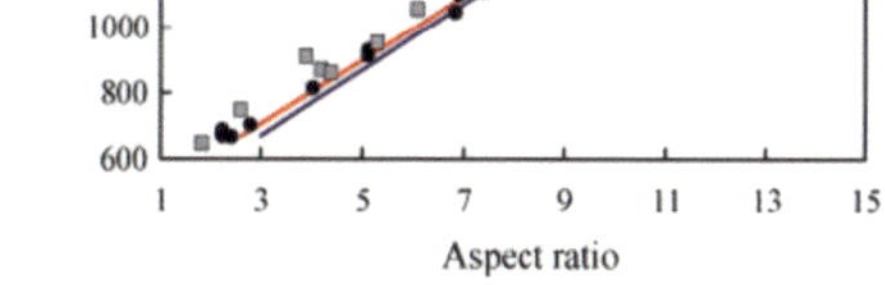

Fig. 10. Longitudinal surface plasmon peak (nm) versus the aspect ratio of nanorods. Simulation results using the DDA method (Kelly, 2003) and the corresponding fit (red straight line) and Gans'calculation (blue straight line). Experimental data from the work (gray squares). Experimental data from our study (black circles) (Park, 2006).

It is well known though that the plasmon resonance is very sensitive to change in the dielectric constant of the medium, and in case of mixed solvents or in sensing applications, this effect must be taken into consideration. Theoretically predicted change in optical properties of colloidal gold suspensions expected upon changing medium has been observed experimentally by several groups (Templeton, 1999; Underwood, 1994). For the gold nanorods, the computed longitudinal plasmon peak increases with an increase in the dielectric constant of medium, as shown in Fig.11. The effect of medium seems more pronounced for longer nanorods, as is evident from the increase in slope observed for higher aspect ratios.

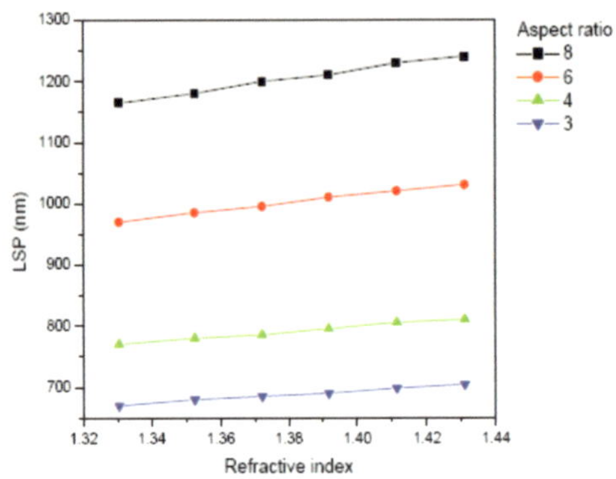

Fig. 11. Calculated LSP as a function of refractive index of medium (Park, 2006).

3.3 Local field enhancements and sensing applications

The electric field is the gradient of potential, and hence using the expression for potential derived earlier, the electric fields inside and outside the sphere are:

$$
\begin{aligned}
E_{in} &= \frac{3\varepsilon_m}{\varepsilon + 2\varepsilon_m} E_0 \\
E_{out} &= E_0 + \frac{3\vec{n}(\vec{n}\cdot\vec{p}) - \vec{p}}{4\pi\varepsilon_0\varepsilon_m} \frac{1}{r^3}
\end{aligned}
\qquad (3)
$$

Resonance in polarizability leads to the resonant enhancement of both the internal and the external dipolar fields. The wavelength at which this resonance occurs depends upon the dielectric function of the metal as well as the medium around it. Since the resonance condition and resulting enhancements of the fields are directly correlated with the shape and size of particle, the basic understanding of this relationship is crucial for their widespread use. The sensitivity of plasmon resonance to the local dielectric environment, implies that any changes within a few nanometers of the particles can be used in say biological or chemical sensing applications (Sharma et al., 2009). For the perfectly spherical particles that can be described by electrostatic approach (Rayleigh limit), only the dipole surface plasmon contributes to the localized enhancement, limiting the overall enhancement achieved. In rod-like particles, highly localized fields can be generated at the tips, providing a much stronger response function for sensing applications. The theoretical and experimental aspects of surface-enhanced Raman scatting and plasmonics based sensing are widely discussed and debated in literature (Willets, 2007; Maier, 2007) and it forms one of the most anticipated applications of non-spherical gold and noble metal particles.

3.4 Color of colloidal dispersions of gold nanorods

Since the color of colloidal gold depends on both the size and shape of the particles, as well as the refractive index of the surrounding medium, it is important to independently account for the color change of gold nanorod suspension due to presence of either nanospheres or any substance that affects the refractive index of the solvent. Since color of the gold sols is traditionally linked to their shape or size, Sharma et al. characterized the dependence of perceived color on shape and dimensions of the nanoparticles using color science. The color was identified by positioning x and y values in the CIE chromaticity diagram.

This visible light region consists of a spectrum of wavelengths, which range from approximately 700 to 400 nm. For the nanorods, the transverse plasmon resonance peak is not quite as sensitive to the change of aspect ratio, as the longitudinal peak, which shows noticeable shifts in the aspect ratio as seen in Fig.12 which shows the UV-vis-NIR spectrum of gold nanorods dispersions. The relatively intensity of transverse peaks shows that mostly nanorods are present, which were obtained by optimizing synthesis and separation techniques. As predicted by theory, the transverse peak blue shifts with an increasing aspect ratio.

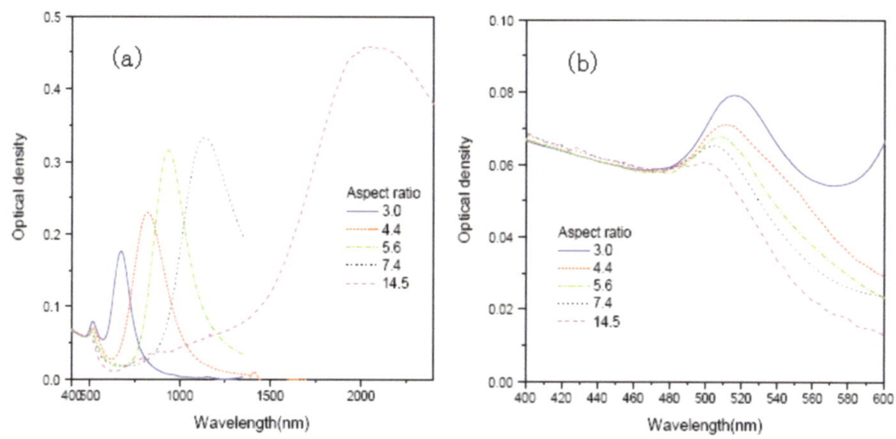

Fig. 12. (a) UV-vis-NIR spectra of dispersions containing gold nanorods with different aspect ratios and (b) transverse peak, showing the blue shift with increase in aspect ratio (Park, 2006).

Fig.13 shows the photograph of the colloidal dispersions of gold nanorods and the color patches simulated using theoretical absorbance data equivalent to the aspect ratio of gold nanorods. The color of solution is basically the same beyond an aspect ratio of around 4. Therefore in a visible region, the dramatic color change cannot be achieved by only changing aspect ratio. But once the longitudinal peak goes beyond 700 nm, (for aspect ratio ~3) the change in peak absorption cannot be detected by the human eye and color of gold nanorod dispersion does not change with further increase in aspect ratio. Therefore the color change could be only observed for relatively short range of aspect ratios. But the tunability of optical properties gold nanorods as a function of aspect ratio provides potentials to use gold nanorods as an optical filter in near infrared region.

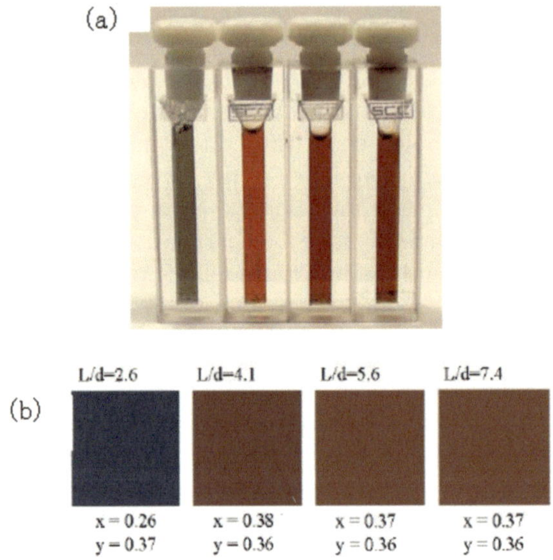

Fig. 13. (a) Photograph of 4 sols of colloidal gold prepared in water. Aspect ratios are 2.6, 4.1, 5.6 and 7.4 (from the left), respectively. (b) The simulated color of dispersion of gold nanorods of different aspect ratio (Park, 2006).

Sharma et al. found that the color in a visible region is rather sensitive to the amount of spherical particles included as byproducts since surface plasmon peak of sphere positions between 500 and 550 nm. Fig.14 shows the color of colloidal dispersion of gold nanorods containing different amount spheres as byproducts. The color changes from purple to brown as the amount of byproducts decreases.

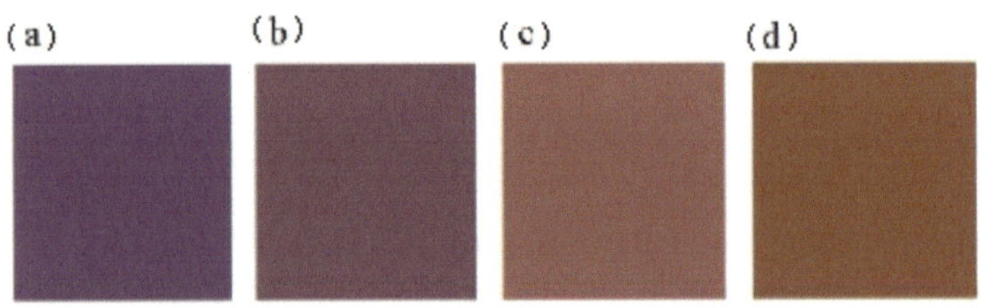

Fig. 14. The color of dispersion of gold nanorods containing different amount of spheres as byproducts: (a) 50 %, (b) 30 %, (c) 10 % and (d) 0 % (Park, 2006).

3.5 Polarization dependent color and absorption in polymer-gold nanocomposite films

The optical properties of gold nanorods are dependent on the state of polarization of incident light, on size and aspect ratio of the particles, and the dielectric properties of the medium. The optical response of a colloidal dispersion of nanorods, as revealed by UV-vis spectroscopy can be thought of as the response from randomly oriented rods. The polarization dependent response of nanorods can be observed by dispersing them in a gel or polymer matrix, and then stretching the matrix uniaxially, thus aligning the dispersed rods. When the incident light is polarized in the direction of stretching or in the direction coinciding with the average orientation of long axis of nanorods, absorbance is dominated by the response due to the longitudinal resonance. As the angle between the stretching direction and polarization of incoming light is increased, the absorbance shows a marked blue shift. Thus the composite films show a marked polarization dependent color and absorption, making them suitable for use as polarization dependent color filters and for other optical applications (Caseri, 2000; Al-Rawashdeh, 1997).

Caseri (Caseri, 2000) presented a very comprehensive historical perspective and discussion of optical properties of polymer/nanoparticle composites. Caseri and co-workers (Caseri, 2000; Dirix, 1999; Dirix, 1999) found that spherical gold nanoparticles can form "pearl necklace type arrays" by aggregating along the stretching direction and produce dichroic filters that have potential application in creating bicolored displays as illustrated in Fig.15. Al-Rawashdeh (Al-Rawashdeh, 1997) studied the linear dichroic properties of polyethylene/gold rods composites and studied how the local field enhancement could make these composite films impacts the infrared absorption of probe molecules attached to the surface of nanorods.

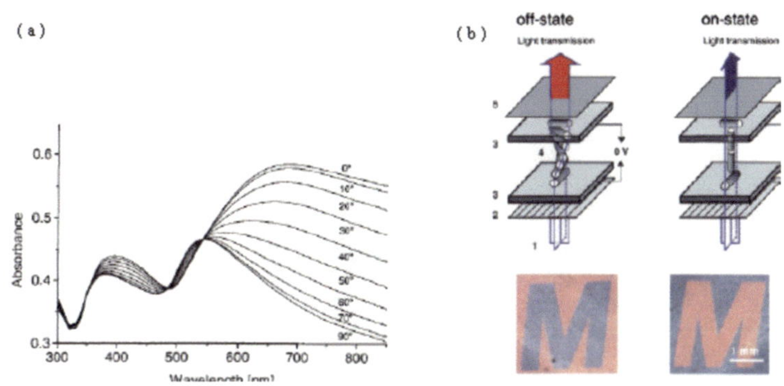

Fig. 15. (a) UV-vis spectra of uniaxially stretched films of high-density polyethylene/gold composites. The angle on spectra indicates the angle between the polarization direction of the incident light and the drawing direction. (b) Twistednematic liquid crystal displays (LCD) equipped with a drawn polyethylene-silver nanocomposite. The "M" represents the on state, the drawing direction is in the picture above parallel and below perpendicularly oriented to the polarizer (Caseri, 2000; Park, 2006).

The transmittance spectra as a function of polarizer angle are shown in Fig.16 for a nanocomposite with gold nanorods of aspect ratio 2.8, and draw ratio of 4 was used for this study. The longitudinal plasmon resonance blue shifts as polarization angle is increased, and the intensity of the peak drops, in accordance with the observations by other groups (Caseri, 2000) (Fig. 17).

Sharma et al. obtained transmittance spectra at different polarizer angles and calculated extinction ratio, E.R. $=10 \log_{10}(T_{\perp}/T_{//})$ [dB] where $T_{\perp}$ and $T_{//}$ are the transmittance perpendicular and parallel to the stretching direction, respectively. Maximum extinction ratio (Park, 2006) is 18 dB at $\lambda = \lambda_{LSP}$ and is comparable to those previously reported in the literature (Matsuda, 2005). The thickness of the film is 50 μm and it has good flexibility.

When the aspect ratio of nanorods is sufficiently large, the LSP shifts to the near-IR region. This indicates that the wavelength region displaying optical dichroism can be shifted from the visible to the near-IR. This enables the fabrication of thin film optical filter that respond to the wavelengths in the near-IR region (Fig.18).

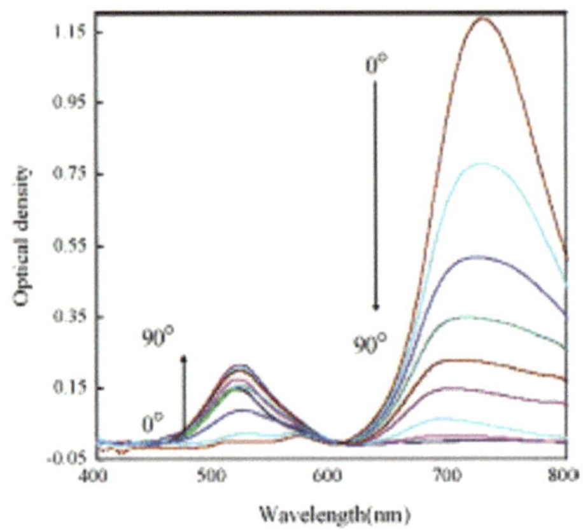

Fig. 16. UV-vis-NIR spectra of PVA/gold nanorods nanocomposites for varying polarization angles L/D of gold Nanorods is 2.8 (Park, 2006).

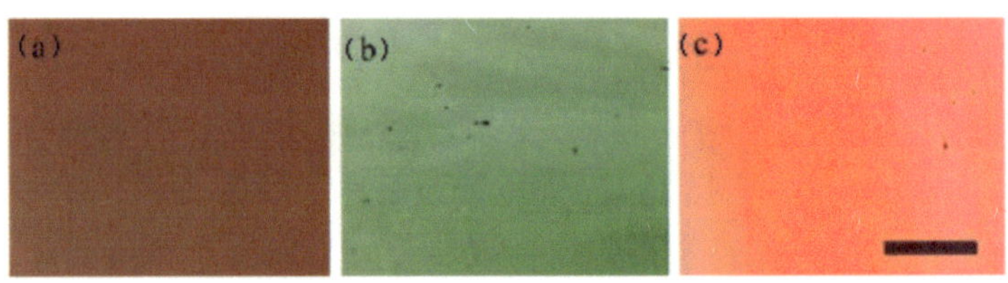

Fig. 17. Optical micrographs of drawn PVA-gold nanocomposites (4 % w/w gold, draw ratio 4): (a) unpolarized, polarization direction, (b) parallel and (c) perpendicular to the drawing direction. Scale bar is 50 mm (Park, 2006).

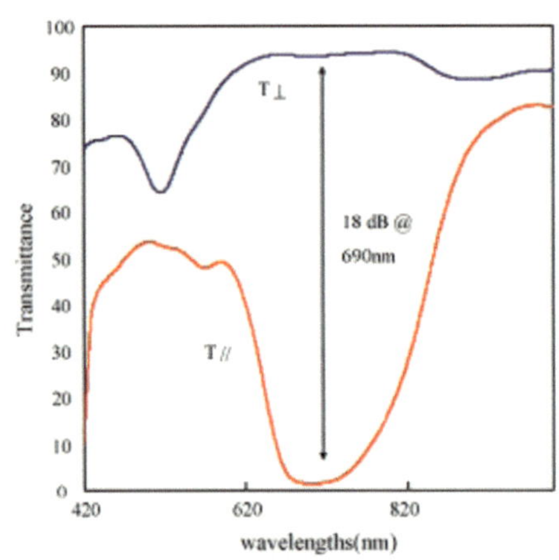

Fig. 18. UV-vis-NIR spectra of PVA/gold nanorods nanocomposites for varying polarization angles L/D of gold Nanorods is 2.8 (Park, 2006).

4. Conclusion

In metal nanomaterial research, the optical properties have been of interest especially because of the applications to medical diagnostics and nanooptics. Gold nanoparticles are attracting great attention due to their unique optical that is dependent on their size and shape. In spherical gold nanoparticles, the plasmon absorption is red shifted with an increase in the diameter of the nanoparticle. Gold nanorods show different color depending on the aspect ratio, which is due to the two intense surface plasmon resonance peaks. The color change provides the opportunity to use gold nanorods as novel optical applications. There have been many applications utilizing this intense color and its tunablity (Pérez-Juste et al., 2005). One of them is in the field of biological system. Nanorods bind to specific cells with greater affinity and one can visualize the conjugated cell using a simple optical microscope due to the enhanced scattering cross section (EI-Sayed, 2005). This is how gold nanorods are used in molecular biosensor for the diagnosis of diseases such as cancer. Nanorods show enhanced fluorescence over bulk metal and nanospheres, due to the large enhancement of the longitudinal plasmon resonance (Eustis, 2005), which will prove to be beneficial in sensory applications. All theses properties make gold nanorod a good candidate for future nanoelectronics, once appropriate techniques allow for the generation of artificial structures in 2D or 3D (Park, 2006).

5. Acknowledgements

We thank Jorge Pérez-Juste , Mohan Srinivasarao and Kyoungweon Park for some contents and ideas of their paper.

6. References

AI-Rawashdeh N., Foss C.A. (1997). UV/visible and infrared spectra of polyethylene/ nanoscopic gold rod composite films: Effects of gold particle size, shape and orientation. *Nanostructured Materials*, Vol.9, No. 1-8, (May 1998), pp. 383-386, ISSN 0965-9773

Bohren C.F., Huffman D.R. (1998). *Absorption and Scattering of Light by Small Particles*. John Wiley & Sons, Inc., ISBN 0471293407, New York, USA

Boyes E.D., Gai P.L. (1997). Environment high resolution resolution electron microscopy and applications to chemical science. *Ultramicroscopy*. Vol.67, No. 1-4, (June 1997), pp. 219- 232, ISSN 0304-3991

Busbee B.D., Obare S.O., Murphy C.J. (2003). An Improved Synthesis of High-Aspect -Ratio Gold Nanorods. *Adv. Mater.*, Vol.15, No.5, (March 2003), pp. 414–416, ISSN 1521-4095

Carrot G., Valmalette J.C., Plummer C.J.G., Scholz S.M., Dutta J., Hofmann H., Hilborn J.G.(1998). Gold nanoparticle synthesis in graft copolymer micelles. *Colloid Polym. Sci.*, Vol.276, No.10, (June 1998), pp. 853-859, ISSN 1435-1536

Caseri W. (2000). Nanocomposites of polymers and metals or semiconductors: Historical background and optical properties. *Macromolecular Rapid Communications*, Vol.21, No.11, (July 2000), pp. 705-722, ISSN 1022-1336

Cepak V.M., Martin C.R.. (1998). Preparation and Stability of Template-Synthesized Metal Nanorod Sols in Organic Solvents. *J. Phys. Chem. B*, vol.102, No.49, (October 1998), pp. 9985–9990, ISSN 1520-6106

Chang S.-S., Shih C.W., Chen C.D., Lai W.C., Wang C.R.C. (1999). The shape transition of gold nanorods. *Langmuir*, Vol.15, No.3, (Februry 1999), pp. 701-709, ISSN 0743-7463

Dirix Y., Bastiaansen C., Caseri W., Smith P. (1999). Oriented pearl-necklace arrays of metallic nanoparticles in polymers: A new route toward polarization-dependent color filters. *Advanced Materials*, Vol.11, No. 3, (March 1999), pp. 223-227, ISSN 1521-4095

Dirix Y., Darribere C., Heffels W., Bastiaansen C., Caseri W., Smith P. (1999). Optically Anisotropic Polyethylene-Gold Nanocomposites. *Applied Optics*, Vol.38, No.31, (November 1999), ISSN 0003-6935

EI-Sayed I.H., Huang X., El-Sayed M.A.. (2005). Surface Plasmon Resonance Scatteringand Absorption of anti-EGFR Antibody Conjugated Gold Nanoparticles in Cancer Diagnostics: Applications in Oral Cancer. *Nano Lett.*, vol.5, No.5, (January 2005), pp. 829-834, ISSN 1530-6984

Eustis S., El-Sayed M.A.. (2005). Aspect Ratio Dependence of the Enhanced Fluorescence Intensity of Gold Nanorods: Experimental and Simulation Study. *J. Phys. Chem. B*, vol.109, No.34, (September 2005), pp. 16350-16356, ISSN 1520-6106

Foss Jr. C.A., Hornyak G.L., Stockert J.A., Martin C.R.. (1992). Optical properties of composite membranes containing arrays of nanoscopic gold cylinders. *J. Phys.Chem.*, vol.96, No.19, (September 1992), pp. 7497-7499, ISSN 0022-3654

Gai P.L. (1998). Direct probing of gas molecule–solid catalyst interactions on the atomic scale. *Adv. Mater*, Vol.10, No.15, (January 1999), pp. 1259-1263, ISSN 1521-4095

Gans R. (1912). Über die Form ultramikroskopischer Goldteilchen. *Annalen Der Physik*, Vol.342, No.5, pp. 881-900, ISSN 1521-3889

Gao J., Bender C.M., Murphy C.J.. (2003). Dependence of Gold Nanorod Aspect Ratio on the Nature of the Directing Surfactant in Aqueous Solution. *Langmuir*, Vol.19, No.21, (August 2003), pp. 9065-9070, ISSN 0743-7463

Hulteen J.C., Martin C.R.. (1997). A general template-based method for the preparation of nanomaterials. *J. Mater. Chem.*, No.7, pp. 1075-1087, ISSN 0959-9428

Jana N.R., Gearheart L., Murphy C.J. (2001). Evidence for seed-mediated nucleation in the chemical reduction of gold salts to gold nanoparticles. *Chem. Mater.*, Vol.13, No.7, (June 2001), pp.2313-2322, ISSN 0897-4756

Jana N.R., Gearheart L., Murphy C.J.. (2001). Wet Chemical Synthesis of High Aspect Ratio Cylindrical Gold Nanorods. *J. Phys. Chem. B*, Vol.105, No.19, (April 2001), pp. 4065–4067, ISSN 1520-6106

Jana N.R., Gearheart L., Obare S.O., Murphy C.J. (2002). Anisotropic chemical reactivity of gold spheroids and nanorods. *Langmuir*, Vol.18, No.3, pp.922-927, ISSN 0743-7463

Jana N.R., L. Gearheart, C.J. Murphy. (2003). Seed-Mediated Growth Approach for Shape-Controlled Synthesis of Spheroidal and Rod-like Gold Nanoparticles Using a Surfactant Template. *Adv. Mater.*, Vol.13, No.18, (September 2003), pp. 1389-1393, ISSN 1521-4095

Jirage K.B., Hulteen J.C., Martin C.R.. (1997). Nanotubule-Based Molecular-Filtration Membranes. *Science*, Vol.278, No.5338, (October 1997), pp. 655-658, ISSN 0036-8075

Johnson C.J., DujardiE. n, Davis S.A., Murphy C.J., Mann S.(2002). Growth and form of gold nanorods prepared by seed mediated, surfactant-directed synthesis. *J. Mater. Chem.*, Vol.12, No.6, (March 2002), pp.1765-1770, ISSN 0959-9428

Johnson P.B., Christy R.W. (1972). Optical Constants of the Noble Metals. *Physics Review B*, Vol.6, No.12, (December 1972), pp. 4370-4379, ISSN 0556-2805

Kelly, K. L.; Coronado, E.; Zhao, L.L.; Schatz, G.C. (2003). The optical properties of metal nanoparticles: The influence of size, shape, and dielectric environment. *J. Phys. Chem. B*, vol.107, No.3, (December 2002), pp. 668-677, ISSN 1520-6106

Kelly K.L., Lazarides A.A., Schatz G.C. (2001). Computational electromagnetics of metal nanoparticles and their aggregates. *Computing in Science & Engineering*, Vol.3, No.4, (July 2001), pp. 67-73, ISSN 1521-9615

Kelly K.L., Coronado E., Zhao L.L., Schatz G.C.(2003). The optical properties of metal nanoparticles: The influence of size, shape, and dielectric environment. *J. Phys. Chem. B*, Vol.107, No.3, (December 2002), pp. 668-677, ISSN 1520-6160

Kim F., Song J.H., Yang P., Am J. (2002). Photochemical Synthesis of Gold Nanorods. *Chem. Soc.*, Vol.124, No.48, (Novermber 2002), pp.14316-14317, ISSN 0002-7863

Link S., El-Sayed M.A. (1999). Size and Temperature Dependence of the Plasmon Absorption of Colloidal Gold Nanoparticles. *J. Phys. Chem. B*, Vol.103, No.21, (May 1999), pp. 4212-4217, ISSN 1520-6106

Maier S.A. (2007). Plasmonics: Fundamentals and Applications. ISBN 978-0387-33150-8 In: *Springer*, Bath, UK

Martin C.R.. (1994). A Membrane-Based Synthetic Approach. *Science*, vol.266, No.5193, (December 1994), pp. 1961-1966, ISSN 0036-8075

Martin C.R.. (1996). Membrane-Based Synthesis of Nanomaterials. *Chem. Mater.*, vol.8, No.8, (August 1996), pp. 1739–1746, ISSN 0036-8075

Matsuda S., Yasuda Y., Ando S. (2005). Fabrication of polyimide-blend thin films containing uniformly oriented silver nanorods and their use as flexible, linear polarizers. *Advanced Materials*, Vol. 17, No.18, (September 2005), pp. 2221-2224, ISSN 1521-4095

Murphy C.J., San T.K., Gole A.M., Orendorff C.J., Gao J.X., Gou L., Hunyadi S.E., Li T. (2005). Anisotropic metal nanoparticles: Synthesis, assembly, and optical applications. *J. Phys. Chem. B*, Vol.109, No.29, (July 2005), pp.13857-13870, ISSN 1520-6106

Nikoobakht B., El-Sayed M.A.(2001). Evidence for bilayer assembly of cationic surfactants on the surface of gold nanorods. *Langmuir*, Vol.17, No.20, (September 2001), pp. 6368-6374, ISSN 0743-7463

Pal T., De S., Jana N.R., Pradhan N., Mandal R., Pal A., Beezer A.E., Mitchell J.C. (1998). *Langmuir*, Vol. 14, No.17, (August 1998), pp.4724-4730, ISSN 0743-7463

Park K. (2006). *Synthesis, Characterization, and Self-Assembly of Size Tunable Gold Nanorods*. In: Doctor of Philosophy, School of Polymer, Textile and Fiber Engineering, Georgia Institute of Technology, Atlanta, USA, December 2006

Pérez-Juste, J.; Pastoriza-Santos, I.; Liz-Marzán, L.M.; Mulvaney, P. (2005). Gold nanorods: Synthesis, characterization and applications. *Coordination Chemistry Reviews*, 2005, vol.249, No.17-18, pp. 1870-1901, ISSN 0010-8545

Pérez-Juste J., Liz-Marz´an L.M., Carnie S., Chan D.Y.C., Mulvaney P. (2004). Electric-Field-Directed Growth of Gold Nanorods in Aqueous Surfactant Solutions. *Adv. Funct. Mater.*, Vol.14, No.6, (June 2004), pp.571-579, ISSN 1616-3028

Pérez-Juste J., Correa-Duarte M.A., Liz-Marz´an L.M. (2004). Silica gels with tailored, gold nanorod-driven optical functionalities. *Appl. Surf. Sci.*, Vol.226, No.1, (March 2004), pp.137-143, ISSN 0169-4332

Sharma V., Park K., Srinivasarao M. (2009). Colloidal dispersion of gold nanorods: Historical background, optical properties, seed-mediated synthesis, shape separation and self-assembly. *Materials Science & Engineering*, Vol.65, No.1-3, (April 2009), pp. 1-38, ISSN 0921-5093

Taub N., Krichevski O., Markovich G. (2003). Growth of Gold Nanorods on Surfaces. *J. Phys. Chem. B*, Vol.107, No.42, (September 2003), pp.11579-11582, ISSN 1520-6106

Templeton A.C., Pietron J.J., Murray R.W., Mulvaney P. (2000). Solvent Refractive Index and Core Charge Influences on the Surface Plasmon Absorbance of Alkanethiolate Monolayer-Protected Gold Clusters. *J. Phys. Chem. B*, Vol.104, No.3, (December 1999), pp. 564-570, ISSN 1520-6106

Underwood S., Mulvaney P. (1994). Effect of the Solution Refractive Index on the Color of Gold Colloids. *Langmuir*, Vol.10, No.10, (October 1994), pp. 3427-3430, ISSN 0743-7463

van der Zande B. M.I., Boehmer M.R., Fokkink L.G.J., Schoenenberger C.. (2000). Colloidal dispersions of gold rods: Synthesis and optical properties. *Langmuir*, Vol.16, No.2, pp. 451-458, ISSN 0743-7463

Wei Z., Mieszawska A.J., Zamborini F.P. (2004). Synthesis and manipulation of high aspect ratio gold nanorods grown directly on surfaces. *Langmuir*, Vol.20, No.11, (April 2004), pp. 4322-4326, ISSN 0743-7463

Willets K.A., Van Duyne R.P. (2007). Localized Surface Plasmon Resonance Spectroscopy and Sensing. *Annual Review of Physical Chemistry*, Vol.58, (May 2007), pp. 267-297, ISSN 0066-426X

Yu Y.Y., Chang S. S., Lee C.L., Wang C.R.C.. (1997). Gold Nanorods: Electrochemical Synthesis and Optical Properties. *J. Phys. Chem. B*, Vol.101, No. 34, (August 1997), pp. 6661-6664, ISSN 1520-6106

9

The Controlled Growth of Long AlN Nanorods and *In Situ* Investigation on Their Field Emission Properties

Fei Liu*, Lifang Li, Zanjia Su, Shaozhi Deng, Jun Chen and Ningsheng Xu
GuangDong Province Key Laboratory of Display Material and Technology,
School of Physics and Engineering,
Sun Yat-sen University, Guangzhou,
People's Republic of China

1. Introduction

Wide bandgap semiconductor nanostructures have been the research focus in recent years because of their unique physical and chemical properties and low electron affinity, which benefits for tunnel emission (Geis et al., 1991; Zhirnov et al., 1997; Kang et al., 2001; Liu et al., 2009). Among them, AlN nanostructures should deserve paid much attention due to their high melting-point (> 2300 °C), high thermal conductivity (K ~ 320 W/m·k), large exciton binding energy and strong endurance to harsh environment (Davis, 1991; Nicolaescu et al., 1994; Sheppard et al., 1990; Ponthieu et al., 1991). There have emerged many synthesis methods to fabricate different morphology of AlN nanostructures, such as nanocone, nanorod, and nanorods (Liu et al., 2009; Zhao et al., 2004; Liu et al., 2004; Tang et al., 2005; Shi et al., 2005, 2006; Wu et al., 2003; Paul et al., 2008). But for actual device applications of AlN nanostructures, there still exist many technique questions, which need to be solved as soon as possible. Firstly, the controlled growth of large area AlN nanostructures with uniform morphology is very difficult because of which cares about the uniformity of their physical properties in devices. Secondly, systemic investigation on the field emission (FE) properties of AlN nanostructures is not enough, which has fallen behind the development of the preparing method. Thirdly, it is unknown to us all that what factors take effect on their FE behaviors and how to find optimal growth conditions for their device applications. So developing a suitable way to controllably prepare AlN nanostructures and investigate on their FE properties in detail is essential for promoting their progress in FE area.

Chemical vapor deposition (CVD) technique is an effective way to fabricate high density nanostructures with uniform morphology. Moreover, it has some merits in comparison with other ways, for example low-cost and easy to realize the controlled growth. By CVD technique, it is reported that different morphologies of AlN nanostructures (nanocones, nanorods, nanocraters and ultra-long nanorods) have been successfully fabricated on the substrate. Their formation mechanisms are respectively proposed for different morphology of nanostructures. In addition, their FE properties are investigated in detail.

---

* Corresponding Author

2. Experimental

Different mass ratios of Al powders (99.99 %) to Fe_2O_3 powders (99.99 %) were used as source materials. Fe_3O_4 nanoparticles were synthesized as the catalysts of the AlN nanocones, nanorods and nanocraters by high temperature solution phase reaction (Yang et al., 2003; He et al., 2001), which were spread over the substrate. The CVD system has been described in our recent works (Liu et al., 2004, 2005, 2008; Cao et al., 2003), as shown in Fig.1.

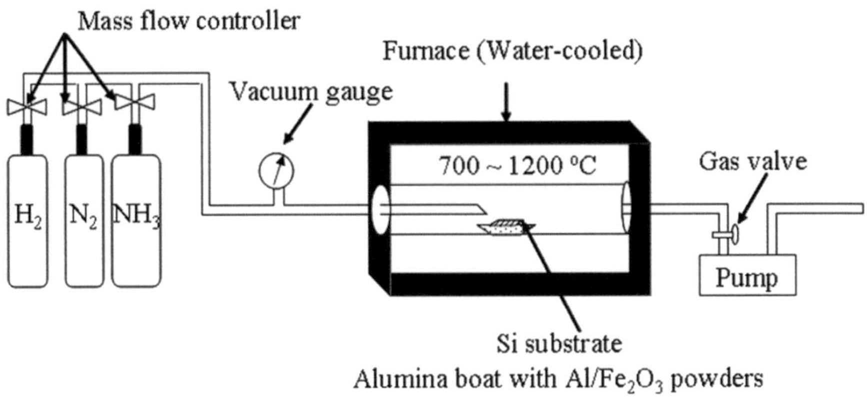

Fig. 1. A scheme of the chemical vapor deposition system in this experiment.

The reaction boat was put in the central region and the substrate was placed above the boat. A two-step increasing temperature method was used to synthesize AlN nanostructures. When the temperature arrived at 400 °C, the reaction vessel was maintained here for 20 ~ 30 min in the carrier gas. In this step, the only difference between the growth of the nanorods and those of other nanostructures is that the carrier gas is H_2 instead of N_2. The base pressure of the chamber was better than 5 Pa, and the flow rate of N_2 or H_2 was kept at 200 sccm. When the temperature of central region was raised to the reaction temperature, and the mixed gas consisted of N_2 and NH_3 was introduced into the chamber. The flow rate ratio of N_2 to NH_3 was fixed at 200: 5 sccm ~ 200: 20 sccm and the reaction pressure was kept at 10 Torr in this procedure. The reaction temperature was ranging from 700 °C to 1200 °C and kept 1-8 hours for the synthesis of different morphologies of AlN nanostructures. When the chamber was cooled to room temperature, a grey white film was found on the substrate.

A field-emission type scanning electron microscope (XL-SFEG, FEI Corp.) was used to observe the morphologies of AlN nanostructures. Transmission electron microscopy (Tecnai-20, PHILIPS) and high-resolution transmission electron microscopy (Tecnai F20, FEI Corp.) were used to obtain the crystalline structure of the nanostructures, respectively. Field emission (FE) properties and work function of AlN nanostructures were performed on the Field emission analysis system and Omicron VT-AFM system equipped with ultraviolet photoelectron spectroscope (UPS), respectively.

3. Results and discussion

By adjusting the synthesis conditions, controlled growth of different morphologies of single-crystal AlN nanostructures has been successfully realized. XRD (X-ray diffraction) technique was applied on four kinds of samples to confirm their chemical compositions. Typical XRD patterns of these four samples are provided in Fig. 2. It is found that there are four AlN characteristic peaks existing in these patterns, which correspond to the data of the Joint Committee for Powder Diffraction Standards (JCPDS) card No. 25-1133. Moreover, these four kinds of AlN nanostructures are found to be well-crystallized. It is also observed that Si (220) peak presenting in these patterns, which should come from the substrate. From these patterns, we can conclude that the alignment of all four samples is not very good because there still have other peaks than the growth direction of [001] existing in the patterns. Based on these patterns, it comes to a conclusion that all of the nanostructures in different morphology are AlN phase with a wurtzite structure.

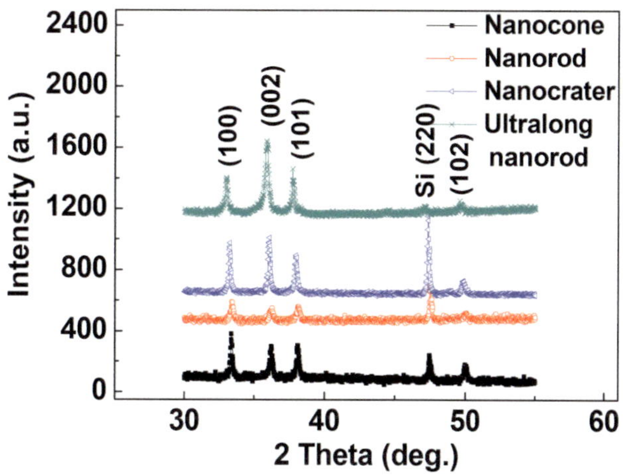

Fig. 2. Typical XRD patterns of AlN nanostructures with different morphologies.

The morphologies of four kinds of samples (nanocone, nanorod, nanocrater and ultra-long nanorod) are shown in Fig. 3. It is seen in Fig. 3a that the AlN nanocones have sharp tips and stand vertically to the substrate. From their high resolution SEM (scanning electron microscope) image (Fig. 3b), the nanocones are observed to have a length of about 2 μm and a mean radius at the top of 40 nm as well as they have smooth surface and high growth density. Fig. 3c shows the SEM image of AlN nanorod arrays in large area. These nanorods are perpendicular to the substrate and well-aligned. It is observed in Fig. 3d that the top of the nanorods seem to be well-facet hexagon and have uniform diameter. The length of the nanorods is about 2 micrometers and their averaged diameter is about 150 nanometers. The perspective view of AlN nanocrater arrays is provided in Fig. 3e. One can see that these AlN nanostructures grow perpendicularly to the silicon substrate and have uniform shapes. They are composed of many surrounded AlN nanocones with the same morphology and have crater-like shapes, so we denote them as nanocraters. In Fig. 3f they are seen have an outer diameter of 400 nm and a length of 2 μm. The top-view and side-view images of ultra-long AlN nanorod are respectively provided in Figs. 3g and 3h. The ultra-long nanorods are seen to have a mean diameter of 100 nm and an averaged length of 50 μm. Their diameter gradually decreases along their growth direction and lie to the substrate with an angle of 70°. Among these four samples, it is obviously that the ultra-long AlN nanorod array has the highest aspect ratio (500) and the lowest growth density ($3 \times 10^{8}/cm^{2}$). Moreover, it is also found that the AlN nanostructures are nearly in vertical arrays whether they are in what morphology, which will benefit for their FE applications.

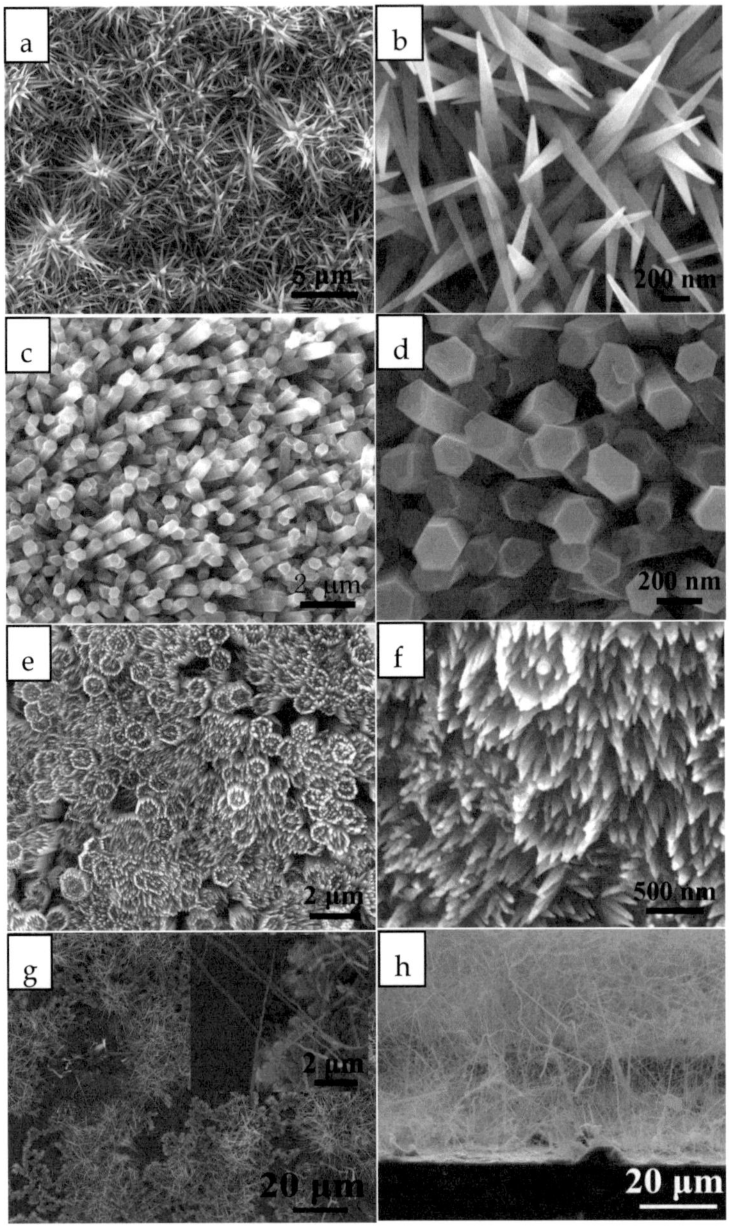

Fig. 3. (a, b) Low-resolution and high-resolution SEM images of the AlN nanocones, respectively. (c, d) Typical SEM images of the AlN nanorods in low magnification and high magnification, respectively. (e, f) The low magnification and high magnification images of the AlN nanocraters, respectively. (g, h) Top-view and side-view images of the ultra-long of nanorods.

To confirm the crystalline characteristics of AlN nanostructures in different morphology, TEM (transmission electron microscopy) technique was performed on these samples. Typical TEM image of AlN nanocone is shown in Fig. 4a. It is seen that the diameter of the nanocone is gradually decreasing from the bottom to the top, which is in agreement with our SEM results in Fig. 3a. The corresponding selected area electron diffraction (SAED) pattern is in the inset, in which one can see that the diffraction spots are clear and sharp. By identifying their clear diffraction spots, these nanocones are proved to have a perfect single crystalline AlN structure and have a growth direction of [001]. And the AlN nanorods have a uniform diameter through the whole length, as shown Fig. 4c. Further SAED (the inset) and HRTEM (Fig. 4d) characterizations indicate that they are also AlN single crystalline structures and grow along [001] direction, which is similar with the nanocones. Figs. 4e and 4f provide the general TEM image of AlN nanocraters, respectively. It is seen that the nanocrater is composed of several surrounded nanocones, as shown in Fig. 4e. Typical SAED pattern (the inset) and HRTEM image (Fig. 4f) of a nanocone (white ellipse) are provided to investigate their crystalline structures. It is noted that every nanocone appears the same diffraction spots and HRTEM (high resolution transmission electron microscopy) images, which proves that they all grow along the [001]. The ultra-long AlN nanorods also exhibit the same crystalline structure and growth direction, as indicated in Fig. 4g and 4h. So it comes to a conclusion no matter what morphology these AlN nanostructures belong to, all of them are perfect single crystals with a growth orientation of [001].

Energy dispersive x-ray (EDX) characterization is performed on different morphologies of AlN nanostructures. These nanostructures have resembled EDX spectrum, so only a typical spectrum is given here. It is clear that the Al L_1, L_2 and L_3-edge peaks (corresponding to its inner energy levels) are existed in Fig. 5a. From the inset, the characteristic peak of N k-edge can be also found in the spectrum. Moreover, Gatan EELS analysis also shows that the total content of element Al and N is over 98 %, thus it can be confirmed that these nanostructures are pure AlN single crystals whether they are in any morphology.

To further comprehend the effect of the experimental parameters to the morphology of AlN nanostructures and understand the growth mechanism for different nanostructures, the detailed synthesis conditions are provided in Table 1. As for the growth of AlN nanocone, nanorod and nanocrater, Fe_3O_4 nanoparticles are used as the catalyst. So we preferred the vapor-liquid-solid (VLS) mechanism to illustrate the formation of AlN nanostructures. Because not any catalysts are found to exist at the top of these three nanostructures, it is proposed that the growth model of these nanostructures is the bottom growth model. As explained in the reference, the tight bonding force may exist between the catalyst and the substrate at the beginning of the growth (Fan et al., 1999). The formation mechanism of three morphologies of AlN nanostructures may be attributed to the diffusion mediated growth mechanism (Shi et al., 2005). Based on this model (Shi et al., 2005), AlN precursors have different diffuse length for different growth planes, and these growth planes should have different growth rate at different temperature at the same time. The anisotropy of the growth plane at different temperature will lead to the formation of nanorod or nanocone, i.e., high temperature (>900 ^{o}C) benefits for the synthesis of AlN nanorods, as shown in Table 1.

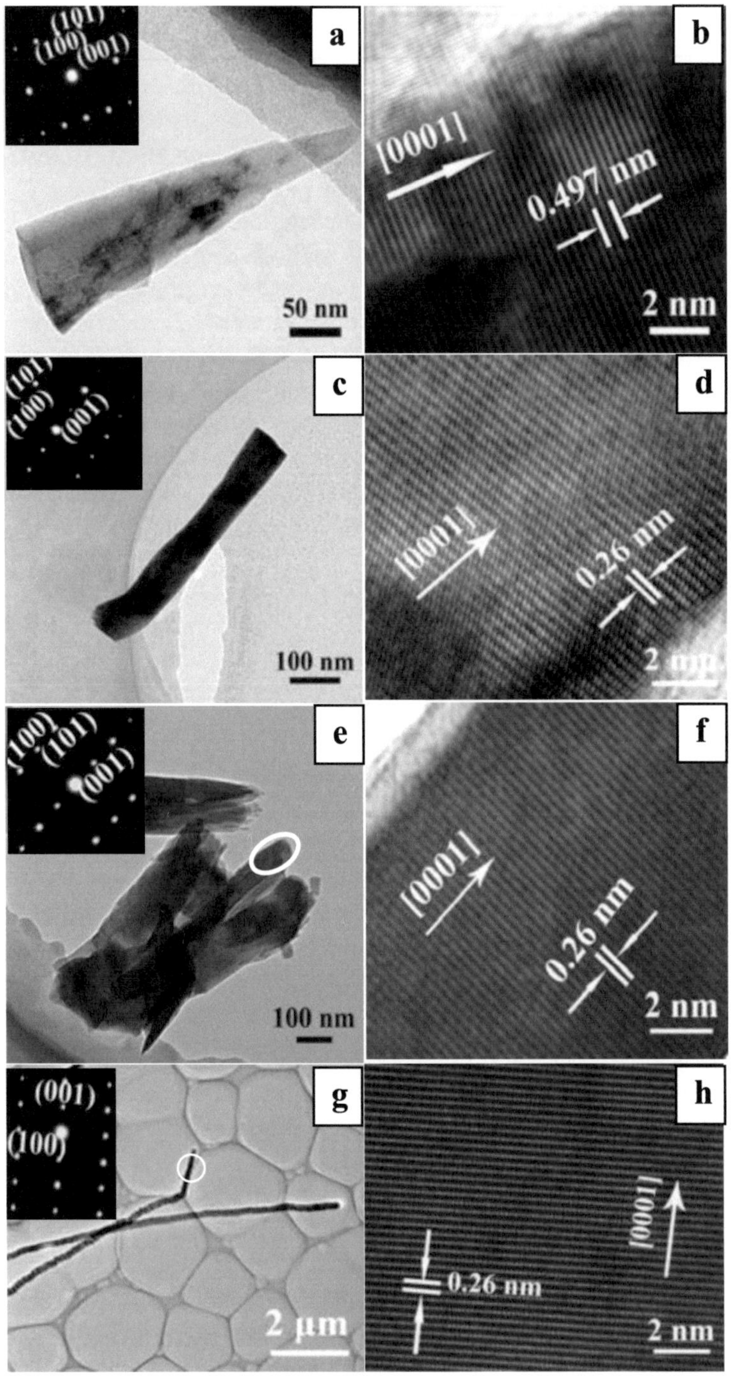

Fig. 4. Typical TEM and HRTEM images of four kinds of AlN nanostructures (a, b) AlN nanocones; (c, d) AlN nanorods; (e, f) AlN nanocraters. (g, h) ultra-long AlN nanorods. The inset is their corresponding SAED pattern and the white circle in Fig. 4e and 4g respectively corresponds to the region in the HRTEM image of Fig. 4f and 4h.

Sample \ Experimental conditions	Source material ratio (Al : Fe_2O_3)	Carrier gas	Growth temperature (°C)	Length (μm)	Outer diameter (nm)	Growth density (/cm²)
Nanocone	0.2 g: 0 g	N_2, NH_3	700-800	2	40	2.2x10⁹
Nanorod	0.2 g: 0 g	N_2, NH_3	900-1000	2	150	1.5 x10⁹
Nanocrater	0.2 g: 0 g	N_2, NH_3	800-900	2	400	3 x10⁸
Ultra-long nanorod	0.6 g : 0.2 g	N_2, NH_3, H_2	1000-1200	50	100	3.2x10⁷

Table 1. The list of morphological characteristics and experimental conditions for different morphologies of AlN nanostructures.

When slightly increasing the growth temperature (still lower than the temperature of the nanorod) at the same molecular ratio of the reaction gas (NH_3/N_2) of the nanocone, the AlN nanocraters can be synthesized. More AlN vapors can dissolve in the catalyst solutions at the growth stage of nanocrater than the formation of AlN nanocones at this situation. Subsequently more AlN seeds can separate from oversaturated solutions because of the overmuch supplement of the vapors in the following stage. Because the concentration of the catalyst is too high, the density of the catalyst becomes too high and the distance between the catalysts turns short accordingly. At the driving of high temperature, the neighbor catalysts migrate to gather together to form larger precursors for the new seeds in synthesizing new nanostructures, which should be the most energy favorable structures. Finally, the nanocraters are fabricated from these gathered precursors with the progress of the reaction.

Although the VLS mechanism may be used to illustrate the formation of the nanocone, nanorod and nanocrater, it is not suitable for the ultra-long nanorods because no catalyst is used in their synthesis process. So the self-catalyzing Vapor-Liquid-Solid (VLS) mechanism (Liu et al., 2004; Z. L. Wang et al., 2003) is proposed to explain the growth of ultra-long AlN nanorod. At the initial stage, the Al powders were evaporated to Al atoms at melting point of about 650 °C and transferred to the region of the substrate by the carrier gas. Subsequently, the transportation Al atoms deposited on the substrate with lower temperature and formed a continuous Al film. At the function of H_2 gas, the continuous film gradually separated into Al nanoparticle film, as described in Ref. (Yao et al., 2008; Jung et al., 2001.) When the temperature was increased to 1000 °C, NH_3 gas was introduced into the vacuum chamber and decomposed into N atoms. At high temperature (>1000 °C), Al nanoparticle precursors turned liquid droplets on the substrate. So the foreign N atoms reacted with the Al liquid droplets and formed the AlN precursors as the seeds of the ultra-long nanorod in the solutions. AlN seeds can continuously separate from oversaturated solutions because of the overmuch supplement of AlN precursors. Finally, with the progress of the reaction, high density of ultra-long AlN nanorod arrays was synthesized on the substrate. And the steric overcrowding (Liu et al., 2004, 2005, 2008; Cao et al., 2003) for these nanostructures is considered as the possible aligned mechanism of AlN nanostructure arrays.

As summarized in Table 1, the growth conditions will influence their density. Through a series of designed experiments, it is found that the growth temperature is the determinate factor for the formation of different morphology of AlN nanostructures. Moreover, the total mass of the source materials has effect on the growth density of the nanostructures. It is observed that the lower mass of raw materials will induce lower density of nanostructure. It is also seen that the ultra-long Al nanorod is hard to be synthesized when the mass ratio of Al powders to Fe_2O_3 powders is lower than 8:1. In addition, the pressure of the carrier gas also takes effects on the synthesis of these nanostructures. If the growth pressure is adjusted to be lower than 100 Pa, only AlN thin film can be found on the substrate.

The surface electron affinity is one of important parameters determining the field emission property of cathode nanomaterials. It is known that AlN nanostructures may have a little positive or negative affinity on (001) crystalline plane, which make them as one of the promising cathode materials in future. The Ultraviolet Photoemission Spectroscopy (UPS) technique is applied to measure the electron affinity of the AlN nanostructures before the beginning of FE measurements. Because these nanostructures have similar UPS spectrum, only a representative spectrum of the sample is given in Fig. 5b. To confirm that the detector can collect all emission electrons from the valence band, a negative 5 voltage was applied to the sample. The whole spectral width ($W_{spectra}$) of the sample can be calculated by linearly extrapolating the emission onset edges to zero intensity at both the low kinetic energy cutoff and the high kinetic end, which are equal to the energy difference between the deepest level in the valence band and the edge of the valence band for semiconductor. According to this method, its width is determined to be about 14.3 eV. The surface electron affinity χ of a sample may be obtained using the equation given below (Benjamin et al., 1996):

$$\chi = h\nu - E_g - W_{spectra} \tag{1}$$

where E_g is the energy band gap, and hv=21.2 eV, which is the radiation energy of He I line. The energy band gap of AlN nanostructures can be assumed to be the same as that of bulk AlN, i.e., 6.28 eV (Edgar et al., 1999). Then the χ values are deduced to be 0.62 eV, which suggests the electron affinity of the as-prepared AlN nanostructures is low enough for electron emission at low voltage operations in FE area.

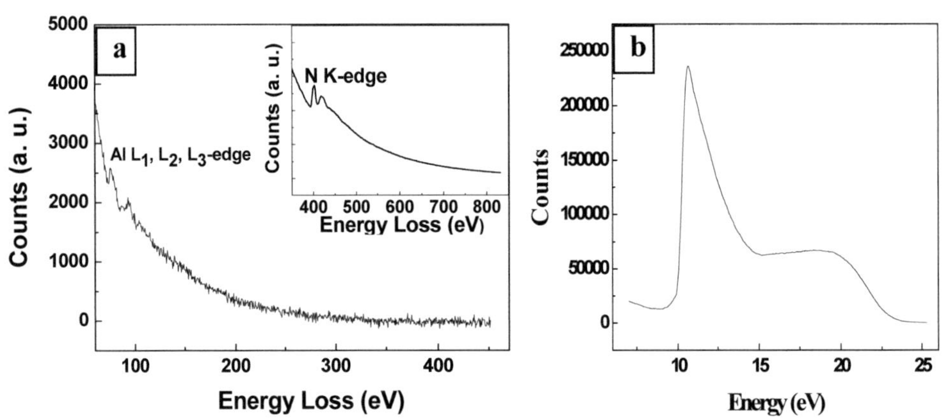

Fig. 5. (a) Representative EELS spectrum of the AlN nanostructures. (b) Their UPS spectrum.

Whether these synthesized AlN nanostructures are promising cold cathode nanomaterials needs to be further valued by their FE properties. Their field emission measurements are carried out in the field emission analysis system, whose base pressure is lower than $3x10^{-7}$

Pa. Transparent anode method was used in the measurements to obtain the FE behaviors of the total film in different morphology. These three samples were applied as the cathodes and the indium-tin-oxide (ITO) glass was used as anode. The area of the samples was around 1x1 cm² and the measurement distance between the anode and the cathode is about 400 μm. The J-E curves and FN plots of the nanocones, the nanorods, the nanocraters and the ultra-long nanorods are shown in Fig. 6a and 6b, respectively. By comparing their FE

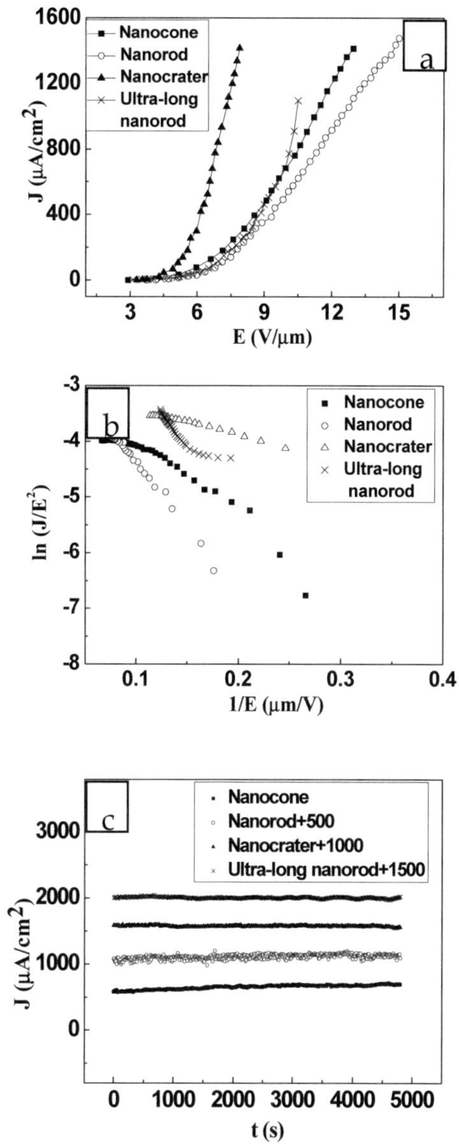

Fig. 6. (a) The field current density-electric field (J-E) curves of the AlN nanostructures in different morphology; (b) Their corresponding FN plots. (c) The stability curves of the AlN nanostructures in different morphology in field emission process.

behaviors, it is clear that the AlN nanocraters have the best FE performance with a turn-on field (defined as the field at 10 μA/cm²) of 3.9 V/μm and a threshold field (defined as the field at 1 mA/cm²) of 7.2 V/μm. It is also found that the FE properties of the AlN nanorods are better, those of the nanocone are worse, and those of the ultra-long AlN nanorods are the worst among them, as provided in Table 2. But even for the AlN ultra-long nanorods with the worst FE performance, their FE behaviors are comparable with many cathode nanomaterials, such as ZnO, WO_3 and CuO. In addition, all of their FN plot exhibits a nonlinearity relationship within the range.

What leads to their different FE performance for different nanostructures is an important question, which is intensively related with their applications. We proposed the following interpretations to analyze the possible mechanism. According to the Metal-Insulator-Vacuum (MIV) theory (Latham & Xu, 1995), the microscopic field E_{Local} acting at the tip of microtrusion is proportional to the enhancement factor β, which is given by

$$E_{Local} = \beta E = \beta \frac{V}{d} \qquad (2)$$

In this expression, V is the applied voltage, d is the distance between the anode and the cathode, and E is the applied macroscopic field. And the enhancement factor β can be expressed as (Fursey & Vorontsov-vel'yaminov, 1967)

$$\beta = \frac{\left(\lambda^2 - 1\right)^{1.5}}{\lambda \ln\left[\lambda + \left(\lambda^2 - 1\right)^{\frac{1}{2}}\right] - \left(\lambda^2 - 1\right)^{\frac{1}{2}}} \qquad (3)$$

where $\lambda = l / r$ is the aspect ratio, i.e., the length of the nanostructure versus its top radius. As discussed in the above theory, the nanostructure with the large aspect ratio should have high β value because they have more strong ability of amplification macroscopic field than other nanostructures at the same applied field. So the nanorod, the nanocones and the nanocraters should have better FE behaviors than the ultra-long nanorods according to this theory. But as observed in Table 2, the FE properties of the ultra-long nanorod with the highest aspect ratio (500) are worse than those of the nanocrater with lower aspect ratio (50), which suggests that this theory can't be directly used to explain the phenomena.

Then at this situation, another discrepancy existed in these nanostructures must be considered due to the screen effects (Bonard et al., 2001; Wang & Tong, 1996), which is their growth density. According to this theory, the surface of the nanostructured film will be very close to a plane when the growth density of the nanostructure is very high. So when the field is applied between the anode and the cathode, it will prevent the macroscopic electrical field from penetrating into the nanostructures film (Bonard et al., 2001), which in versus lowers the local field amplification existing among the nanostructures and make electron emission hard to occur. Based on the classical Fowler-Nordheim (FN) theory (Fowler & Nordheim, 1928), the relationship between current density J and applied field E should be described as follows:

$$J = \eta a \left(\frac{\beta^2 E^2}{\varphi} \right) \exp\left(\frac{-B\varphi^{\frac{3}{2}}}{\beta E} \right), \tag{4}$$

Where a=1.54x10^{-6} (AV^{-2}eV), B=6.83x10^{9} (Vm^{-1}eV^{-3}/2), and φ is the work function of the cathode emitters, which adopts 3.7 eV for AlN (Edgar et al., 1999). So the enhancement

FE Properties / Morphology	Aspect ratio	Turn-on field at 10 μA (V/μm)	Threshold field at 1 mA (V/μm)	β	s (field screen parameter)	Emission Instability (2 h)
Nanocone	50	4.2	11.3	1740	0.087	9 %
Nanorod	23	4.9	12.4	1250	0.234	7 %
Nanocrater	50	3.9	7.2	2140	0.2675	4 %
Ultra-long Nanorod	500	4.5	10.4	1997	0.250	6 %

Table 2 The list of comparisons on the field emission parameters of the as-prepared AlN nanostructures in different morphology.

factor β values of the AlN nanostructures can be calculated from the slopes of their FN plots, which reflect the amplification ability of macroscopic field. From Table 2, it is observed that the obtained β values are ranging from 1250 to 2140, which suggests that AlN nanostructures are one of the most promising FE materials in future. It can be observed in Table 2 that the nanorod and the nanocrater with lower density have higher field enhancement factor β than the nanocones with higher density. Combined the nonlinearly (low and high electric field)exhibited in their FN plots with the screen effect, Filip model was more appropriate for these samples (Filip et al., 2001), which can be expressed as follows:

$$E_{local} = s\frac{V}{r} + (1-s)\frac{V}{d}, \tag{5}$$

Where E_{local} is the local field nearby the emitters when considering the screen effect, r is the radium of the tip of the emitter, s is the screen effect parameter, V is the applied voltage and d is the distance between the anode and cathode, which is 400 μm in our measurements. Combined equation (2) with (5) and considered the d/r value is far higher than 1 in our experiment, the field enhancement factor can be derived as:

$$\beta \approx 1 + s\frac{d}{r} \Rightarrow s = \frac{(\beta-1)r}{d}, \tag{6}$$

Substitute typical d value, r value and β value obtained for four samples, their corresponding field effect parameter s can be obtained, which is indicated in Table 2. It is clear that the smaller the s value, and the greater the role that screening effect plays in the actual field emission process. So it is concluded that the field screen effect of the nanocone is the most distinguished among these four nanostructures because of its smallest s value (0.087). Therefore, the nanocraters should have the best FE behaviors and the nanorods should have better FE behaviors among these four nanostructures based on this rule, which is in well agreement with our observed results in Fig. 6. Moreover, the nonlinearity of the FN plots of four samples can also be explained by Filip model (Filip et al., 2001). It is also proposed that the morphology and the growth density of the AlN nanostructures determine their FE behaviors to a great extent.

The emission stability of nanostructures is another question need to be paid much attention for their actual application in FE area, and Fig. 6c shows the stability curves of the AlN nanostructures in different morphology. Because the stability appearance at high working current is more important than at low current for application, the emission current was kept at 500 μA at the beginning of the measurement whether they are in what morphology. The whole measurement lasted for 2 hours. It is seen in Fig. 6c that the nanocraters are found to possess the most stable emission performance with a current fluctuation lower than 4 % and the nanocones are the worst (9 %), as indicated in Table 2. Other nanostructures also exhibit excellent field emission stability, which is good enough for the potential application in field emission area. In combined with all the field emission behaviors of these four AlN nanostructures, we can draw a conclusion that the AlN nanocrater arrays have the best FE performance whether from the turn-on field or the stability, which should have the most promising future in FE applications.

Then, what to do for further inproving their field emission properties? We should firstly find the important factor, which strongly affects their emission performance. In order to confirm this, measurement on individual AlN ultra-long nanorod is very essential because many physical information of AlN nanostructure is concealed by the continuous film due to some effects. Few reports so far are concerned with the characterization of individual single AlN ultra-long nanorods because the AlN nanostructures are often fabricated in high density, which is hard to carry out the in-situ measurement on the physical properties of individual nanostructures. Since we have mastered the controlled growth tecnique of the AlN ultra-long nanorod in growth density, measurement on the physical properties of individual nanostructure becomes feasible. The measurement was carried out in a modified SEM system, which has been depicted in detail in our previous papers (Liu et al., 2008, 2010a, 2010b; She et al., 2007). Figs. 7a and 7b provides the SEM image of the measurement during the electric conductivity process and during field emission process, respectively. It is seen from Fig. 7a that the W probe tightly contacts to individual ultra-long AlN nanorod on its end. Individual ultra-long nanorod has a length of about 50 μm and a diameter of about 100 nm, which is consistent with the observed images in Fig. 2. One can see in Fig. 7b that the distance between the W probe and individual nanorod was kept at about 2 μm through field emission measurement.

Representative curve of the electrical transportation property of individual ultra-long AlN nanorod is shown in Fig. 8a. Here only a typical I-V curve of a single nanostructure is given because their electrical conductivity is similar, having the same order (in the range from $2 \times 10^{-4} - 7 \times 10^{-4}\ \Omega^{-1}cm^{-1}$). From the shape of the I-V curve of Fig. 8a, it can be concluded that the electric contact in conductivity measurement should be Schottky barrier rather than the ohmic contact because it is nonlinear and asymmetrical. We use the following descriptions to illutrate this phenomena. The contact resistance $R_{Contact}$ in measurements can be divided into two types, in which one is the contact resistance R_1 between the nanorod and the substrate and another is the contact resistance R_2 between the nanorod and the W probe, which can be written as:

$$R_{Contact} = R_1 + R_2 \tag{7}$$

So the total resistance R_{Total} consists of the contact resistance $R_{Contact}$ and the intrinsic resistance R_{AlN} of AlN nanorod, which is as follows:

$$. R_{Total} = R_{AlN} + R_{Contact} = R_{AlN} + R_1 + R_2 \tag{8}$$

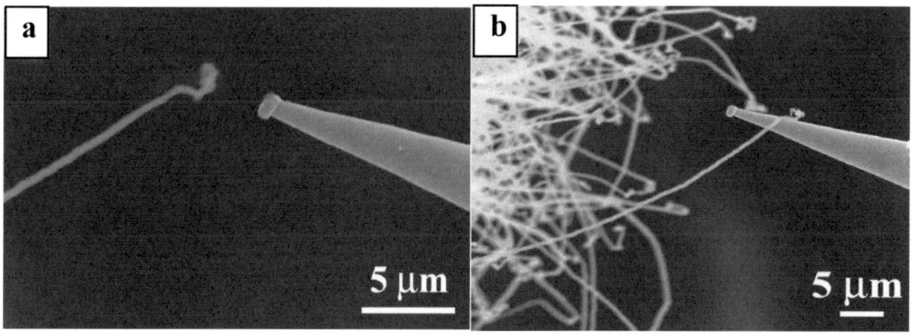

Fig. 7. The SEM images of individual AlN nanorod and the tungsten probe (a) during the conductivity measurement, and (b) during the field emission measurement.

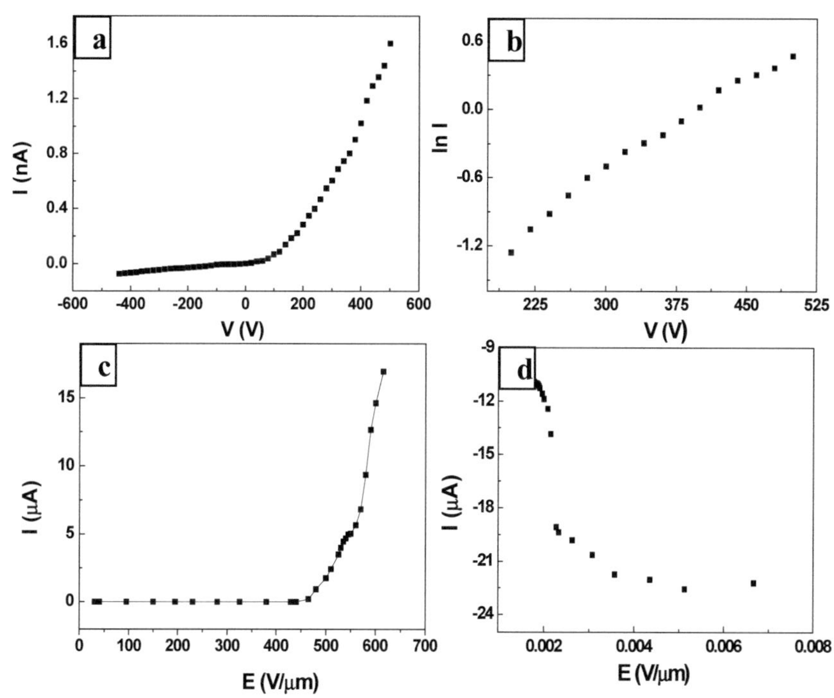

Fig. 8. (a) Representative electrical transport curve of single AlN nanorod. (b) The lnI versus V curve in conductivity measurment. (c) The field emission I-E curves of single nanorod. (d) Its corresponding FN plots.

The work function of W, AlN and n-doped Si is usually 4.5 eV, 3.8 eV and 4.2 eV, respectively. At high growth-temperature over 1100 °C, the AlN clusters and n-doped Si will be alloyed together, which suggests that their contact resistance R_1 is very low close to ohmic contact after the growth. But for R_2, high contact resistance will occur because their large discrepancy of the work function between W probe and individual AlN nanorod, which results in the formation of Schottky barrier. In this situation, R_2 determines the contact resistance $R_{Contact}$. Based on the thermionic emission theory (Heo et al., 2004; Sze et al., 1971), the reverse current of an ideal Schottky barrier should be at a very low level because the barrier is often elevated to a very high value at reverse voltage, so the reverse current can be neligible in contrast to their positive current. The positive current will exponentially increase with the positive voltage applied on the Schottky diode, followed by the equation (Sze et al., 1971)

$$J(V,\varphi_b) = A^{*}T^{2}\exp\left(-\frac{\varphi_b}{kT}\right)\exp\left(\frac{qV}{nkT}\right)\times\left[1-\exp\left(-\frac{qV}{kT}\right)\right] \quad (9)$$

, where $A^*=4\pi m^*qk^2/h^3$ is the Richardson constant of the semiconductor, φ_b is the effective height of Schottky barrier, k is the Boltzmann constant, q is the magnitude of electron charge, n is usually adopted to be 1 for ideal Schottky diode, T is the absolute temperature. According to this equation, the linear behavior of lnI versus V should be observed at positive voltage. It is clearly seen in Fig. 8b that lnI has a linear relation with V, which is in well agreement with the proposed theoretical results. It suggests that the electron transportation process of measurement on individual nanorod should obey the thermionic emission model, which proves our explaination on Schottky barrier is reasonable. Moreover, the electrical conductivity can be obtained by approximately calculating the slope of I-V curve at high voltage. It can be understood that the intrinsic resistance R_{AlN} dominates the total resistance R_{Total} because the Schottky barrier at positive voltage will rapidly decreased to a very low value with the the increasing of the applied voltage, which is far lower than R_{AlN}. So based on this way, the intrinsic conductivity of AlN nanorod is deduced to be $2.7\times10^{-4}\ \Omega^{-1}cm^{-1}$, which is almost 10^2 times larger than with that ($8\times10^{-6}\ \Omega^{-1}cm^{-1}$) of other researcher group (Zheng et al., 2008). The relatively higher conductivity should be originated from the doping of element Si and Fe, which incorporated into the engergy gap of AlN to be a shallow impuries energy level. Element Si may come from the Si substrate or quartz tube, and element Fe may come from Fe_2O_3 in source powders, which possibly diffused into AlN nanorods during the high-temperature growth process. Therefore, the intrinsic electric conductivity of ultra-long AlN nanorod is overestimated in this measurement.

Fig. 8c gives typical field emission current versus applied field (J-E) curve of a single ultra-long AlN nanorod, and its corresponding Fowler-Nordheim (FN) plot is indicated in Fig. 8d. From Fig. 8d, individual AlN nanorod is observed to have a mean 1 nA field (defined as the electric field when the emission current is 1 nA) of 440 V/μm and 1 μA field (defined as the electric field when the emission current is 1 μA) of 480 V/μm. The FE performance of the as-synthesized ultra-long AlN nanorod is better than that of AlN nanocones (500 nA at 1000 V//μm) in recent report (Bonald et al., 2001), which should come from the conductivity of individual nanorod on our synthesis process is higher than that in other researher groups due to the doping of element Fe and Si. It is also found that the field emission current exhibits a increase tendency with the applied field and doesn't reach the saturation until the applied field arrives at 800 V/μm, which proposes AlN nanorod has a high endurance ability to large emission current due to its high thermal conductivity. The FN plots of individual AlN nanorod is seen to be nonlinear. We attributed the nonlinear behavior of the FN plots to be the drecrease of the contact resistance R_2 with the increase of the temperature followed by high applied field, because of which induces higher effective field appearing between cathode and anode. Even though it is found from Fig. 8c that the FE property of individual AlN nanorod is not very good, in comparison with the others such as boron nanotubes and nanowires (Liu et al., 2008, 2010a, 2010b), $W_{18}O_{49}$ (Li et al., 2009) etc, as we reported early, and that this finding is in consistent with the findings reported above for the FE performance of AlN nanorod films. The relatively worse FE performance of a single AlN nanorod is strongly related to its low intrinsic electrical conductivity, as may be explained in the following. It is obvious that the intrinsic conductivity of individual AlN nanorod in our experiment is lower than that of many nanomaterials with excellent FE properties, such as CNTs (1 - 2 x 10^3 $\Omega^{-1}cm^{-1}$) (Dai et al., 1996), $W_{18}O_{49}$ nanorod (10^{-1} - 10^{-2} $\Omega^{-1}cm^{-1}$) (Li et al., 2009), ZnO (2 - 4 x 10^{-2} $\Omega^{-1}cm^{-1}$) (She et al., 2008) and boron nanotube (10 - 20 $\Omega^{-1}cm^{-1}$) (Liu et al., 2010). So when an equal voltage is applied, more voltage drops on the nanorod's intrinsic reistance rather than falls across the vacuum gap between nanorod and anode probe, which leads to a lower effective field at the tip of the nanorod. Thus, relatively higher turn-on and threshold field is needed for the AlN nanorod in order to realize the tunnel effects, which results in their not very good FE performance. The relatively worse FE properties of individual nanorod must naturally take effect on the emission performance of their corresponding film. It is known that the element-doping can effectively improve the electric conductivity of individual nanostructures by forming a impurities energy level in their energy gap, which can provide enough electrons for tunnel current at low applied field. Once the electric conductivity of AlN nanostructures can be enhanced to a high level, they should have a more promising future in FE devices.

4. Conclusions

Different morphology of AlN nanostructure arrays have been successfully synthesized by CVD technique. They are confirmed to have perfect single crystalline AlN structures with a growth direction of [001]. The VLS and Self-catalyzing VLS mechanism is respectively used to explain the formation mechanism for these four nanostructures. And the AlN nanocrater arrays are found to have the best FE properties among four kinds of nanostructures, which have a turn-on field of 3.9 V/μm and a threshold field of 7.2 V/μm. Moreover, measurements on individual ultra-long nanorods show that they have a lower electric conductivity (2.7 x 10^{-4} $\Omega^{-1}cm^{-1}$) and relative worse FE performance (1 nA field of 440 V/μm and 1 μA field of 480 V/μm) than CNTs. The former leads to their FE performance. All of the synthesized AlN nanostructure arrays exhibit good FE performance, which suggests that they should be one of the most promising cold cathode nanomaterials in future.

5. Acknowledgements

The authors are thankful for the support from the National Basic Research Program of China (973 Program, Grant No. 2007CB935501, 2010CB327703, Grant No. 50802117, 51072237, 50725206,), the National Joint Science Fund with Guangdong Province (Grant No. U0634002, U0734003), the Foundation of Education Ministry of China (Grant No. 20070558063 and

2009-30000-3161452), the Science and Technology Department of Guangdong Province, the Education Department of Guangdong Province, and the Science and Technology Department of Guangzhou City.

6. References

Benjamin, M. C.; Bremser, M. D.; Jr., T. W. Weeks., King, S. W.; Davis, R. F.; Nemanich, R. J. (1996). UV Photoemission Study of Heteroepitaxial AlGaN Films Grown on 6H-SiC, *Appl. Surf. Sci.*, Vol. 104/105, No. , (September 1996) , pp. 455-460, ISBN 0169-4332

Bonard, J. M.; Weiss, N.; Kind, H.; Stöckli, T.; Forró, L.; Kern, K.; Châtelain, A. (2001). Tuning the Field Emission Properties of Patterned Carbon Nanotube Films, *Adv. Mater.*, Vol. 13, No. 3, (February 2001), pp. 184-188, ISBN. 0935-9648

Cao, P. J.; Gu, Y. S.; Liu, H. W.; Shen, F.; Wang, Y. G.; Zhang, Q. F.; Wu, J. L.; Gao, H. J. (2003). High-Density Aligned Carbon Nanotubes with Uniform Diameters, *J. Mater. Res.*, Vol. 18, No. 7, (July 2003), pp. 1686-1690, ISBN 0022-2402

Davis, R. F., (1991). III-V Nitrides for Electronic and Optoelectronic Applications, *Proceeding of the IEEE.*, Vol. 79, No. 5, (May 1991), pp. 702-712, ISBN 0018-9219

Dai, H.; Wong, E. W.; Lieber, C. M. (1996). Probing Electrical Transportation Properties in Nanomaterials: Conductivity of Individual Carbon Nanotubes, *Science*, Vol. 272, No. 5261, (April 1996), pp 523-526, ISBN 0036-8075

Edgar, J. H., Strite, S.; Akasaki, I.; Amano, H.; Wetzel, C. (1999). Properties, Processing and Applications of Gallium Nitride and Related Semiconductors, INSPEC., ISBN 978-0-86341-775-7, Part A, London, UK

Fan, S. S.; Chapline, M. G.; Franklin, N. R., Tombler, T. W.; Cassell, A. M. ; Dai, H. J. (1999). Self-Oriented Regular Arrays of Carbon Nanotubes and Their Field Emission Properties, *Science*, Vol. 283, No. 5401, (January 1999), pp. 512-514, ISBN 0036-8075

Fan, Z. Y.; Wang, D. W.; Chang, P. C.; Tseng, W. Y.; Lu, J. G. (2004). ZnO nanowire Field-Effect Transistor and Oxygen Sensing Property, *Appl. Phys. Lett.*, Vol. 85, No. 24, (December 2004), pp. 5923-5925, ISBN 0003-6951

Filip, V.; Nicolaescu, D.; Tanemura, M.; Okuyama, F. (2001). Modeling the Electron Field Emission from Carbon Nanotube Films, *Ultramicroscopy*, Vol. 89, No. 1-3, (October 2001), pp. 39-49, ISBN 0304-3991

Fowler, R. H. & Nordheim, L. W. (1928). Electron Emission in Intense Electric Fields, *Proceeding of the Royal Society of London*, Ser. A, Vol. 119, No. 781, (May 1928), pp. 173-181

Fursey, G. N. & Vorontsov-Vel'yaminov, P. N. (1967). Qualitative Model of Initiation of Vacuum Arc II. Field Emission Mechanism of Vacuum Arc Onset, Sov. Phys: Tech. Phys., Vol. 12, No. 5, (May 1967), pp. 1377-1385, ISBN 0038-5662

Geis, M. W.; Efremov, N. N.; Woodhouse, J. D.; McAleese, M. D.; Marchywka, M.; Socker, D. G.; Hochedez, J. F. (1991). Diamond Cold Cathode, *IEEE Electron Device Lett.*, Vol. 12, No. 8, (August 1991), pp. 456-459, ISBN 0741-3106

He, S. T.; Yao, J. N.; Jiang, P.; Shi, D. X.; Zhang, H. X.; Xie, S. S.; Pang, S. J.; Gao, H. J. (2001). Formation of Silver Nanoparticles and Self-Assembled Two-Dimensional Ordered Superlattice, *Langmuir*, Vol. 17, No. 5, (February 2001), pp. 1571-1575, ISBN 0743-7643

Jung, M.; Eun, K. Y.; Lee, J. K.; Baik, Y. J.; Lee, K. R.; Park, J. W. (2001). Growth of Carbon Nanotubes by Chemical Vapor Deposition, *Diamond Relat. Mater.*, Vol. 10, No. 3-7, (March 2001), pp 1235-1240, ISBN 0925-9635

Kang, D.; Zhirnov, V. V.; Sanwald, R. C.; Hren, J. J.; Cuomo, J. J. (2001). Field Emission from Ultrathin Coatings of AlN on Mo Emitters, *J. Vac. Sci. Technol. B*, Vol. 19, No. 1, (January 2001), pp. 50-54, ISBN 1071-1023

Latham, R. V. & Xu, N. S. (1995). High Voltage Vacuum Insulation, Academic, ISBN 0-12-437175-2 , London, UK

Li, Z. L.; Liu, F.; Xu, N. S. (2009). Improving field-emission uniformity of large-area $W_{18}O_{49}$ nanowire films by electrical treatment, J. Vac. Sci & Technol. B, Vol. 27, No. 6, (December 2009), pp. 2420-2425, ISBN 1071-1023

Liu, C.; Hu, Z.; Wu, Q.; Wang, X. Z.; Chen, Y.; Sang, H.; Zhu, J. M.; Deng, S. Z.; Xu, N. S. (2004). Vapor−Solid Growth and Characterization of Aluminum Nitride Nanocones, *J. Am. Chem. Soc.*, Vol. 127, No. 4, (January 2004), pp. 1318-1322, ISBN 0002-7863

Liu, F.; Cao, P. J.; Zhang, H. R.; Li, J. Q.; Gao, H. J. (2004). Controlled Self-Assembled Nanoaeroplanes, Nanoflowers and Tetrapod Networks of Zinc Oxide, *Nanotechnology*, Vol. 15, No. 8, (August 2004), pp. 949-952, ISBN 0957-4484

Liu, F.; Cao, P. J.; Zhang, H. R.; Shen, C. M.; Wang, Z.; Li, J. Q.; Gao, H. J. (2005). Well Aligned ZnO Nanorods and Nanorods Prepared Without Catalysts, *J. Cryst. Growth*, Vol. 274, No 1-2. , (January 2005), pp. 126-131, ISBN 0022-0248

Liu, F.; Tian, J. F.; Bao, L. H.; Yang,T. Z.; Shen, C. M., Lai, X. Y.; Xiao, Z. M.; Xie, W. G.; Deng, S. Z.; Chen, J.; She, J. C.; Xu, N. S.; Gao, H. J. (2008). Fabrication of Vertically Aligned Single Crystalline Boron Nanowire Arrays and Investigation on Their Field Emission Behaviors, *Adv. Mater.*, Vol. 20, No. 13, (July 2008), pp. 2609-2615, ISBN 0935-9648

Liu, F.; Su, Z. J.; Liang, W. J.; Mo, F. Y.; Li, L.; Deng, S. Z.; Chen, J.; Xu, N. S. (2009). Controlled Growth and Field Emission Investigation of Vertically Aligned AlN Nanostructures in Different Morphology, *Chin. Phys. B*, Vol. 5, No. 18, (May 2009), pp. 2016, ISBN 1674-1056

Liu, F. ; Su, Z. J.; Li, L; Mo, F. Y.; Jin, S. Y.; Deng, S. Z.; Chen, J; Shen, C. M.; Gao, H. J.; Xu, N. S. (2010). Effect of Contact Mode on the Electrical Transport and Field Emission Performance of Individual Boron Nanowire, *Adv. Funct. Mater.*, Vol. 20, No. 12, (June 2010), pp. 1994-2003, ISBN 1616-301X

Liu, F.; Shen C. M.; Su, Z. J.; Ding, X. L.; Deng, S. Z; Chen, J; Xu, N. S.; Gao, H. J. (2010). Metal-like Single Crystalline Boron Nanotubes and Their Electric Transport and Field Emission Properties, *J. Mater. Chem.* , Vol. 20, No. 11, (January 2010), pp. 2197-2205, ISBN 0022-2461

Nicolaescu, I. V.; Tardos, G.; Riman, R. E. (2005). Thermogravimetric Determination of Carbon, Nitrogen, and Oxygen in Aluminum Nitride, J. Am. Ceram. Soc., Vol. 77, No. 9, (March 2005), pp. 2265-2272, ISBN 0002-7820

Paul, R. K.; Lee, K. H.; Lee, B. T.; Song, H. Y. (2008). Formation of AlN Nanowires Using Al Powder, *Mater. Chem. Phys.*, Vol. 112, No. 2, (December 2008), pp. 562-565, ISBN 0254-0584

Ponthieu, E.; Grange, P.; Delmon, B.; Lonnoy, L.; Leclercq, L.; Bechara, R.; Grimblot, J. (1991). Proposal of A Composition Model for Commercial AlN Powders, *J. Eur. Ceram. Soc.*, Vol. 8, No. 4, (April 1991), pp. 233-241, ISBN 0955-2219

She, J. C.; An, S.; Deng, S. Z.; Chen, J.; Xiao, Z. M.; Zhou, J.; Xu, N. S. (2007). Laser Welding a Single Tungsten Oxide Nanotip on a Handable Tungsten Wire: A Demonstration of Laser-welding Nanoassembly, *Appl. Phys. Lett.*, Vol. 90, No. 7, (February 2007), pp. 073103-1-3, ISBN0003-6951

She, J. C.; Xiao. Z. M.; Yang, Y. H.; Deng, S. Z.; Chen, J.; Yang, G. W.; Xu, N. S (2008). Correlation Between Resistance and Field Emission Performance of Individual ZnO One-dimensional Nanostructures, *ACS Nano*, Vol. 2, No. 10, (September 2008), pp. 2015-2022, ISBN 1936-0851

Sheppard, L. M. (1990). Aluminum Nitride: A Versatile but Challenging Material, *Am. Ceram. Soc. Bull.*, Vol. 69, No. 11, (June 1990), pp. 1801-1812, ISBN 0002-7812

Shi, S. C.; Chen, C. F.; Chattopadhyay, S.; Lan, Z. H.; Chen, K. H.; Chen, L. C. (2005). Growth of Single-Crystalline Wurtzite Aluminum Nitride Nanotips with A Self-Selective Apex Angle, *Adv. Func. Mater.*, Vol. 15, No. 5, (April 2005), pp. 781-786., ISBN 1616-301X

Shi, S. C.; Chattopadhyay, S.; Chen, C. F.; Chen, K. H.; Chen, L. C. (2006). Structural Evolution of AlN Nano-Structures: Nanotips and Nanorods, *Chem. Phys. Lett.*, Vol. 418, No. 1-3, (January 2006), pp. 152-157, ISBN 0009-2614

Sze, S. M.; Coleman, D. J.; Loya, A. (1971). Current Transport in Metal-semiconductor-metal Structures, *Solid-State Electron.*, Vol. 14, No. 12, (June 1971) pp. 1209-1218, ISBN 0038-1101

Tang, Y. B.; Cong, H. T.; Chen, Z. G.; Cheng, H. M. (2005). An Array of Eiffel-Tower-Shape AlN Nanotips and Its Field Emission Properties, *Appl. Phys. Lett.*, Vol. 86, No. 23, (June 2005), pp. 233104 1-3, ISBN 0003-6951

Wang, B. P. & Tong, L. S. (1996). A Study of the Optimum Field Emitter Shape for Vacuum Electronics Applications, *Appl. Surf. Sci.*, Vol. 94/95, (May 1996), pp. 101-, ISBN 0169-4332

Wang, Z. L.; Wang, X. Y.; Zuo, J. M. (2003). Induced Growth of Asymmetric Nanocantilever Arrays on Polar Surfaces, *Phys. Rev. Lett.*, Vol. 91, No. 18, (October 2003), pp. 185502-185505, ISBN 0031-9007

Wu, Q.; Hu, Z.; Wang, X. Z.; Lu, Y. N.; Huo, K. F.; Deng, S. Z.; Xu, N. S.; Shen, B.; Zhang, R.; Chen, Y. (2003) Extended Vapor–Liquid–Solid Growth and Field Emission Properties of Aluminium Nitride Nanowires, *J. Mater. Chem.*, Vol. 13, No. 8, (June 2003), pp. 2024-2027, ISBN 0959-9428

Yang, H. T.; Shen, C. M.; Su, Y. K.; Yang, T. Z.; Gao. H. J.; Wang, Y. G. (2003). Self-Assembly and Magnetic Properties of Cobalt Nanoparticles, *Appl. Phys. Lett.*, Vol. 82, No. 26, (June 2003), pp. 4729-4731, ISBN 0003-6951

Yao, R. H.; She, J. C.; Xu, N. S.; Deng, S. Z.; Chen, J. (2008). Self-Assembly of Au-Ag Alloy Nanoparticles by Thermal Annealing, *J. Nanosci. Nanotechnol.*, Vol. 8, No. 7, (July 2008), pp. 3487-3492, ISBN 1533-4880

Zhao, Q.; Xu, J.; Xu, X. Y.; Yu, D. P. (2004). Field Emission from AlN Nanoneedle Arrays, *Appl. Phys. Lett.*, Vol. 85, No. 22, (November 2004), pp. 5331-5333, ISBN 0003-6951

Zheng, J.; Yang, Y.; Yu, B.; Song, X. B.; Li, X. G. (2008). [0001] Aluminum Nitride One-dimensional Nanostructures: Synthesis, Structure Evolution, and Electrical Properties, *Acs Nano*, Vol. 2, No. 1, (January 2008), pp. 134-142, ISBN 1936-0851

Zhirnov, V. V.; Wojak, G. J.; Choi, W. B.; Cuomo, J. J.; Hren, J. J. (1997). Wide Band Gap Materials for Field Emission Devices, *J. Vac. Sci. Technol. A*, Vol. 15, No. 3, (May 1997), pp. 1733-1738, ISBN 0734-2101

10

Manipulation of Nanorods on Elastic Substrate, Modeling and Analysis

A. H. Fereidoon[1], M. Moradi[1] and S. Sadeghzadeh[2,*]
[1]*Department of Mechanical Engineering, Semnan University, Semnan,*
[2]*Department of Mechanical Engineering,*
Iran University of Science and technology (IUST), Tehran,
Iran

1. Introduction

Nanorods are one of the nano-scale structures in nanotechnology that may be synthesized of metal or semiconductor materials. Nanorod range of applications is different, ranging from display technology (reflection of the nanorod can be changed by applying electric field) to build Micro Electro Mechanical systems (MEMS).

Nano-manipulation, or controlled positioning at nanoscale, is the first step towards fabrication and assembly purposes. In this process, the Atomic Force Microscope (AFM) tip makes contact with nanoparticles and pushes them on the substrate. The AFM deflections during pushing task can be sensed and recorded using photodiode and optical methods.

This chapter of the book reviews manipulation of nanorods in various functional conditions. This review includes:

- Nanorod applications
- Different types of nanorods
- Applications of nanorods kinesiology in different technology
- Different methods of manipulation of Nanorods.
- Classification of effective parameters in manipulation of Nanorods
- Effective forces and available theory in nanorod modeling
- Various theories to model nanorod
- Appropriate strategy for manipulation of Nanorods and efficient algorithm
- A complete sample modeling and motion simulation
- The basic variables and their effects in manipulation

1.1 Nanorod applications, applications of nanorods kinesiology in different technology

The applications of nanorods are diverse, ranging from display technologies (the reflectivity of the rods can be changed by changing their orientation with an applied electric field) to

*Corresponding Author

MEMS. Nanorods based on semiconducting materials have also been investigated for application as energy harvesting and light emitting devices (wikipedia website).

Prominent among them is in the use in display technologies. By changing the orientation of the nanorods with respect to an applied electric field, the reflectivity of the rods can be altered, resulting in superior displays. Picture quality can be improved radically. Each picture element, known as pixel, is composed of a sharp-tipped device of the scale of a few nano meters. Such TVs, known as field emission TVs, are brighter as the pixels can glow better in every colour they take up as they pass through a small potential gap at high currents, emitting electrons at the same time.

Nanorod-based flexible, thin-film computers can revolutionize the retail industry, enabling customers to checkout easily without the hassles of having to pay cash (articleworld website).

1.2 Different types of nanorod

There are several demands to use manipulation of Nanorods based on how the nanowire is placed on the sample substrate (fig 1a-c). Also, different kinds of nanorod such as biological, metallic, etc exist in nature and industry. Rigid particles can be moved without considerable deflection; flexible particles can be moved with considerable deflection; but soft particles may be damaged when pushing force exceeds the yield stress (Fig. 1).

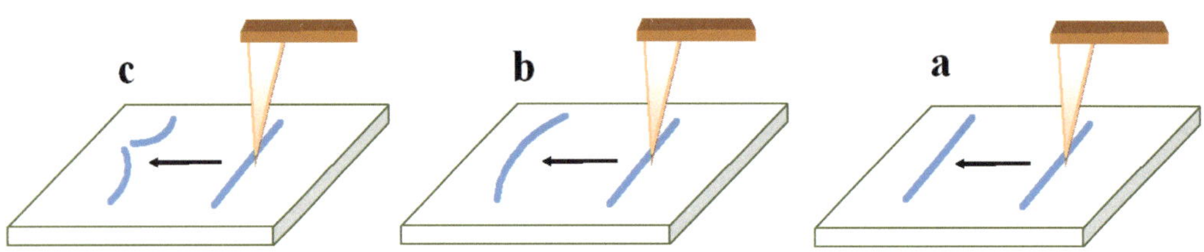

Fig. 1. Three expected results: a- rigid nanorod, b- flexible nanorod, c- soft nanorod

1.3 Different methods of manipulation of nanorods

Generally, there are two methods for nanorod and nanowire manilulation. These are Pick/Place and Push/Pull. Figures 2 and 3 show strategies for Pick/Place and Push/Pull manipulation, respectively.

2. Manipulation of nanorods

2.1 Classification of effective parameters on manipulation of nanorods

In the AFM, there are five effective parts in manipulation and each of them has different parameters (Table1). Several processes can be designed composing these parts and their parameters.

2.2 Effective forces in manipulation of nanorods

There exist various nano forces in the AFM based nano-manipulation with a micro probe. However, what are main forces and how they work to remain not very clear (Tian et al.,

2007). Based on the recent researches (Bhushan, 2005; Israelachvili, 1991) and considering effective factors such as humidity and electrostatic charge, the crucial nano forces can be summarized as van der Waals, repulsive contact force and friction (three basic nano forces). Furthermore, the capillary force aroused by humidity or biological substrates, where, the electrostatic force caused by the electrostatic charge. Based on their effect in nano-manipulation, these forces can be categorized into attractive, repulsive and frictional forces (Tian et al., 2007) . The nano forces between tip and particle can be described as shown in figure 4.

General Factors	1	2	3	4
Manipulation Task	Push/pull	Pick/Place	Cutting	Bending/ Buckling
AFM Specification	Contact/Non-contact/ Tapping Mode	Wet/Dry Environment	Cantilever/ Gripper	Rectangular/ V Shaped Cantilever
Particle	Sphere/Rod/Tube	Dimensions	Rigid/Elastic	Metallic/ Biologic
Substrate	Smooth/Rough	Rigid/Elastic	Metallic/ Biologic	Solid/ Fluid
Process Dynamics	Dominant Forces	2D/3D	Straight/Curved Path	Constant Velocity/Acceleration

Table 1. General effective parts and parameters in manipulation process (Moradi et al., 2011)

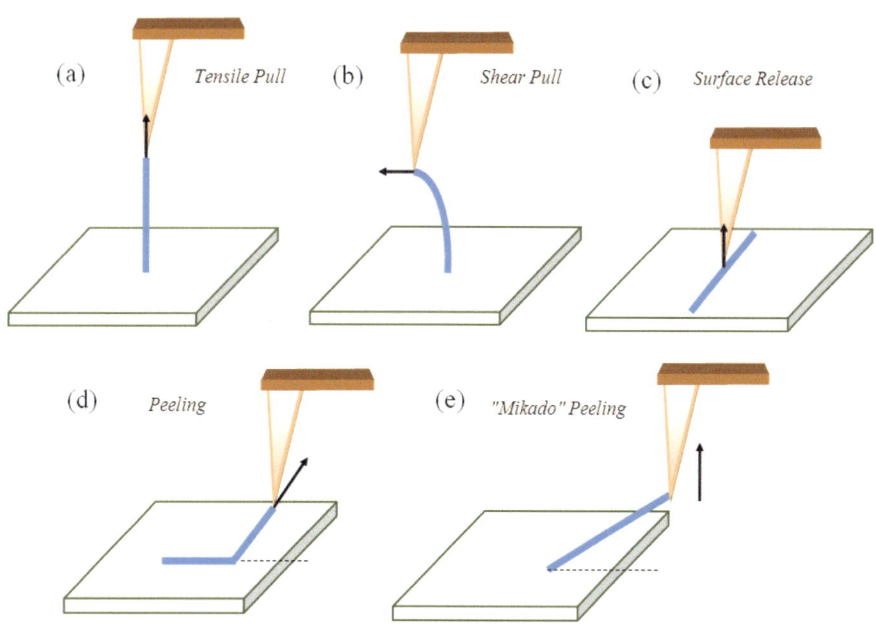

Fig. 2. Strategies for picking up a nanorod. Picking nanorods can be done by pulling along (a) or sideways (b) to the axis of a standing nanorod. Nanorods lying on substrates can beare attractiv ich consistforces,tractive van der Waals force, capillar are repulsive forces composed of repulsive contact force, repulsive van der Waals force and repulsive electrostatic force, respectively (Sitti & Hashimoto, 2000).

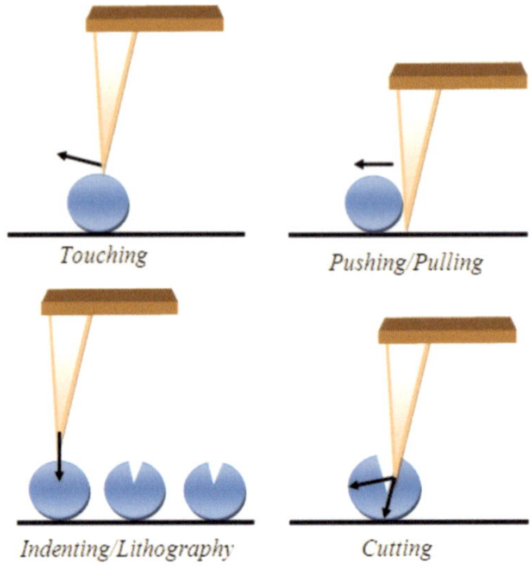

Fig. 3. Possible mechanical push/pull task using AFM probe

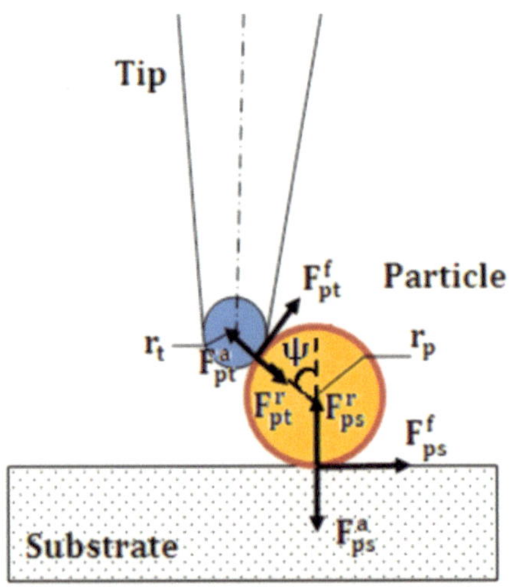

Fig. 4. The nano forces among tip, particle and substrate (Sitti & Hashimoto, 2000)

Where subscripts *p*, *t* and *s* correspond to probe, tip and substrate and superscripts *f*, *r* and *a* correspond to the friction, repulsive and attractive forces, respectively. F_{ts}^{f} and F_{pt}^{f} are rictional forces, F_{ps}^{a} and F_{pt}^{a} are attractiv ich consistforces,tractive van der Waals force, capillary force and attractive electrostatic force, and F_{ps}^{r} and F_{pt}^{r} are repulsive forces composed of repulsive contact force, repulsive van der Waals force and repulsive electrostatic force, respectively (Sitti & Hashimoto, 2000).

Like gravitational force, van der Waals forces exist for every material in any ambient condition. These forces originate from electromagnetic forces between two dipoles and depend on material types, separation distances and object geometry (Bhushan, 2005) . For spherical tip-flat surface, the van der Waals force is as

$$f(h)_{wdv}=-\frac{A_H R_t}{6h^2} \quad (1)$$

Where A_H, R_t and h are Hamaker constant (about 10^{-19}), tip radius and distance between tip and substrate, respectively. The minus sign indicates attractive force and the plus sign indicates repulsive one (Serafin & Gewirth, 1997). When the tip approaches to the substrate, reaches to a region of mechanical instability that the force gradient of the potential F_{pt}^{f} exceeds the spring constant of the cantilever (Serafin & Gewirth, 1997)

$$\frac{df(h)_{wdv}}{dh}=K_Z \quad (2)$$

At this instability, the probe will jump into contact with the surface with a characteristic "snap-in" distance, d_s as

$$d_s=\left(\frac{A_H R_t}{3K_Z}\right)^{1/3} \quad (3)$$

Snap in substrate phenomena using photodiode data can be detected (Serafin & Gewirth, 1997).

The water layers on the surfaces of probe, particle, or substrate result in the adhesion force. A liquid bridge occurs between the tip surfaces at close contact (Sitti & Hashimoto, 2000). The adhesion force between a non-deformable spherical particle of radius R_t and a flat surface in an atmosphere containing a condensable vapor is

$$F_S=4\pi R_t(\gamma_{LV}\cos\varepsilon+\gamma_{SL}) \quad 4)$$

Where ε is the contact angle, the first term is due to the Laplace pressure of the meniscus (γ_{LV} : Liquid-Vapor surface energy) and the second one is due to t: Solid-Liquid surface energy) (Israelachvili, 1991).

As there will be some electrical charge accumulated on the surface of particles or the tip, the particle is prone to adhere on the tip and manipulation may be failed. Since the particles are not picked up, the electrostatic force between the particle and the substrate is not important. However, after pushing, the charge on the particle is transferred to the tip which can cause an electrostatic force. Electrostatic force between tip and substrate will be as

$$F_e=\kappa R_t Z e^{\kappa h} \quad (5)$$

Where is the Debye length, Z is the characteristic parameter of the tip - particle and h is distance (Bhushan, 2005) . In actual experiment condition, probe and substrate can be grounded to release the electrostatic charge for minimizing electrostatic force and also the experiment condition can be kept dry to minimize the capillary force. Thus, the crucial forces between the tip and the substrate are mainly van der Waals, Friction and repulsive contact force (Tian et al., 2007) .

Contact force causes the indention on contact surfaces, which is considerable in nano-scale and affect the manipulation. Several models like Hertz, Johnson–Kendall–Roberts (JKR),

Maugis–Dugdale (MD) has been utilized as the continuum mechanic's approaches in nanoscale (Bhushan, 2005) . The Hertz model only takes consideration of mechanical deformation under external force. However, the surface adhesion force becomes significant in the contact of micro and nano objects, which must be taken into consideration. Using JKR model (Bhushan, 2005) , the adhesion force (F_{adh}) and the surface energy (W_{adh}) are related as

$$F_{adh} = -\frac{3}{2}\pi R_t W_{adh} \tag{6}$$

In JKR theory, the contact radius was found to be

$$a^3 = \frac{R}{K}(P_0 + 3\pi RW + \sqrt{6\pi RWP_0 + (3\pi RW)^2} \tag{7}$$

Where P_0, R, a and K are normal force, equivalent radius, contact radius and equivalent elastic modulus of particle- surface, respectively (Israelachvili, 1991). R and K are as

$$K = \frac{4}{3}[\frac{1-v_1^2}{E_1} + \frac{1-v_2^2}{E_2}]^{-1} \tag{8}$$

$$R = \frac{R_1 R_2}{R_1 + R_2} \tag{9}$$

Where v_i and E_i are poisson ratio And modulus of elasticity in (8). R_1 and R_2 are radius of curvature of particle and surface in (9), respectively. Contact area has an important effect in nanorod friction and so in motion behavior. Friction in nanoscale can be calculated as

$$f = \mu F + S_c A \tag{10}$$

Where μ, F, S_c and A are friction coefficient, normal force, critical shear stress and contact area (Bhushan, 2005) .

3. Theories for nanorod modeling

Common approaches to model cylindrical elastic system are Euler-Bernoulli and Timoshenko beam theory (Falvo et al., 1999; Falvo et al., 1997; Wu et al., 2010; Hsu et al., 2008). Falvo et al. studied behavior of a TMV virus using Euler- Bernoulli beam. In another work, they modeled CNTs as an Euler- Bernoulli beam to study rolling and sliding in nanoscale. Also, Hsu et al. used Timoshenko beam theory on the elastic substrate to calculate natural frequencies and mode shapes of CNTs (Hsu et al., 2008). Wu et al. manipulated CNTs with various diameters using a cantilever tip of the AFM to investigate the motion properties. They used Euler- Bernoulli theory to model flexible behavior of one-dimensional nanomaterials on a structured surface (Wu et al., 2010). In the present study, at the first time, manipulation process is considered nanorod as an Euler- Bernoulli beam on the elastic substrate (Fig. 5).

Euler- Bernoulli equation is as

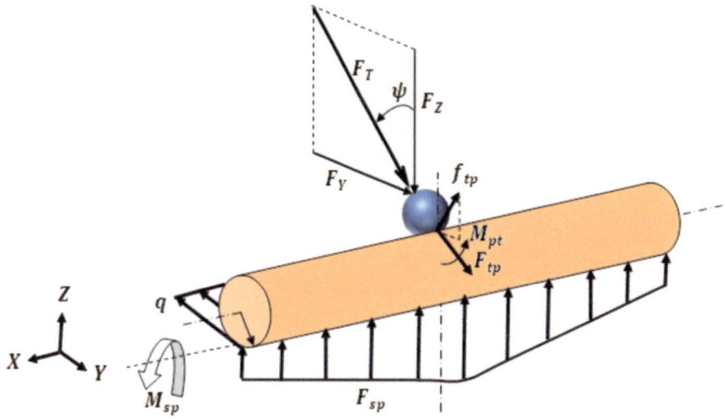

Fig. 5. Free-body-diagram of nanorod and corresponding parameters

$$EI\frac{\partial^4 y}{\partial x^4} = -K_s y \tag{11}$$

Where y, K_s, E and I are deflection in a plane perpendicular to the substrate, elastic constant of substrate- nanorod interface, nanorod modulus of elasticity and moment of inertia. General solution of equation is

$$y = e^{-\beta x}\left(C_1 \cos\beta x + C_2 \sin\beta x\right) + e^{\beta x}\left(C_3 \cos\beta x + C_4 \sin\beta x\right), \beta = \sqrt[4]{\frac{K_s}{4EI}} \tag{12}$$

Due to fourth order equation, we need four boundary conditions (Boresi et al., 1993). These boundary conditions can be defined as pushing force in center of nanorod and finiteness of nanorod length. So, boundary conditions in finite beam are

$$\begin{cases} x \to \pm\frac{L}{2} \Rightarrow M = EI\frac{d^2 y}{dx^2} = 0 \\ x = 0 \Rightarrow \Theta = \frac{dy}{dx} = 0 \\ x \to 0^+ \Rightarrow V = -EI\frac{d^3 y}{dx^3} = -\frac{P}{2} \end{cases} \tag{13}$$

Where Θ, P, V and M are deflection angle, normal pushing force, shear force and moment. Using definition $a = \frac{P}{8EI\beta^3}$, exact solution will be (Boresi et al., 1993)

$$y = \frac{2ae^{\beta(L-x)}}{1-e^{2\beta L}}[-\left(1+e^{\beta L}\right)\cos\beta x + (1-e^{\beta L})\sin\beta x] + \frac{2ae^{\beta x}}{1-e^{2\beta L}}[-\left(1+e^{\beta L}\right)\cos\beta x + (1-e^{\beta L})\sin\beta x] \tag{14}$$

Mechanical model of an elastic nanorod under pushing force is illustrated in figure 6, schematically. In this figure, A and B are the ends of the sample; T is the push point. Unlike prior works, q changes along the nanorod length as friction force. Which q in each section can be obtained as

$$q(x) = \frac{\mu F}{L} + \frac{S_c A(x)}{L} \tag{15}$$

Where μ, S_c , L and A are friction coefficient, critical shear stress, length of nanorod and contact area. In each instance, contact area using JKR model calculates and adds to (15). According to contact area along the length, this relation improves accuracy of the model. So, deflection deviation along length can be obtained as (Boresi et al., 1993)

$$\begin{cases} U = -\dfrac{q(x)x^2}{24EI}\left(4Lx + x^2 + 6L^2\right), -\dfrac{L}{2} < x < 0 \\ U = +\dfrac{q(x)x^2}{24EI}\left(4Lx - x^2 - 6L^2\right), 0 < x < \dfrac{L}{2} \end{cases} \tag{16}$$

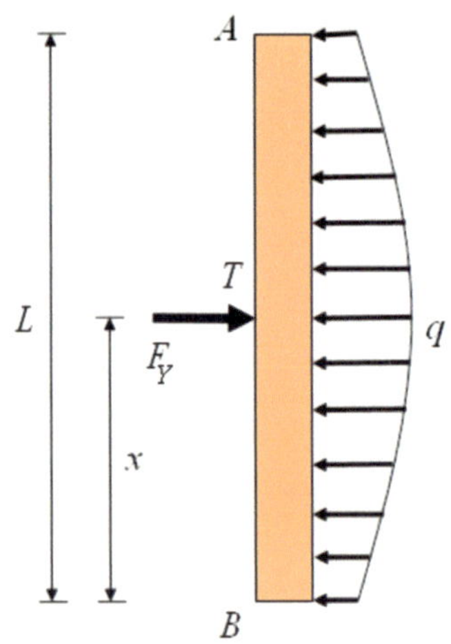

Fig. 6. Free-body-diagram of nanorod along path

In this problem, both the substrate and the nanorod are stationary at the beginning. Then, the probe moves with constant velocity and pushing force causes deflection along the path.

4. Strategy and algorithm for manipulation of Nanorods

The manipulation process cannot be observed in real time. During the pushing of the objects, imaging is not possible because imaging and manipulation tools are same. As a solution, surface and targeted particles could be imaged before and after manipulation. Using the obtained images, relative position of particles to the basic reference point can be determined (Requicha, 1999). Due to lack of real time images using the force feedback data during the for proper manipulation is crucial.

Two methods can be considered for pushing nanoparticles in constant speed: (1) Moving the substrate while AFM probe is in contact with particle; (2) AFM probe tip moves and pushes the targeted particle on the immobile substrate. Dynamic results for both methods would be the same (Korayem & Zakeri, 2008; Tafazzoli et al., 2005). The first method is used in this paper where the probe forces F_T acting between the tip and the particle is kept constant during nanoparticles movement on substrate (Fig. 7).

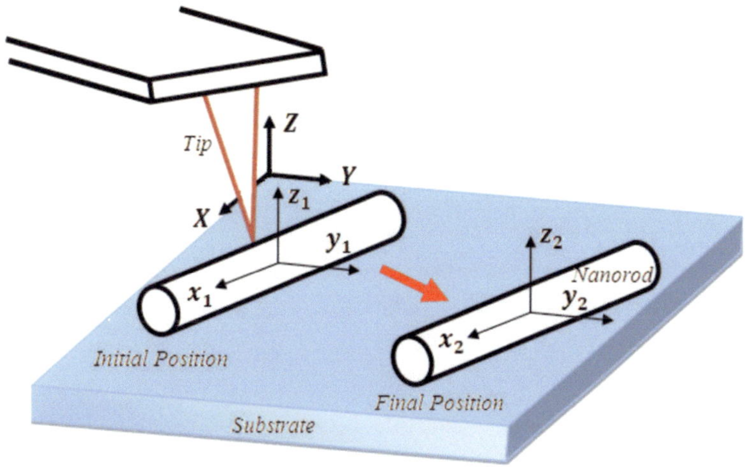

Fig. 7. AFM tip moves nanorod on substrate (Moradi et al., 2010)

4.1 Manipulation strategy

As mentioned above, at the beginning of the process both the substrate and particle are in the stationary state. Then, the probe moves down to approach substrate (Auto Parking). Van der Waals force increases until snap in instability point, respectively. In this point, the tip jumps to the substrate. This phenomenon can be detected using photodiode data. Then the tip starts to move upward. Deflection in the cantilever increases until pulling force overcame attraction force. According to the adhesion force between the tip and substrate, retraction force is more than attraction force. Next, the tip moves to reach the desired particle, horizontally. Also, van der Waals force between the tip and the particle increases until snap in particle. Then, the substrate motion follows and pushing force on particle increases. The tip may be crossed the particle and the process being failed. To ensure the desired contact, a small normal preload, F_{z0} is exerting by providing normal deflection offset, Z_{P0} on the AFM probe. Then, substrate moves in constant velocity and particle sticks to that and moves with substrate. Lateral motion of particle assists to increase pushing force, *FT*. Finally, pushing force reaches to the critical force required to overcome adhesion forces between particle/substrate. The particle motion with the substrate stops, when particle reached to desired position. At this time, the suggested behavior will be expected by particle depending on dynamic mode diagrams of particle. The probe moves upward and goes to the initial reference position when process completes (Fig. 8).

4.2 Manipulation algorithm

A new algorithm is presented to manipulate nanorods automatically (Fig. 9). As mentioned in manipulation strategy, both actuator and sensor in the AFM as nanorobots are same. Due to the image- manipulation- image cycle of the process for proper manipulation this algorithm has three parts; imaging before motion, manipulation and imaging after manipulation (Requicha, 1999).

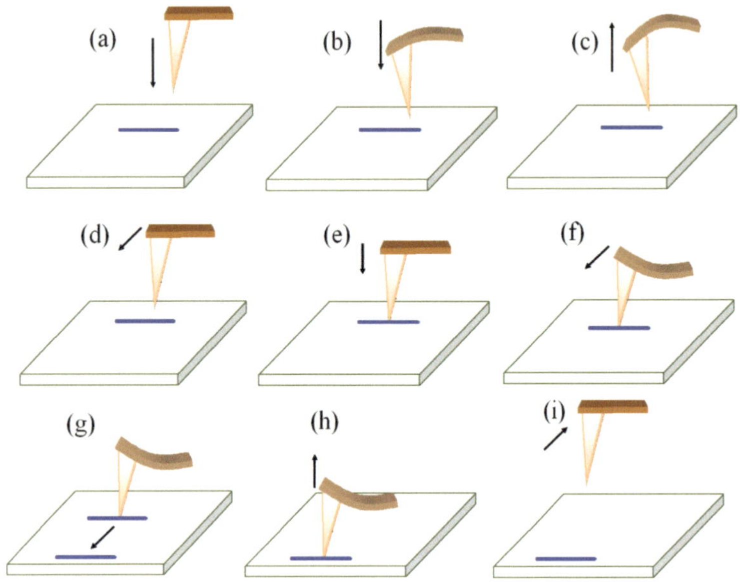

Fig. 8. Nano-manipulation strategy using the AFM: a) Auto parking, b) Snap in substrate, c) Pull out substrate, d) Approach to nanorod, e) Snap in nanorod, f) Offset in Z direction, g) Pushing, h) Pull out nanorod, k) Going to Reference point (Moradi et al., 2011)

In manipulation before motion, the AFM in the non-contact mode starts imaging and specifies particle geometries. User can choose particles and their desired position to the operation. Then the AFM detects center of particles. If the center of particles placed in the desired position, the process will be finish. Otherwise, second part of process will be start. In the second part, after contact between the probe- particle, stage moves with defined constant velocity. This motion makes the cantilever bend and increase pushing force. As mentioned above, pushing force affects on deformation, depression and indention of particle and substrate. According to the presented model these changes can be obtained and damage condition can be checked for different kinds of nanorods. If primary pushing force don't damage nanorod, the pushing should be follow to desired place.

Stage turn down when the particle pushing take place. In third part, the AFM scans the path in line backward and controls achievement. If the particle had been placed in desired position with minimum deviation, the process completed. It may some deviation occur due to different nonlinearities such as drift, creep and hysteresis in the process. When there is deviation, the AFM goes to the second part and process continues until the particle reaches to its destination.

5. Complete modelling

The AFM nanorobot has a cantilever as a manipulator. This cantilever probe consists of a connected canonical tip. The AFM is modelled as a linear spring to account for normal deflection in z direction, and a torsional spring to record lateral twisting of the probe. Spring coefficients of the springs in expression (17) (Tafazzoli et al., 2005) are a function of the geometry and the mechanical properties of the AFM probe.

$$K_z = \frac{Ewt^3}{4L^3}, K_\theta = \frac{Ewt^3}{6L(1+\upsilon)} \tag{17}$$

Where L, w, and t are the length, width, and thickness of the cantilever. E and ν are the young's modulus and poison's ratio of the probe, respectively.

Spring force/moment is linear product of the spring constant and deflection/twisting. Spring force and moment (F_z, M_θ), shear force (F_υ), normal (F_Z) and lateral tip force (F_Y), and tip pushing force (F_T) are depicted during lateral movement of the particle in AFM tip free body diagram (Fig. 10).

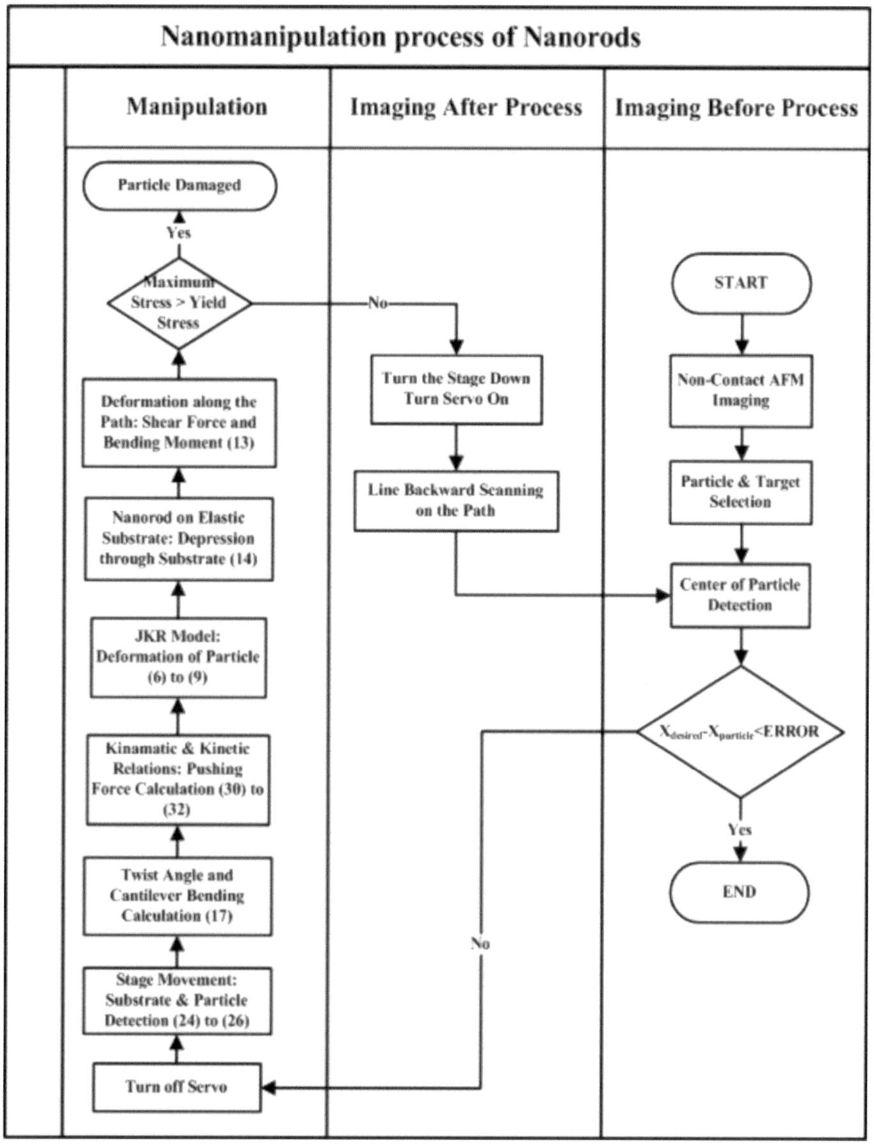

Fig. 9. Manipulation of Nanorods algorithm

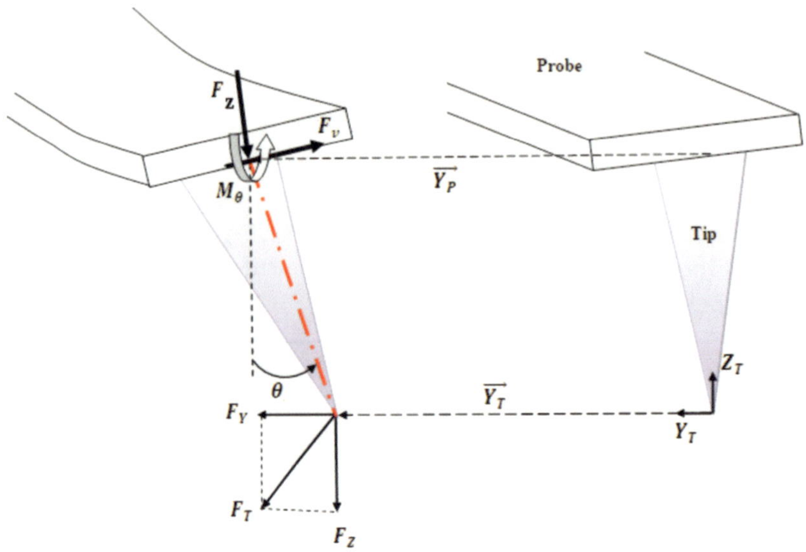

Fig. 10. Free body diagram of the AFM tip and corresponding parameters

Since the nano-manipulation task is implemented by the very front part of the tip which is very small compared with the whole probe tip body, the forces applied on tip can be viewed as applied on the tip apex (Sitti & Hashimoto, 2000).

Van der Waals force between the cantilever tip and the flat surface is as

$$f_1(h) = -\frac{A_H R_t}{6h^2} \tag{18}$$

Where A_H, R_t and h are Hamaker constant (about 10^{-19}), tip radius and distance between tip and substrate, respectively. The minus sign indicates attractive force and the plus sign indicates repulsive one (Dong & Mao, 2005). When tip approaches to substrate, reaches to a region of mechanical instability that the force gradient of the potential exceeds the spring constant of the cantilever (Dong & Mao, 2005)

$$\frac{df(h)_{wdv}}{dh} = K_Z \tag{19}$$

At this instability, the probe will jump into contact with the surface with a characteristic "snap-in" distance, d_s as

$$d_s = \left(\frac{A_H R_t}{3K_Z}\right)^{1/3} \tag{20}$$

Snap in substrate phenomena using photodiode data can be detected (Dong & Mao, 2005). Then the probe moves upward until spring force equals to pull out force (21).

$$f_2(h) = -\frac{3}{2}\pi R_t W_{adh} = -K_Z Z_P \tag{21}$$

Now, the probe approaches to the nanorod, horizontally. Van der Waals force between the tip and the nanorod increases when the distance decreases (Serafin & Gewirth, 1997). This force and snap in point are presented in (22) and (23).

$$f_3(h) = -\frac{A_H R_{pt}^{1/2}}{8\sqrt{2}h^{5/2}}, R_{pt} = \frac{R_p R_t}{R_p + R_t} \tag{22}$$

$$d_s = \left(\frac{5A_H R_{pt}^{1/2}}{16\sqrt{2}\mathrm{H}K_\theta}\right)^{\frac{2}{7}} \tag{23}$$

Where H is the tip height. To ensure the contact with the particle a small normal deflection offset is exerted on the AFM probe. The positioning stage is moved with a constant velocity. Assuming this velocity to be small, the quasi-static assumption can be considered to be valid ($\frac{dy_T^2}{dt^2} = 0$) (Korayem & Zakeri, 2008; Tafazzoli et al., 2005). Thus, the lateral position, velocity and acceleration of the tip in Y direction can be defined as relations (24) to (26).

$$y_T = y_P - H\sin\theta \tag{24}$$

$$\dot{y}_T = \dot{y}_P - H\dot{\theta}\cos\theta \tag{25}$$

$$\ddot{y}_T = 0 - H\ddot{\theta}\cos\theta + H\dot{\theta}^2\sin\theta \Rightarrow \ddot{\theta} = \dot{\theta}^2\tan\theta \tag{26}$$

Where P corresponds to the probe and T corresponds to the tip; θ represents the twisting angle of the probe. To determine the dynamics of the system, the normal deflection of the probe (Z_P) is critical.

$$Z_P = Z_T - H\cos\theta; Z_T = Z_{P0} + H = \text{cte.} \tag{27}$$

$$\dot{Z}_P = H\dot{\theta}\sin\theta \tag{28}$$

$$\ddot{Z}_P = H\ddot{\theta}\sin\theta + H\dot{\theta}^2\cos\theta \tag{29}$$

Using Newton-Euler method, forces on the AFM tip can be derived as

$$F_Z = \left(\frac{I_P\ddot{\theta} - M_\theta}{H}\right)\sin\theta + (F_z - \frac{m}{2}\ddot{z}_P)\cos^2\theta \tag{30}$$

$$F_v = \frac{1}{\sin\theta}\left(F_Z - F_z + \frac{m}{2}\ddot{z}_P\right) \tag{31}$$

$$F_Y = F_v\cos\theta \tag{32}$$

Where I_p is the AFM tip moment of inertia through its rigid contact with the probe, and m is the mass of the AFM tip. Forces expressions (30) to (32) would determine the pushing force and pushing angle of the nanoparticle.

$$F_T = \sqrt{F_Y^2 + F_Z^2}, \quad \psi = \tan^{-1}(F_Y / F_Z) \tag{33}$$

Where ψ is the pushing force angle. Normal (F) and frictional forces (f) on the particle (Fig. 2) using the pushing force, are derived as (34) to (37)

$$f_t = F_T \cos\zeta, \quad \zeta = \psi - \varphi - \frac{\pi}{2} \tag{34}$$

$$F_t = -F_T \sin\zeta \tag{35}$$

$$f_s = F_T \sin\psi \tag{36}$$

$$F_s = F_T \cos\psi \tag{37}$$

Where t and s corresponds to the AFM tip and the substrate; φ is contact angle. Normal forces (F_t, F_s) would cause deformations on the tip-nanorod and the nanorod- substrate interfaces, respectively.

According to the increase in the lateral pushing force and the contact deformation, frictional force grows and reaches to its critical value. Expressions (38) to (40) indicate sliding and rolling conditions on the tip and the substrate (Tafazzoli et al., 2005).

$$F_s > \mu_{ss} F_s + S_{ss} A_s \tag{38}$$

$$F_t > \mu_{ts} F_t + S_{ts} A_t \tag{39}$$

$$(f_s + f_t) R_p > (M_s + M_t) \quad where \;\; M_t = \mu_{tr} F_t + S_{tr} A_t \tag{40}$$

Where s and r correspond sliding and rolling. So, μ_{tr} indicates tip friction coefficient in the rolling mode. Using expressions (34) to (39) and simplifying relations (38) to (40), critical forces for sliding on the tip (41), sliding on the substrate (42) and rolling the nanorod (43) in Y directions are obtained as

$$F^*_{t-Sliding} = \frac{S_{ts} A_t}{\sin\zeta - \mu_{ts} \cos\zeta} \tag{41}$$

$$F^*_{s-Sliding} = \frac{S_{ss} A_s}{\sin\psi - \mu_{ss} \cos\psi} \tag{42}$$

$$F^*_{s-Rolling} = \frac{S_{sr} (A_s + A_t)}{R_p(\sin\psi + \cos\zeta) - \mu_{sr}(\sin\zeta - \cos\psi)} \tag{43}$$

6. Simulations

Aim of this section is simulation of nanorods and studies its behavior during motion. Critical force, critical time, maximum deflection and safety factor are some of the most important parameters of the process. The Si AFM probe is used to simulate nanorod. Geometrical and mechanical properties of the probe are summarized in Table 2 (Israelachvili, 1991). Due to few available experiments with complete quantitative parameters of materials, the reference 21 is selected for simulation. In this reference, complete parameters for polystyrene particles are presented. Polystyrene is a hard and solid polymer with specified tribological properties (Table 3) (Israelachvili, 1991). Recently, Dong and Mao (Requicha, 1999) have reported the synthesis and characterization of polystyrene nanorods. After 3 hours reaction time, they produced nanorods with length = 1 μm and width = 60–85 nm (Requicha, 1999).

Polystyrene nanoparticles are pushed on transparent glass slide (ITO glass) substrate using the Si AFM probe. This model is verified by using available theoretical and experimental results. Then polystyrene nanorod is pushed to study the process. In this simulations, polystyrene nanorod with diameter Rp=85 nm and different length are pushed on the ITO glass substrate that moves with 5nm/s constant velocity. Contact mechanics and tribological parameters can be obtained experimentally for different materials which are in contact (Table 3). The constant friction coefficients for static and dynamic movement of the nanoparticle on the substrate are μs=0.8, μd=0.7, respectively. Shear strength is assumed to be constant on the both contact surfaces between the particle/substrate and the tip/substrate. Surface energy between the nanorod and the tip/substrate is ω=0.1 J/m2 and contact angle is φ =45 (Israelachvili, 1991).

Geometrical parameters					Mechanical properties			
L(μm)	W(μm)	t(μm)	H(μm)	R_t(nm)	E(GPa)	υ	G(GPa)	ρ(kg/m^3
225	48	1	12	20	169	0.27	66.54	2330

Table 2. Geometrical and mechanical parameters of the AFM cantilever

Friction Coefficient		Shear Strength	
μ_s	μ_r (nm)	S_s (MPa)	S_r (Pa.m)
0.8	80	28	28

Table 3. Tribological parameters between nanorod/substrate (Israelachvili, 1991)

7. Results and discussions

As mentioned in Modeling based on AFM Section, the manipulation process modeling is presented considering various forces and nano-scale features. A polystyrene nanorod with 85 nm diameter and 1 μm length is selected to push on the ITO glass substrate.

Initially, the tip has 50 nm distances from the nanorod in both vertical and longitudinal directions. Since the stage moves with a constant velocity (5 nm/s), the probe moves down to reach 1.5 nm in heights (instability region) after 9.7 seconds (Fig. 11). In this region, the probe snaps in the substrate. Jump to the surface can be detected using registered photodiode data. Then, the cantilever moves up to overcome the adhesion forces. The probe motion will continue to reach 16 nm heights (pull out). Then, the cantilever moves up 73 nm after 11.4 seconds to create a proper contact angle. After that, it moves 90 nm after 9 seconds to contact with the desired nanorod, horizontally. The tip jumps to the particle when the tip-particle distance reaches to 10 nm. The probe is in contact with the nanorod and the necessary conditions for pushing are provided (Moradi et al., 2011).

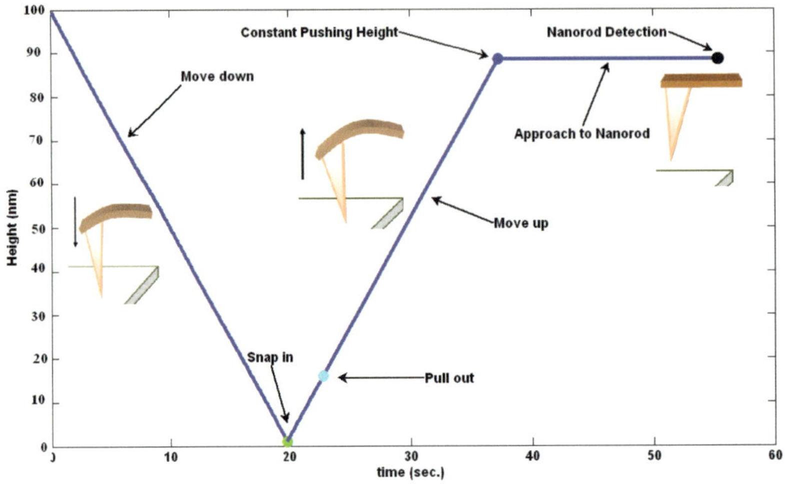

Fig. 11. Height & Time During approach/retract to the surface (Moradi et al., 2011)

The cantilever may pass over the nanorod and losses it during the pushing process. To prevent unpredicted effects and to ensure the desired contact angle of 45 degrees, initial offset in Z direction should be considered. Here 50 nm have been applied to the model. As shown in figure 12, initial pushing force is larger than critical sliding force on the tip. The substrate motion increases the pushing force to reach the critical sliding force. At critical time (0.17 seconds), the critical force (about 27.8 nN) overcomes the critical adhesion and friction forces (Table 4). The nanorod is stationary up to critical time. In the critical time, it begins sliding on the substrate. Unlike micro particles, for the nanorods, the sliding is dominant dynamic mode (Moradi et al., 2011).

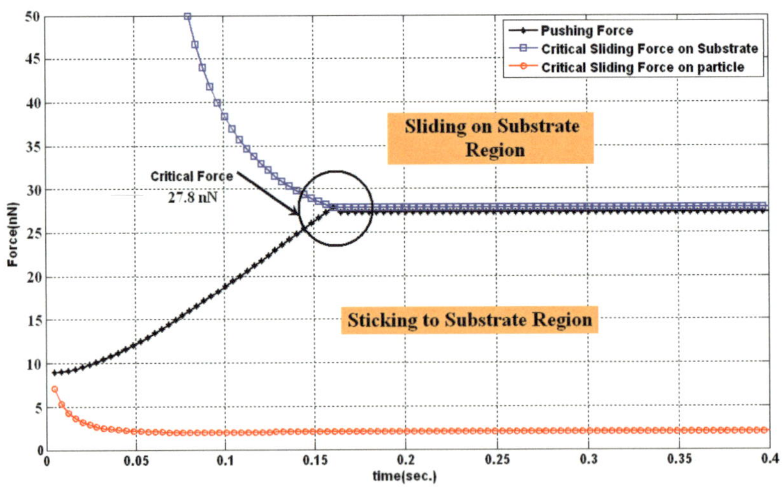

Fig. 12. Critical force to sliding the polystyrene nanorod on the ITO glass substrate (Moradi et al., 2011)

Here based on von mises yield criteria, particle damage condition is checked in each instance. Operation may be failed when safety factor being less than one. Shear force and moment variation along length are used to calculate safety factor in critical time. Using common elastic relations (Dong & Mao, 2005), the Safety factor for this process is estimated about 1.01 that shows safe pushing process of nanorod to a desired position. Complete nanorod pushing procedure consists of initial position, maximum deflection of the nanorod at critical time, the nanorod motion in each instance and the final desired position respectively (Fig. 13). This process takes about 50 seconds, totally (Moradi et al., 2010, 2011).

No	Process Output	Value
1	Critical Force (nN)	27.8
2	Critical Time (Sec.)	0.161
3	Maximum Deflection (nm)	75
4	Maximum Indention (nm)	0.033
5	Process Safety Factor	1.01
6	Adhesion Energy (10^{-16}J)	1.38
7	Potential Energy (10^{-16}J)	10.4

Table 4. Results of process simulation

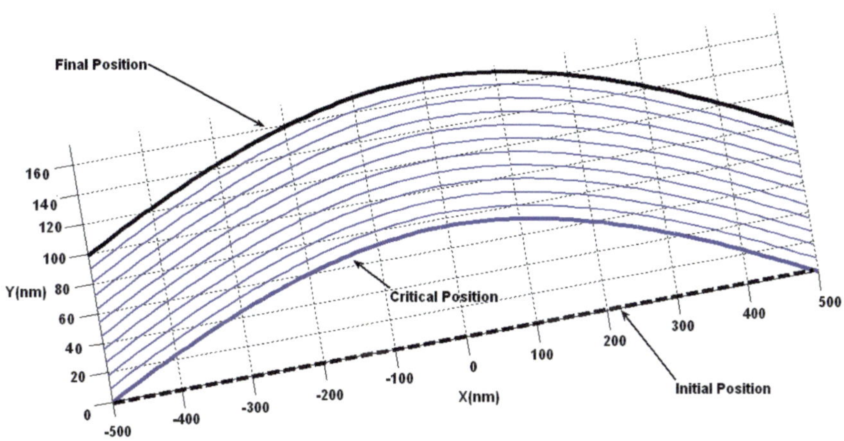

Fig. 13. The polystyrene nanorod pushing procedure to a desired position (Moradi et al., 2010, 2011)

7.1 After manipulation

When nanorod reached to desired position, tip move upward and goes to reference position. There is a question that based on adhesive force, if nanorod remain deflected or no. To answer this question, first we calculate adhesion energy using JKR theory (44) and potential energy (45) in deflected position. Then, these values are compared. Nanorod relaxes to its straight shape If adhesion energy is less than potential energy else it remain deflected. Figure (14) shows these energy values for different aspect ratio. As shown in this figure, when aspect ratio is less than 6, nanorod remain deflected after tip removal. But for higher aspect ratio, adhesion force can't keep it deflected. So, nanorod starts to relax to its straight position.

$$U_{adhesion} = 2\int_{-L/2}^{+L/2} (2\omega)b(x)dx \tag{44}$$

$$U_{Potential} = \int_{-L/2}^{+L/2} F_T x dx \tag{45}$$

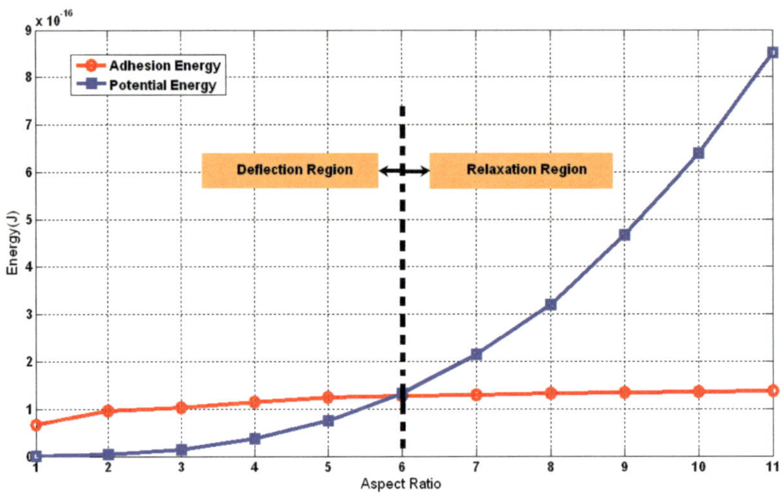

Fig. 14. Aspect ratio effects on adhesion and potential energy

7.2 Nanorod length effect analysis

The effect of aspect ratio on the process is one of the most important issues in manipulation of Nanorods. To study this parameter, a lot of simulations are repeated for different aspect ratio with constant radius. For aspect ratios 1:1 to 1:11, effects on the process are shown in Fig. 15. As shown in Fig. 15, for very short length (1:1 aspect ratio), deflection is very small

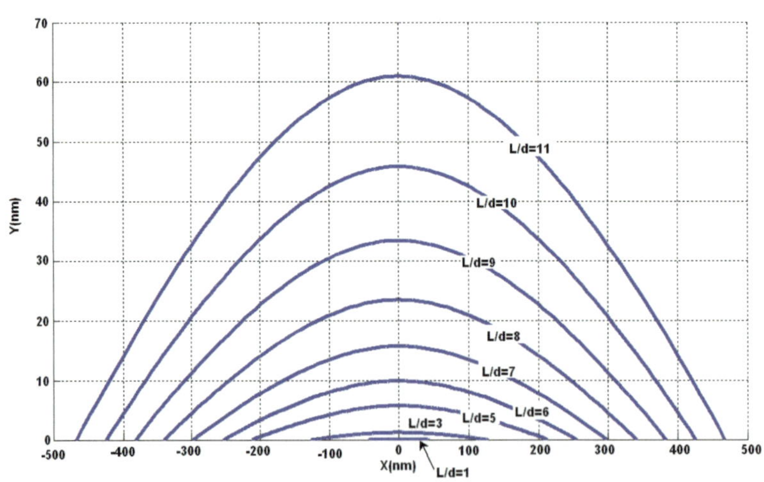

Fig. 15. Aspect ratio effects on deflection and motion of a nanorod (Moradi et al., 2010)

and negligible. The length increment causes the deflection to be more considerable. The maximum deflection is observed at 61 nm in 1:11 aspect ratios. The purpose of manipulation was 100 nm displacements in the straight path, but the longest nanorod pushed 61 nm because of the critical deflection. So, for fabrication in the nanoscale, deflection should be considered before process setting. To ensure a safe process, yield criteria, maximum shear and deflection should be considered. Using Von Mises theory, the safety factor is calculated for each one. When the aspect ratio is greater than 1:11.76, the nanorod approaches the failure region and may be damaged. Owing to the adhesion force, increasing the aspect ratio causes critical force and time increment, respectively (Moradi et al., 2010).

Also, based on presented model, for different aspect ratio pushing force and maximum deflection are obtained. As shown in figure 16, increase in aspect ratio increase these two outputs respectively.

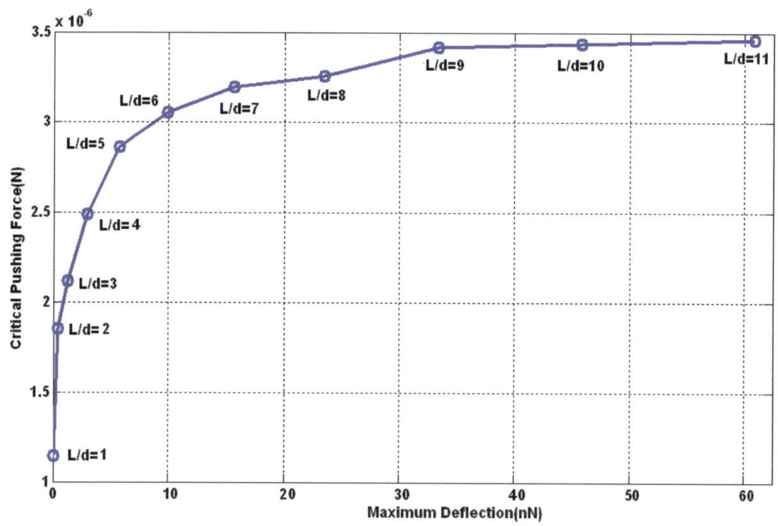

Fig. 16. Aspect ratio effects on Critical pushing force and maximum deflection

9. Conclusion

Using accurate model in the nano scale, physical phenomena of nanotechnology can be understood. However, this model can be used in accurate prediction of manipulation of Nanorods, control of process, optimization of parameters and user interface system development for SPMs. In addition, it can be a suitable tool in automation of nano assembly and nano manufacturing.

It obviously has been shown that after manipulation, nanorod relaxes to its straight shape If adhesion energy is less than potential energy else it remain deflected. When aspect ratio is less than 6, nanorod remain deflected after tip removal. But for higher aspect ratio, adhesion force can't keep it deflected. So, nanorod starts to relax to its straight position. In addition, in this book chapter, using Von Mises theory, the safety factor is calculated for various criterions. When the aspect ratio is greater than 1:11.76, the nanorod approaches the failure region and may be damaged. Owing to the adhesion force, increasing the aspect ratio causes critical force and time increment, respectively

Also, based on presented model, for different aspect ratio pushing force and maximum deflection are obtained. As shown in figures, increase in aspect ratio increase these two outputs respectively.

10. References

http://www.en.wikipedia.org/wiki/Nanorod

http://www.articleworld.org/index.php/Nanorod

K. Mølhave, T. Wich, A. Kortschack, "Bøggild, P. Pick-and-place nano-manipulation using microfabricated grippers", Nanotechnology 2006, 17, 2434-2441.

M. Sitti, "Survey of nano-manipulation systems", In Proc. IEEE Nanotechnology Conference, 2001, Maui, HI, 75–80.

Moradi M., Fereidon A. H., Sadeghzadeh S.,"Dynamic modeling for nano-manipulation of polystyrene nanorod by atomic force microscope", Scientia Iranica 2011, 18 (3), 808–815.

X. Tian et al., "A Study on Theoretical Nano Forces in AFM Based Nano-manipulation", Proc. of the 2nd IEEE Int. Conf. on Nano/Micro Engineered and Molecular Systems, 2007.

B. Bhushan, "Nanotribology and Nanomechanics: an introduction", 2005, Springer-Verlag.

J.N. Israelachvili, "*Intermolecular and surface forces*", 1991, Academic Press, London.

M. Sitti, H. Hashimoto, "Controlled Pushing Of Nanoparticles: Modeling And Experiments", Proc. of IEEE/ASME Trans. on Mechatronics, Vol. 5, No. 2, (2000).

J.M. Serafin, A.A. Gewirth, "Measurement of Adhesion Force to Determine Surface Composition in an Electrochemical Environment", J. Phys. Chem. B, 101, (1997) 10833-10838.

M.R. Falvo et al., "Nanometre-scale rolling and sliding of carbon nanotubes", Nature, 397 (1999) 236-8.

M.R. Falvo et al., "Manipulation of Individual Viruses: Friction and Mechanical Properties", Biophysical J., Vol. 72, (1997) 1396-1403.

S. Wu et al., "Manipulation and behavior modeling of one-dimensional nanomaterials on a structured surface", App. Surface Sci., (2010).

J.C. Hsu, R.P. Chang, W.J. Chang, "Resonance Frequency of Chiral SWNT using Timoshenko Beam Theory", Phys. Letters A, 372, (2008) 2757-2759.

A.P. Boresi, O.M. Sidebottom, F.B. Seely, J.O. Smith, "*Advanced Mechanics of Materials*", 1993, John Wiley and Sons.

A.G. Requicha, "*Nanorobotics: Handbook of Industrial Robotics*", 2nd ed., 1999, Wiley, pp 199–210.

A. Tafazzoli, M. Sitti, "Dynamic modes of nanoparticle motion during nano probe based manipulation", Proc. Of IMECE'04, 2004.

M.H. Korayem, M. Zakeri, "Sensitivity analysis of nanoparticles pushing critical conditions in 2-D controlled nano-manipulation based on AFM", Int. J. Adv. Manuf. Technol. , (2008).

A. Tafazzoli, C. Pawashe, M. Sitti, "Atomic force microscope based two-dimensional assembly of micro/ nanoparticles", (ISATP 2005) 6th IEEE Int. Symp., 230 – 235, 2005.

Moradi M., Fereidon A. H., Sadeghzadeh S., "Aspect Ratio and Dimension Effects on Nonorod Manipulation by Atomic Force Microscope (AFM)", Micro Nano Letters, 5, 5, 324–327, 2010.

J. Dong, G. Mao, "Polystyrene nanorod formation in $C_{12}E_5$ hemimicelle thin film templates", Colloid Polym. Sci. (2005) 284: 340–345.

11

Recent Developments in the Synthesis of Metal-Tipped Semiconductor Nanorods

Sabyasachi Chakrabortty and Yinthai Chan
Department of Chemistry, National University of Singapore, Singapore

1. Introduction

Semiconductor nanocrystals (NCs), commonly known as quantum dots (QDs), have received great attention over the last two decades due to their, unique size and shape-dependent optoelectronic properties, as well as their flexible surface chemistry. While efforts to produce colloidal NCs date back to the pioneering work of Rossetti et al at Bell Labs[1] and Ekimov et al at the Vavilov State Optical Institute[2] in the early 1980's, the ability to obtain monodisperse spherical and highly crystalline NCs remained largely elusive until the introduction of the hot injection method by Murray et al. in 1993.[3] In this method, organometallic precursors of the semiconductor material are rapidly injected into an organic solvent at an elevated temperature under inert conditions. This results in a rapid cooling of the reaction mixture, effectively separating the nucleation and growth phases of the intended semiconductor NCs. Despite the addition of organic surface-capping groups, as-synthesized core NCs typically suffer from poor surface passivation and possess surface trap states. These surface trap states result in fast non-radiative relaxation pathways for photogenerated charge carriers, thus leading to reduced fluorescence quantum yields (QYs) typically on the order of ~10-20%. In order to improve the fluorescence efficiency as well as the photostability of semiconductor NCs, growth of an inorganic shell (typically a wider bandgap semiconductor) is generally adopted. One of the earliest and most widely used techniques for the overcoating of semiconductor NCs even today was introduced by Hines et al. in 1996,[4] where precursors of the semiconductor shell material are added dropwise to a relatively dilute solution of NC cores at temperatures sufficiently low to prevent homogeneous nucleation of the precursors or Ostwald ripening of the NC cores. Growth of the semiconductor shell can lead to effective surface passivation of the core NC, leading to near unity QYs in the case of CdSe/CdS core-shell NCs,[5] although QYs in the range of 50-70% are more common. The synthetic development of various II-VI, IV-VI and III-V colloidal semiconductor NCs have been reported to date,[3,6,7] and have led to intense research efforts in the study of their fundamental optoelectronic properties as well as their use in applications in areas as diverse as light-emitting diodes (LEDs),[8] solar cells,[9] and biological imaging.[10]

While early efforts in the synthetic development of semiconductor NCs focused on minimizing polydispersity in single component spherical particles such as the now-ubiquitous CdSe NCs, a more sophisticated understanding of the underlying mechanisms of

NC growth over the last decade or so oversaw the field evolving quickly towards more complicated structures ranging from highly anisotropic shapes to multi-component particles bearing more than one nanoparticle within the same nanostructure.[11,12] Previous reports have suggested that several experimental parameters are important in controlling the size as well as the shape of the nanoparticles, such as the choice of precursors and their respective concentrations, the reaction temperature, and the types of ligands used.[13-17] These parameters play a decisive role in the overall morphology and size dispersity of the nanoparticles ultimately synthesized. To date, semiconductor nanoparticles of a wide variety of different shapes have been reported: rods,[18] tetrapods,[19] stars,[20] cubes,[21] and hyperbranched structures,[22,23] as illustrated in Fig. 1. Owing to the exhibition of strong quantum confinement effects in semiconductor NCs whose dimensions are comparable to their Bohr exciton radius, the shape of such nanostructures is expected to play a significant role in their size-dependent optoelectronic properties. For example, in the case of CdSe nanorods, fluorescence emission is linearly polarized along the long axis of the rod[24] while non-radiative Auger relaxation processes are less pronounced as compared to spherical CdSe NCs. Additionally, it has been suggested that branched semiconductor NCs may be more desirable than their spherical counterparts as the active material in solution-processed photovoltaics due to increased percolation pathways for charge transport.[25] A more recent thrust in the field has been the development of hybrid heterostructures in which different functionalities of individual components may be incorporated into the same nanostructure, thus offering a basis for multifunctional NCs. Hybrid metal-semiconductor nanoheterostructures comprising of a metal nanoparticle in intimate contact with a semiconductor NC serve as a good example of such multi-component systems. Such structures either possess the inherent functionalities of the metal and semiconductor nanoparticle, or unique optoelectronic properties that result from close coupling between the two components.

This chapter traces the development of hybrid metal-semiconductor NCs as a current research trend in the field of colloidal semiconductor NCs. Despite the plethora of examples of different-shaped metal-semiconductor nanostructures reported,[12,26-28] metal-tipped semiconductor nanorods are perhaps the most intensively studied, and will be the focus of the chapter. Different wet-chemical synthetic strategies to fabricate various colloidal semiconductor nanorods and their growth mechanisms are presented, and methods to deposit metal nanoparticles at specific sites on nanorods are discussed. The possible applications of such metal-tipped nanorod structures are then elaborated on in the context of their unique physicochemical properties. Finally, we present a relatively recent technique in which previously inaccessible metals may be specifically deposited onto the tips of semiconductor nanorods via a photoinduced mechanism.

2. Colloidal semiconductor nanorods

An enormous effort has been made in the past decade or so in developing strategies for the synthesis of colloidal semiconductor nanorods in terms of their composition, morphology, size dispersity and surface properties.[11,29-32] Exquisite control over these parameters is undoubtedly very important in terms of gaining a good understanding of the fundamental properties in such systems as well as for their purported applications in solution processed photovoltaics,[33] catalysis,[34] etc. The colloidal synthesis of semiconductor nanorods via the hot injection method is in many respects similar to that of spherical NCs, where the injection of a high concentration of monomers into a solvent at elevated temperatures results in near instantaneous generation of semiconductor nuclei, whereupon the temperature drops significantly and subsequent monomer consumption occurs primarily via the growth of the NCs. Growth into anisotropic structures involves the formation of a preferred crystalline phase in which the adsorbing surfactant molecules specifically bind to individual crystallographic planes[35] as illustrated by a wurtzite-based nanocrystal structure in Fig. 2. This in turn exerts a degree of control over the addition of monomers to specific facets on the growing crystal, resulting in anisotropic growth. Thus it is relatively straightforward to visualize how the exclusive addition of monomers to opposite facets of a spherical construct would eventually result in elongated rod-like structures. The mechanism by which such growth occurs will be elaborated in more detail in the next section which describes the different synthetic strategies to make colloidal semiconductor nanorods of various material compositions.

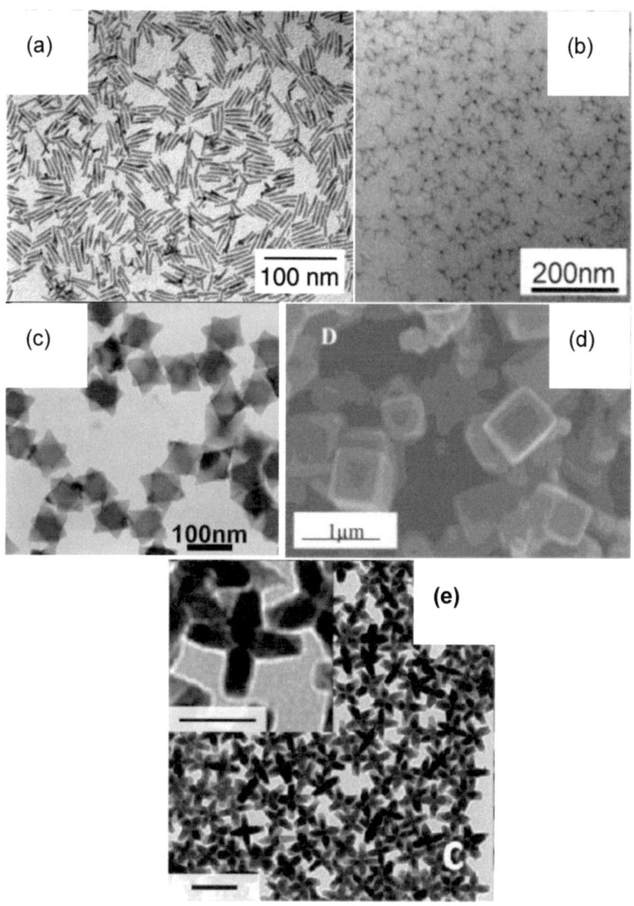

Fig. 1. Examples of colloidal NCs grown with different shapes: TEM images of (a) CdSe nanorods,[18] (b) CdTe tetrapods,[19] (c) PbS Stars,[20] (d) PbSe cubes[21] and (e) hyperbranched CdSe/CdS octapods (scale bar 100 nm). Inset showing one single structure(scale bar 50 nm).[23] *Courtesy: (a) J. Am. Chem. Soc. 2000, 122, 12700-12706; (b) Chem. Mater. 2003, 15, 4300-4308; (c) Adv. Mater. 2006, 18, 359; (d) Nano Lett. 2003, 3, 857-862; (e) Nano Lett. 2010, 10, 3770-3776.*

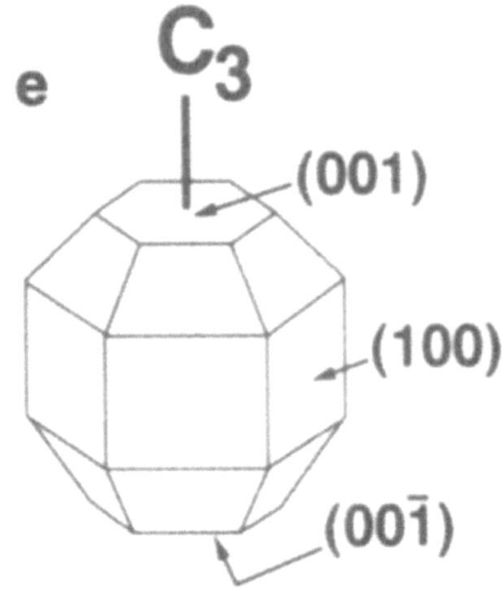

Fig. 2. Model wurtzite nanocrystals showing different facets. *Courtesy: J. Phys. Chem. 1995, 99, 17417-17422.*

2.1 Known mechanisms of growth

2.1.1 Selective adhesion

Surfactant molecules which are able to bind to the facets of semiconductor NCs influence their relative growth rates, resulting in anisotropic growth when particular facets are preferred. The extent of growth anisotropy is governed by the binding affinity of the surfactant molecules present in the reaction mixtures to a particular facet, thus influencing the growth kinetics at those facets. It has also been suggested that the anisotropic growth occurs via kinetically driven processes where the system is flooded by a high concentration of monomers.[13] Under these circumstances, the least passivated facets (and therefore most reactive) will have a tendency to grow faster. In addition to the monomer concentration, the growth times and reaction temperatures are also crucial factors for obtaining rod-like structures of relatively high aspect ratios. Sufficient growth times would be required in order to observe dominant growth long a unique axis while adequately high reaction temperatures facilitate the breakdown of precursors and raise the monomer concentration. In the literature, the anisotropic shape control in binary II-VI semiconductor nanorods are mostly reported on the basis of this selective adhesion technique. Among the II-VI semiconductor nanorods such as ZnS, CdTe, CdSe and CdS,[11,36-38] the latter two are arguably the most well-developed in terms of chemical synthesis. The first wet-chemical synthesis of colloidal semiconductor nanorods (CdSe) via the hot injection method is widely ascribed to Alivisatos, Peng and co-workers in which Cd and Se precursors were injected into a mixture of n-trioctylphosphine oxide (TOPO) and hexylphosphonic acid (HPA) at ~ 360°C.[11] Fig. 3 (a) shows different batches of as synthesized CdSe nanorods obtained through this method. The amount of HPA plays a decisive role for obtaining the rod-like structure as it helps to favor growth at the $(00\bar{1})$ facet of the wurtzite CdSe nanocrystal (see Fig. 2) relative to its other facets.[18] This results in the anisotropic growth of wurtzite (w-) CdSe nanoparticles along its C-axis into elongated rod-like structures, as exemplified by Fig. 3(b).[11] Thus with a judicious combination of surfactant molecules, growth kinetics at the various facets of semiconductor NCs may be tailored such that semiconductor nanorods of different aspect ratios may be reproducibly synthesized. It should be pointed out that the formation of nanorods in a monosurfactant system cannot be satisfactorily explained using this growth model, however, since it relies heavily upon the distinct binding affinities of different surfactant molecules to the facets of the NC.

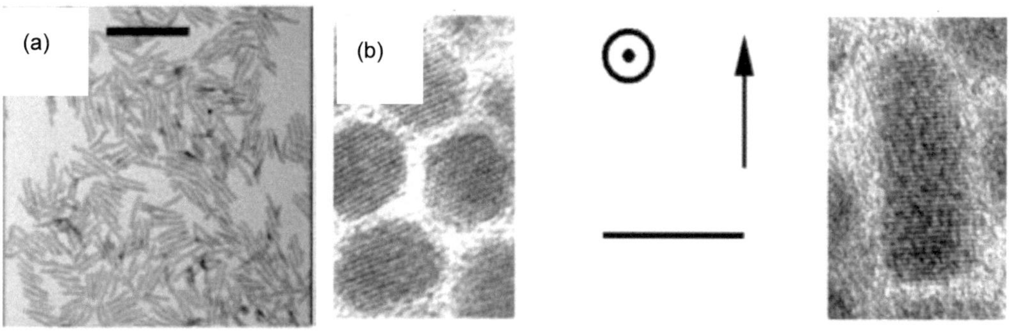

Fig. 3. (a) Low resolution TEM images of colloidal CdSe nanorods (scale bar 50 nm). (b) Showing the growth along C-axis to obtain rod shape from spherical nanpparticle(scale bar 10 nm). *Courtesy: Nature 2000, 404, 59-61.*

2.1.2 Oriented attachment

The oriented attachment growth mechanism involves the formation of nearly isotropic spherical NCs and then subsequent fusion of these NCs preferentially along a specific set of crystallographic facets, ultimately leading to the formation of elongated structures. Thus the size dispersity of the resulting rod-like nanostructures is influenced by that of the initial ensemble of spherical particles. Diametrically located regions on the NCs bearing high energy facets serve as nucleation sites for attachment, and the coalescence of the individual crystals along a certain direction may be energetically favorable as it can reduce the existence of the high energy facets, thus providing the driving force for oriented attachment.[39] The wet-chemical synthesis of colloidal PbSe nanorods with rock-salt structure provides an example of the oriented attachment growth mechanism, where it has been shown that the formation of monodisperse PbSe nanorods, as shown in Fig. 4(a), takes place via the coalescence of spherical PbSe NCs.[29] Analogous to the synthesis of CdSe nanorods, single crystalline PbSe nanorods may be prepared (Fig. 4) via the introduction of a Se precursor to a Pb precursor containing solvent at ~ 170 °C in the presence of oleic acid as a surfactant. The presence of high energy Pb or Se terminated (111) facets in the initial PbSe NC cores were found to be most likely responsible for the oriented attachment into rod-like structures. Use of HRTEM revealed that the synthesized PbSe nanorods are highly crystalline in nature, with no obvious grain boundaries or periodic stacking faults to suggest their formation via the fusion of multiple spherical NCs. In order to prove the oriented attachment mechanism of growth, aliquots at different time intervals were taken from the reaction vessels and carefully characterized via TEM.[29] The formation of coalesced particles at early reaction times provided strong evidence for the oriented attachment mechanism, as depicted in Fig. 4(c).

2.1.3 Seeded growth

While the previous two mechanisms for nanorod formation described were based on the homogeneous nucleation and growth from the breakdown of semiconductor precursors, another method to fabricate monodisperse colloidal semiconductor nanorods using pre-formed spherical semiconductor NCs as seeds was reported independently by Carbone et.al.[30] and Talapin et. al.[31,32] The resulting nanorod structures thus comprise of a spherical core within a rod-like shell, which in their case were CdSe/CdS core-shell nanorods, or

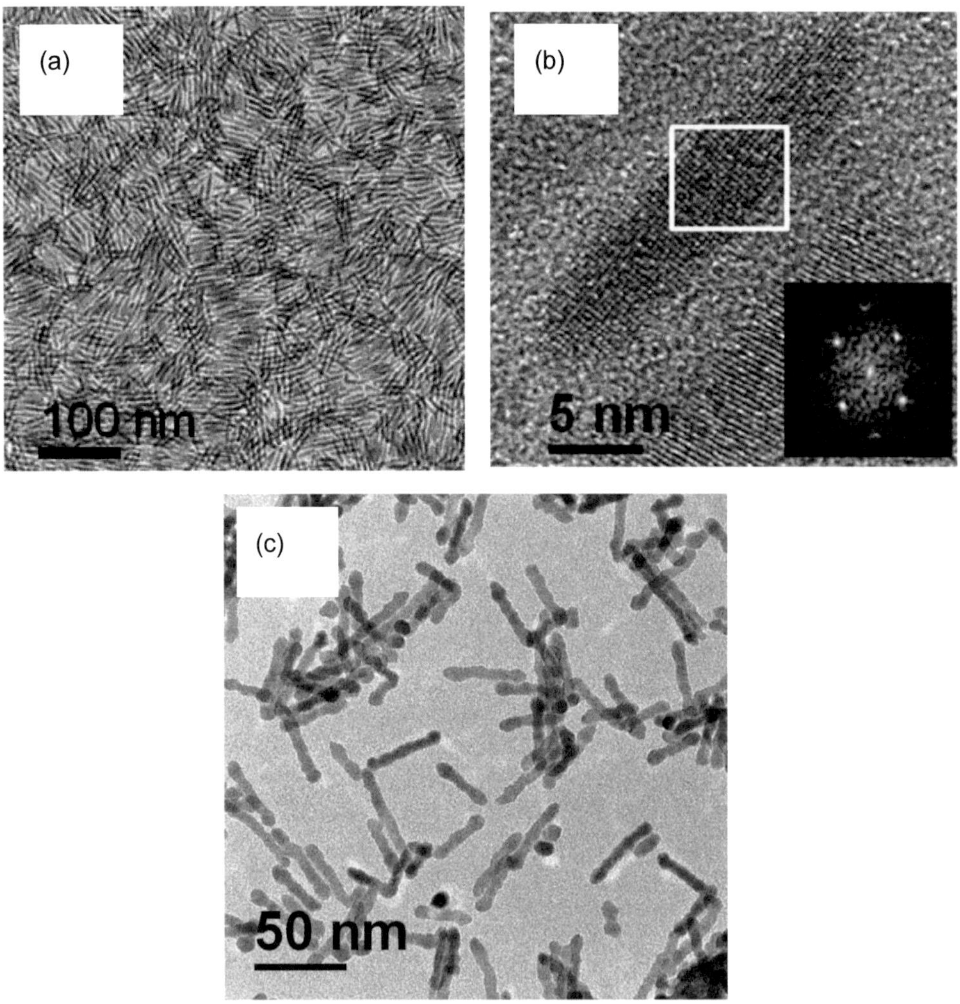

Fig. 4. (a) Low resolution TEM image of PbSe nanorods that formed via an oriented attachment mechanism. (b) It is not readily evident from the High-resolution TEM (HRTEM) image of an individual PbSe nanorod after the synthesis reaction has gone to completion that the formation of the rod occurs through the linear adhesion of a couple of sphere-like PbSe NCs. Inset showing the FFT image. (c) TEM image from an aliquot, taken in the intermediate state of the oriented attachment process. *Courtesy: J. Am. Chem. Soc. 2010, 132, 3909-3913.*

perhaps more widely known as CdSe seeded CdS nanorods. For this seeded growth approach, a mixture of spherical CdSe nanocrystal seeds (separately prepared) and a S precursor was injected at elevated temperatures of ~ 350°C to a Cd precursor containing solvent in the presence of alkyl phosphonic acids as surfactants.[30] As the activation energy barrier for heterogeneous nucleation is generally lower compared with homogeneous nucleation, nearly exclusive heterogeneous nucleation and growth of CdS on the CdSe seed occurs. Owing to the different binding affinities of the alkyl phosphonic acids to the various facets of the CdSe seed and subsequently CdS, continued addition of monomers resulted in a unidirectional growth of the CdS shell, consistent with the growth mechanism described earlier in Section 2.1.1. By obviating the need for homogeneous nucleation, the CdS precursors could be utilized primarily for growth of the shell material on CdSe, thus allowing for very monodisperse rods with a wide range of aspect ratios. Fig. 5 shows transmission electron microscopy (TEM) images of as synthesized CdSe seeded CdS nanorods with aspect ratios ranging from 1:3 to 1:30.[30] The CdSe seed was found to be asymmetrically located along the length of the CdS shell due to the fast growth rate along one axis of the wurtzite seed (i.e., $(00\bar{1})$ facet, see fig. 2) compared to the other, where the preferential binding affinities of different capping groups play the decisive role. Available examples include ZnSe/CdS,[30] ZnTe/CdS,[30] CdSe/CdTe[40] etc. Thus, these numerous varieties of materials composition could be further exploited to advance applications in various fields.

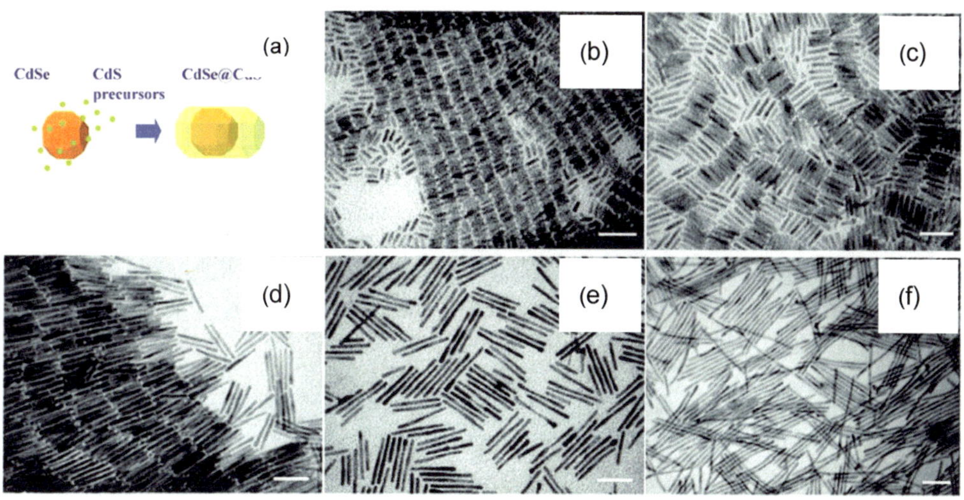

Fig. 5. (a) Schematic illustration of seeded growth approach resulting in the formation of CdSe seeded CdS nanorods. (b) - (f) show CdSe seeded CdS nanorods with aspect ratios ranging from 1:3 to 1:30. *Courtesy: Nano Letters 2007, 7, 2942-2950.*

3. Hybrid metal-semiconductor composites

Multicomponent nanoparticles pave the way to create smart materials by potentially incorporating different functionalities of the individual components within the same nanostructure. Hybrid metal-semiconductor nanostructures are one example of such multicomponent materials, which possibly hold such promise. In these hybrid structures, the metal and its semiconductor nanoparticle counterpart are closely coupled such that novel properties or applications may emerge. For example, the selective growth of metal tips onto semiconductor nanorods or tetrapods offers contact points for self-assembly,[41] good electrical contacts in optoelectronic devices,[42] modification of its nonlinear optical response[43] or impart magnetic functionality.[44] Recently, there has been a fairly large number of examples in the literature where growth of Pt,[27] Au,[26,45] or Co[28,44] onto semiconductor nanorods have been reported.The following section will focus on different synthetic strategies adopted for combining a metal with its semiconductor nanorod counterpart.

3.1 Synthetic strategy

It has been shown that the facets at the tips of nanorods such as CdS or CdSe seeded CdS are different in terms of their reactivity when compared with the facets along the length of the rod. The facet reactivities of the two tips of the nanorod themselves differ, and it may generally be concluded that the distribution of facet reactivities on anisotropic semiconductor NCs is quite wide.[39] The difference in facet reactivities may be exploited to yield selective metal deposition, which may be understood from the standpoint of monomers of the metal precursor exclusively nucleating and growing at the most reactive facets of the semiconductor nanostructure. In terms of thermodynamics, the preferential deposition of a secondary (metal, in this case) particle is governed by minimizing the overall surface and interfacial energy of the system. However, kinetic factors such as reactant diffusion and how labile the surfactant molecules on the semiconductor are play a pivotal role in the fabrication of the hybrid metal-semiconductor nanostructure as well. The next few sections will introduce different techniques that have been reported for the fabrication of hybrid metal-semiconductor nanorods.

3.1.1 Site selective metal deposition

It was previously suggested that the atomic arrangements in the apex facets in semiconductor nanorods are dissimilar compared to their longitudinal sidewalls, which gives rise to different surface energies depending on whether it was the facets at the apex or sides of the nanorod that were being considered. This phenomenon is illustrated well by the wurtzite (w-) lattice, which is generally believed to have chemically non-equivalent surface termination at opposing ends of the structure's growth axis. For example, in w-CdS nanorods, one end is Cd rich whereas the other is S rich.[46] As a result, the corresponding reactivity of the facets at the sulfur- and cadmium-rich ends can differ significantly, which would be expected to lead to dissimilar rates of heterogeneous nucleation and growth of metal clusters at the different ends on the nanorod. Saunders et. al.,[46] previously showed that under anaerobic conditions, Au growth occurs preferentially at the S-rich tip of the CdS nanorod. In contrast, dumbbell-like Au-CdS-Au structures are obtained in the presence of O_2 (as summarized in Scheme 1) due to etching of the Cd rich end, which is expected to increase its surface reactivity and allow for Au nucleation and growth. By taking into account that the reactivities at the sides of the nanorod are very different as compared to its ends, a hierarchical order of free energy barriers to nucleation at the different sites of the nanorod can be anticipated. Under this basis, Chakrabortty et. al. [45] reported a facile Au deposition process to synthesize CdSe seeded CdS nanorods of various Au decorated morphologies by using the concentration of the Au precursor as the only adjustable parameter. Fig. 6(a)-(c) summarizes the different morphologies of Au on the CdS rod surface as concentrations of the Au monomers were increased. This trend of selective Au growth with increasing concentrations of added Au precursor was reproduced at a variety of fixed temperatures (25 -90 °C) and sufficiently long growth times (up to 6 h), which in turn supports the presence of a hierarchical order of reactivities between the facets at the tips and sides of the nanorod.[45] In order to exploit this hierarchy of facet reactivities for the fabrication of nanostructures of higher structural complexity and therefore functionality, sequential exposure to precursors of Au and then Ag was pursued. Thus the deposition of Au to give matchstick-like Au-CdSe/CdS nanostructures was followed by exposure to Ag. Unlike the Au precursors, the Ag precursors undergo cationic exchange with CdS, which results in the formation of Ag_2S nanoparticles at the end of the nanorod that did not bear a Au tip. This resulted in novel "Janus-type" dumbbell structures,[45] where Au is primarily located at one end of the nanorod while Ag_2S is located at the other end as shown in Fig. 6(d). Several other groups have also utilized the fact that the differences in facet reactivities at the sides and tips of semiconductor nanorods can be exploited to yield site selective deposition, and a rapidly growing library of metal-tipped semiconductor nanorod heterostructures (e.g. Co-CdSe/CdS[28] and Pt-CdSe/CdS[27]) have been synthesized.

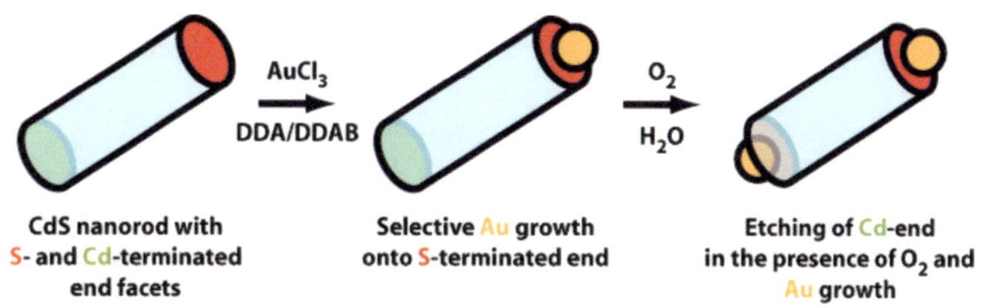

Scheme 1. Schematic of the growth process of Au nanocrystals onto CdS nanorods, where in the absence of O_2 Au deposited on the S-rich facet and in presence Au deposited at both end.. *Courtesy: J. Phys. Chem. B 2006, 110, 25421-25429.*

3.1.2 Light controlled metal deposition

The ground state electronic structure of semiconductors is generally characterized by a valence band that is filled with electrons and a conduction band that is empty. Upon light absorption an electron is promoted from the valence band into the conduction band, leaving behind a hole in the valence band. The photo-generated hole in the valence band and electron the conduction band can react with electron donors and acceptors, respectively, thus allowing the semiconductor surface to act as an oxidation or reduction point. Thus a solution containing metal cations can undergo photoreduction at the semiconductor surface to nucleate and subsequently grow into metal nanoparticles assuming that it is energetically favorable for electron transfer from the semiconductor to the metal to occur. In 2004, Pacholski et. al., [47] demonstrated photoinduced Ag^+ reduction at the surface of ZnO nanorods to obtain Ag-ZnO nanorod composites, as shown in Fig. 7. The use of stabilizer molecules facilitated the deposition of Ag monomers primarily at the end of the long ZnO nanorods, as shown in Fig. 7(b), thus offering a certain degree of control over the site-selectivity of the Ag deposition process. The large, dominant Ag particle growth at a particular region of the nanorod and the absence of smaller clusters of Ag throughout the rod surface strongly suggests that once a small Ag nucleus is formed, the rest of the monomers preferentially add to that nucleus instead of heterogeneously nucleating at other locations on the rod. This photoinduced strategy to facilitate the deposition of metals onto a semiconductor surface was not only exploited to synthesize various metal-semiconductor hybrid structures but also to allow for the growth of very large metal domains. As an example, the average diameter of Au domains on Au-CdSe/CdS nanorods synthesized via a thermal growth process[48] is typically on the order of ~2.0 nm. However, UV irradiation of the Au-CdSe/CdS nanorod causes a transfer of electrons from CdS to the Au particle, allowing for the additional reduction of Au ions in solution at the Au particle surface, thus increasing the size of the Au particle to diameters as large as ~ 15 nm.[48] The holes are left behind in the CdS nanorod, and are eventually transferred to the solvent bath. Fig. 8(a) illustrates the growth of large Au domains in Au-CdSe/CdS nanorods via a photoinduced process. Fig. 8(b) is a TEM image showing the dramatic enlargement of the Au domains on CdSe/CdS nanorods via the photoinduced process described above and may becontrasted with the Au-CdSe/CdS structures shown in Fig. 6(a), which were synthesized via a thermal growth process. The addition of hole scavengers such as EtOH, $CHCl_3$ etc, increase the efficiency of the charge separation by ensuring that there is no buildup of holes in the semiconductor. In summary, this light driven mechanism of getting metals to deposit onto semiconductors may be exploited to achieve more complex hybrid metal-semiconductor nanostructures, as will be elaborated on in a later Section.

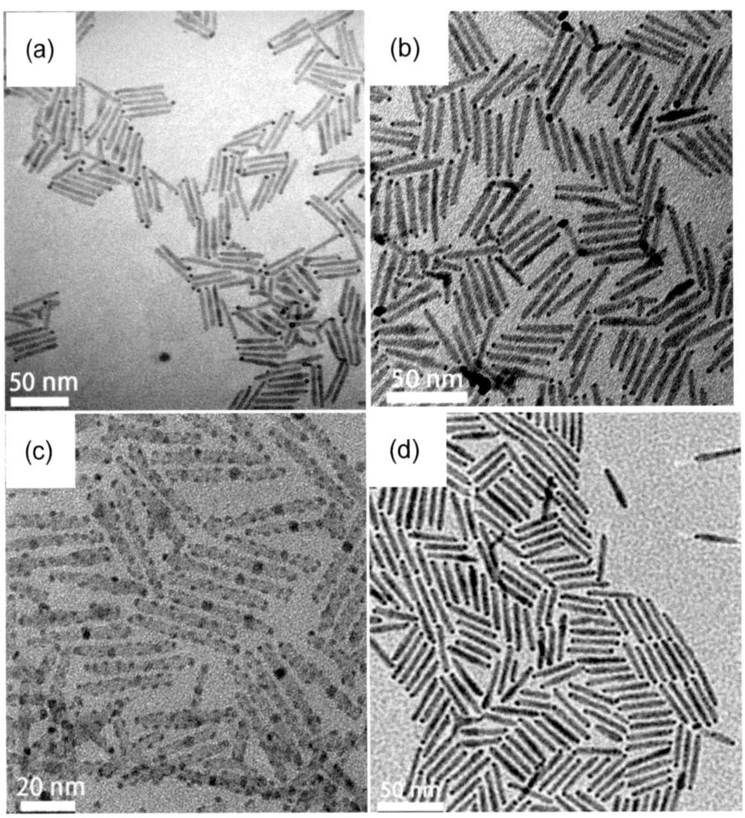

Fig. 6. (a) - (c) are low resolution TEM images showing CdSe seeded CdS nanorods exposed to increasing amounts of Au precursor where the deposition trends proves the hierarchical order of facet reactivities. (d) TEM image of the "Janus-dumbbell" having Au nanoparticle at one end and Ag_2S nanoparticle at the other end. *Courtesy: Angew. Chem. Int. Ed. 2010, 49, 2888-2892.*

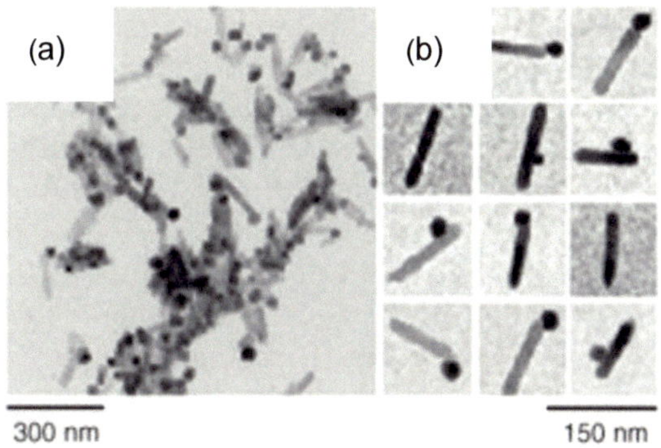

Fig. 7. TEM images of (a) ZnO nanorods with deposited silver particles, (b) gallery of different Ag decorated ZnO surface. *Courtesy: Angew. Chem. Int. Ed. 2004, 43, 4774 –4777.*

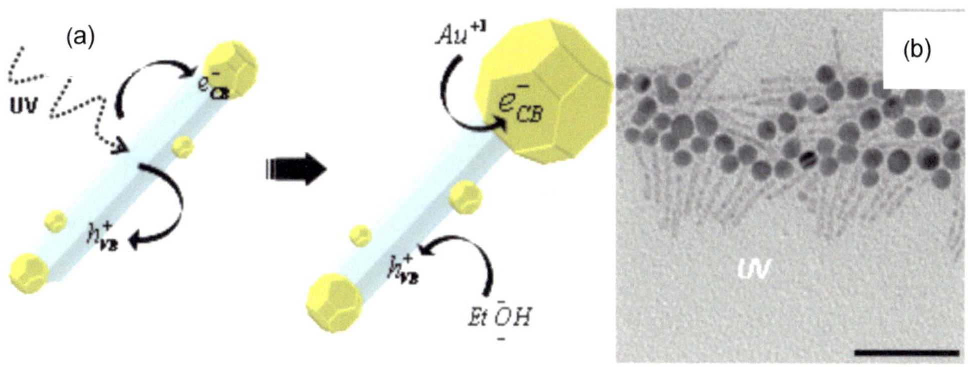

Fig. 8. (a) Schematic, showing the mechanism photo induced large Au growth technique. (b) TEM image of Au-CdS hybrid nanorods formed after UV irradiation having lagre Au domain (scale bar 50 nm). *Courtesy: Nano Lett. 2009, 9, 3710-3714.*

3.1.3 Electrochemical Ostwald-ripening

Ostwald ripening in NCs is a process in which under conditions where all the monomers have been depleted during growth, the smallest and least thermodynamically stable NCs in the ensemble dissolve and re-deposit onto the largest and most thermodynamically stable NCs. While Ostwald ripening in NCs is typically an inter-particle process, it was shown in the case of hybrid Au-CdSe nanorods that the Ostwald ripening of Au occurs via an electrochemical, intra-particle process. While the initial Au deposition was done via synthetic strategies described in Section 3.1.1, continuation of the reaction after all Au monomers had been depleted resulted in the transformation of dumbbell-like Au-CdSe to matchstick-like Au-CdSe nanocomposites,[26] as shown in Fig. 9.

The electrochemical Ostwald ripening process that is relevant to this Au-CdSe nanorod system first involves the dissolution of Au atoms from the smallest Au particles into the solution via an oxidative process. The released electrons are shuttled across the semiconductor nanorod surface by hopping through surface states to the largest Au nanoparticles, where upon reduction of the Au ions in solution occurs.[49] This entire process is summarized in Scheme 2 below. It is evident from Fig. 9(a) that the two Au domains in the dumbbell-like morphology are significantly different in size, and the smaller Au domain possesses a higher surface energy and stronger susceptibility towards oxidation than its larger counterpart, thus providing a driving force for the electrochemical ripening process. The continued dissolution and re-deposition of small Au particles onto the largest ones eventually results in semiconductor nanorods with a single large Au tip.

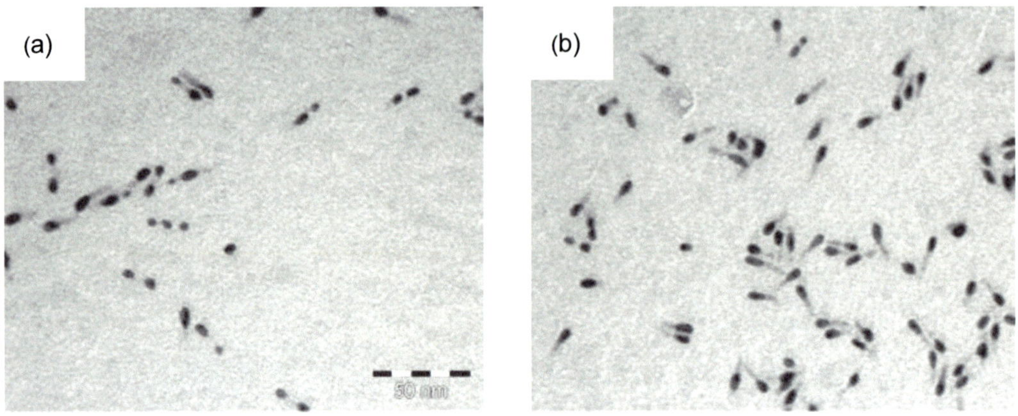

Fig. 9. TEM image of (a) Au-CdSe-Au nanodumbbell and (b) Au-CdSe nano 'nano-bell-tongues.' *Courtesy: Nat. Mater. 2005, 4, 855-863.*

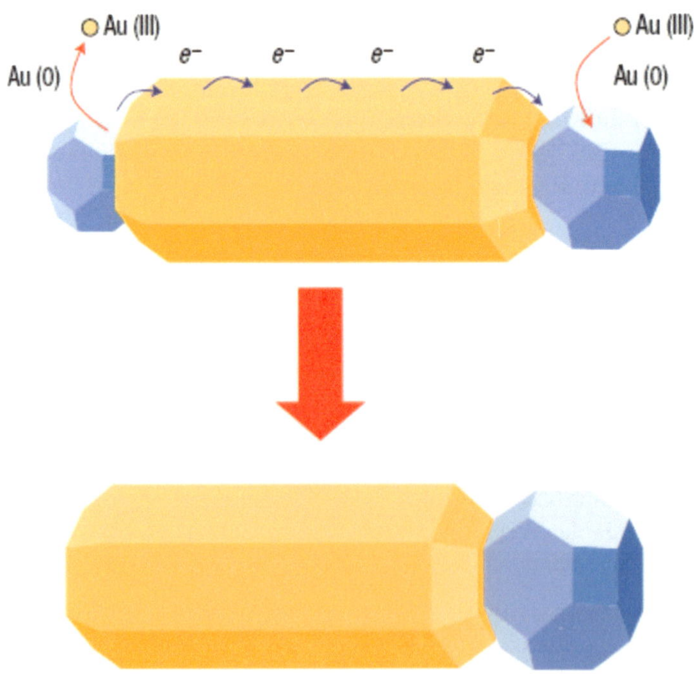

Scheme 2. Schematic of the ripening process in which matchstick-like Au-CdSe nanorods (also known as nano-bell-tongues) are obtained from a dumbbell-like Au-CdSe morphology. *Courtesy: Nat. Mater. 2005, 4, 801-802.*

3.1.4 Surfactant driven metal deposition

Facet-preferential adhesion of ligands or surfactants may also direct the metal deposition process and the resulting morphology of the hybrid metal-semiconductor nanorod structure. An example of such a system is that of magnetic/semiconductor Co decorated TiO_2 nanorods,[50] where careful variation of the amounts of different surface ligands allowed for the derivation of different Co-decorated TiO_2 nanorod morphologies. Through the systematic tuning of different ratios of 1-octanoic acid and oleylamine as strong binding surface capping groups, it was found to be possible to switch from tip-preferential growth to non-selective Co deposition, as summarized in Scheme 3. When the effective concentrations of the ligands are high, they inhibit heterogeneous nucleation on the semiconductor surface, and homogeneous nucleation and growth of Co takes place. A more optimium amount of the surfactants, on the other hand, passivates the sides of the semiconductor surface such that tip selective growth of Co occurs almost exclusively. When exceedingly low amounts of the capping ligands are used, their expectedly insufficient ability to effectively passivate the side facets of the nanorods results in the growth of Co nanoparticles at the tips and throughout the sides of the nanorod. The TEM images corresponding to the various morphologies depicted in Scheme 3 are summarized in Fig. 10. Thus, it is seen from this example that the surfactant can play a very significant role in determining the architecture of anisotropic metal-semiconductor nanostructures.

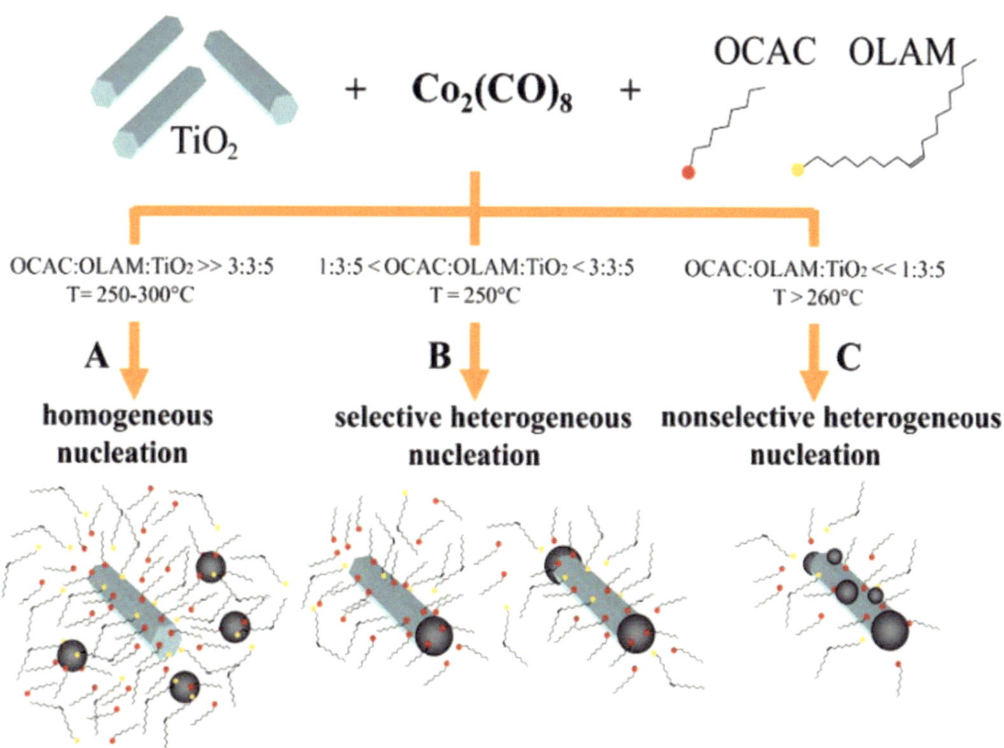

Scheme 3. Schematic for the mechanism of Surfactant-Controlled Co Deposition on TiO_2 nanorod Seeds. *Courtesy: Nano Lett. 2007, 7, 1386-1395.*

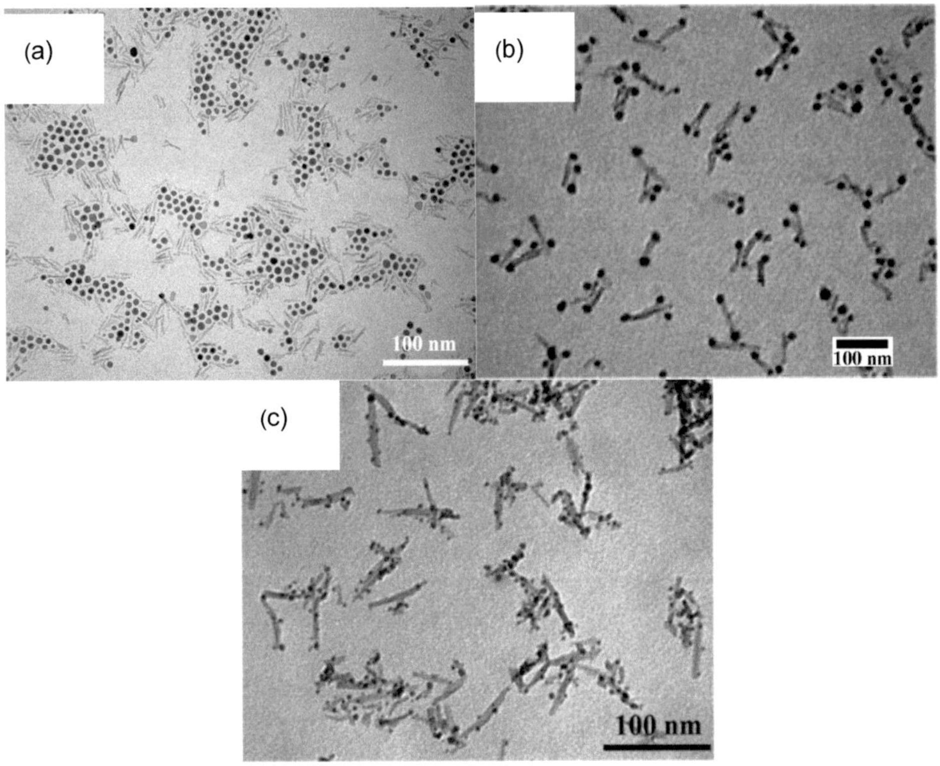

Fig. 10. (a)-(c) Low resolution TEM images corresponding to the various morphologies of Co/TiO_2 hybrid structure, where (a) represents the homogeneous nucleation of Co, (b) shows the tip-preferential growth and (c) illustrate the non-selective Co deposition with TiO_2 nanorods. *Courtesy: Nano Lett. 2007, 7, 1386-1395.*

3.2 Nature of interface

As described earlier, various metal nanoparticles can be specifically deposited onto semiconductor nanorods at different locations such as the tip region or throughout the nanorod surface. While the overall morphology of the hybrid metal-semiconductor nanostructure is undoubtedly important from the standpoint of applications in a more general sense, the nature of the interface between the metal and semiconductor plays a pivotal role for charge transfer processes across the metal-semiconductor interface. In the case of Au-Cd chalcogenide nanorods, which is one of the most studied metal-semiconductor nanorod systems,[51,52] it was found that under the mild chemical synthesis conditions reported by most groups,[45] the Au particle typically does not grow epitaxially on the host semiconductor surface, as illustrated in Fig. 11(a). A study by Figuerola et, al.,[53] investigated the effects of thermal annealing under solvent-less conditions in vacuum for Au nanoparticle decorated CdSe nanorods that were deposited onto a substrate, as shown in Fig. 11(b). Upon heating at 250 °C, all small Au clusters along the sides of the nanorods had disappeared, as evident from Fig. 11(c), while it was noticed that the Au domains at the tips grew significantly larger. This phenomenon cannot be ascribed to the intra-particle electrochemical Ostwald ripening described earlier since the dissolution of Au into its ions cannot take place without the presence of a solvent. In contrast, it has been proposed that the combination of atomic diffusion and cluster diffusion mechanism is responsible for the steady disappearance of Au nanocrystals from the lateral facets of CdSe nanorods. These mechanisms can occur on the substrates and were supported by in situ TEM movies.[53] Closer investigation of the Au-CdSe region via HRTEM (as shown in Fig. 11(d)) revealed a well-defined epitaxial interface between the Au nanoparticle and CdSe nanorod, thus suggesting that thermal annealing may be a viable strategy towards reconstructing the metal-semiconductor interface in wet-chemically synthesized hybrid metal-semiconductor nanostructures. This would certainly be important in optoelectronic applications involving the transport of charges across the metal-semiconductor interface.

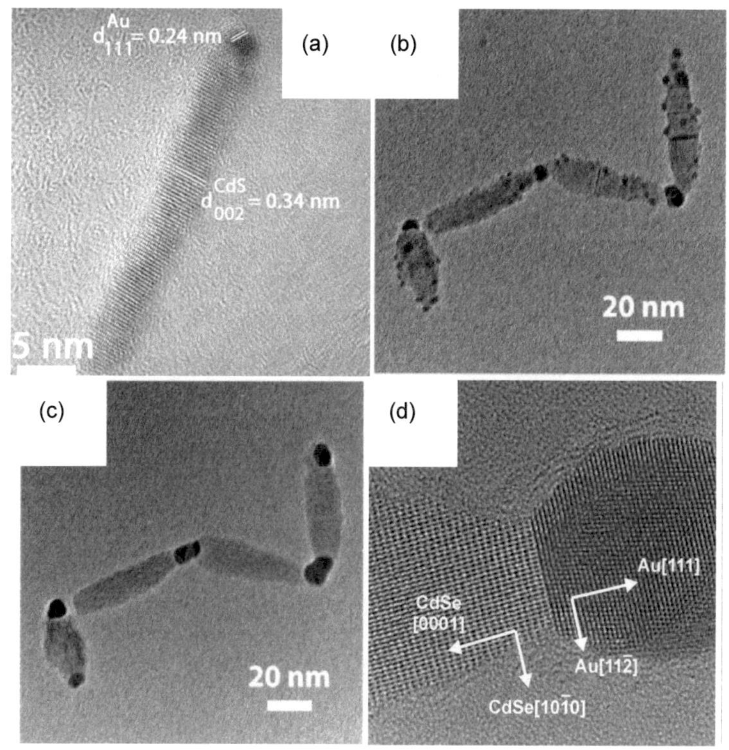

Fig. 11. (a) HRTEM image showing a gold nanoparticle at the apex of the nanorod. (b), (c) TEM images of Au decorated CdSe nanorods heterostructures before and after themmal annealing respectively. (d) HRTEM image of one representative example of epitaxial Au domain with respect to the nanorod. *Courtesy: (a) Angew. Chem. Int. Ed. 2010, 49, 2888-2892; (b),(c),(d) Nano Lett. 2010, 10, 3028-3036.*

3.3 Various applications

As mentioned before, hybrid metal-semiconductor nanostructures can potentially have applications in areas such as directed self-assembly, solution processed optoelectronics and in photocatalysis. Here we describe a few of these applications in order to illustrate the utility of such multicomponent nanostructures. In terms of directed self-assembly processes, the presence of metal tips on semiconductor nanorods provides anchor points for functionalizing ligands with functional groups that differ from the ligands bound to the semiconductor. Zhao et. al. fabricated close-packed films of dumbell-like Au-tipped CdSe nanorods[54] where the rods are aligned parallel to each other due to interactions between alkyl phosphonic acids bound to the CdSe surface, as shown in Fig. 12(b). In addition, the

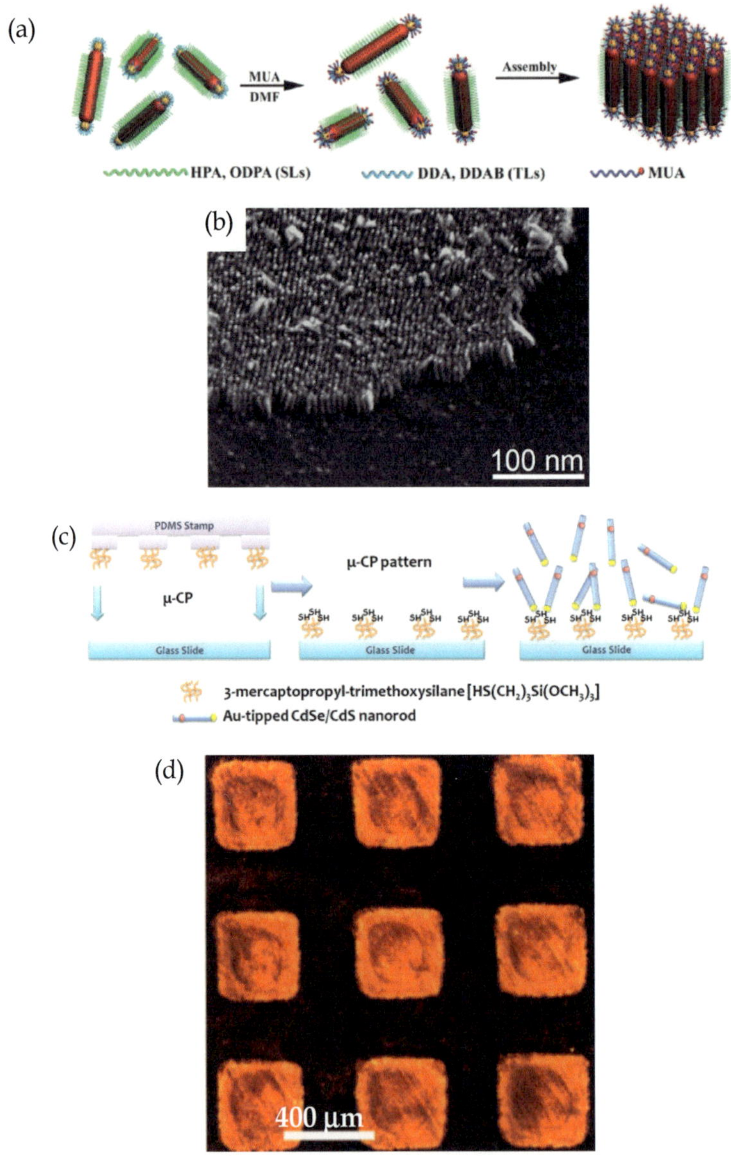

Fig. 12. Examples of the directed assembly of metal-tipped semiconductor nanorods. (a) Schematic illustration of the self-assembly of dumbbell like Au-tipped CdSe NRs (b) SEM image of assembly triggered by interactions between ligands at sides of Au tipped CdSe nanorods. (c) Schematic of directed assembly process with one tipped Au-CdSe/CdS nanorods (d) Showing the fluorescent patterned surfaces on glass substrate. *Courtesy: (a),(b) Nano Lett. 2009, 9, 3077-3081; (c),(d) Small 2011, 7, 2847-2852*

formation of closed-packed films were facilitated by the use of 11-mercaptoundecanoic acid which can conjugate with the Au-tips via the thiol moiety and also help to build hydrogen bonding networks through carboxylic acid groups between them. Recently Chakrabortty et. al. demonstrated that careful engineering of the Au-tipped CdSe seeded CdS nanorod architectures can lead to a significant retention of their fluorescence properties,[55] unlike in the case of Au-CdSe nanorods whose fluorescence are nearly wholly quenched. By patterning a surface with accessible thiol functional groups as illustrated in Fig. 12(c), they were able to show highly specific directed assembly onto the patterned surface via strong Au-S interactions.[56] Fig. 12(d) is a fluorescence microscope image of Au-tipped CdSe seeded CdS nanorods assembled onto the patterned glass slide. The uniformity of the fluorescence pattern formed showcases the fidelity of the directed assembly process. The photocatalytic properties of metal-semiconductor nanorods was also studied by a number of research groups.[57,58] It was suggested that under light irradiation, charge separation of the photogenerated exciton can take place at the interface between the semiconductor and metal depending on the relative position of the Fermi energy level of the metal with respect to the conduction band of the semiconductor. The transfer of these separated charges to surrounding molecules in the solvent thus becomes a basis for photocatalytic, redox-based reactions. One example of such a system is that of dumbbell-like Au-CdSe, where enhanced charge-separation at the metal-semiconductor interface upon light irradiation leads to the highly efficient degradation of methylene blue (Fig. 13(a)-(b)).[57] Given the vast combinations of metals and semiconductor nanomaterials that have already been reported, opportunities for exploring the photocatalytic properties of such systems seem numerous. The use of metal tips as electrical contacts for semiconductor nanorods has also been explored at the single nanorod level by Sheldon et. al.,[42] and it was found that the conductance could be enhanced by a factor of at least 100,000-fold compared to standard CdSe rods (Fig. 13(c)-(d)).

4. Light-induced metal deposition on metal tipped semiconductor nanorods

Expanding the range of metals that can be deposited onto the semiconductor nanorod surface can result in properties previously inaccessible, opening up avenues for newfound applications or enhancing current ones. However, successful deposition of a particular metal on a semiconductor nanostructure depends on the affinity of the metal and its precursor for the semiconductor. Thus exposure of the CdSe seeded CdS nanorods to Pd precursors under the same reaction conditions as those used for Au deposition does not yield any observable Pd deposition onto the semiconductor surface, which can be rationalized by the fact that the Pd-S interaction is much weaker than the Au-S interaction on the CdS surface. This can place a limit on the range of metals that can be attached to a given semiconductor. In order to circumvent this problem, Li et. al.,[59] reported a deposition approach where existing metal tips in a metal-semiconductor nanostructure are used as heterogeneous nucleation sites for the subsequent deposition of a different metal under conditions of UV photo-irradiation Fig. 14 illustrates the resulting structures when Pd precursors are exposed to Au-tipped CdSe seeded CdS nanorods under UV excitation. It is readily observed that Pd can selectively be deposited onto the Au tipped CdSe seeded CdS nanorods, resulting in a significant increase in particle diameter. Interestingly, it was found from HRTEM (High-Resolution Transmission Electron Microscopy) and HAADF- STEM (high angle annular dark field-scanning transmission electron microscopy) data that the Pd forms a homogeneous alloy with the Au particle rather than an expected core-shell structure. The successful deposition of Pd may be understood as follows: under UV excitation, the photogenerated electron migrates to the Au tip, making it more reducing, while the hole is left behind in the CdS rod and is eventually transferred to the solvent bath by hole scavengers. The Pd^{2+} ions are then reduced to Pd^0 upon coming into contact with the Au nanoparticle surface, and end up forming an alloy with Au. This strategy was also extended to the deposition of Fe, which can exhibit magnetic properties even upon oxidation.

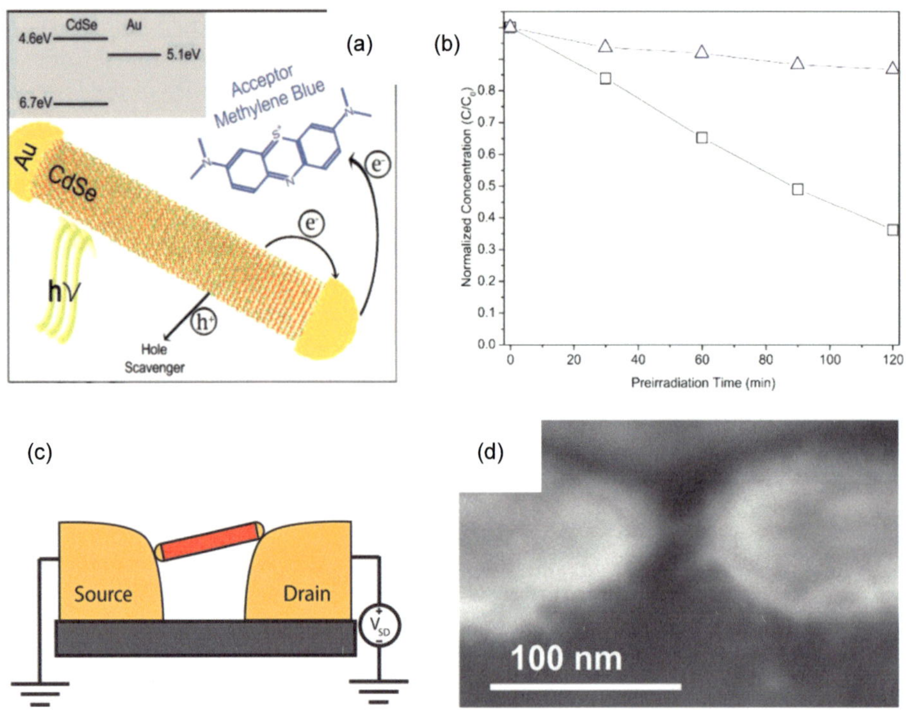

Fig. 13. (a) Scheme of a light-induced charge separation mechanism in Au-tipped CdSe nanorod where the photo-generated electron resides at the gold tip to serve as a reduction point. (b) Comparing the reduction efficiency of CdSe nanorods and gold nanoparticles mixture (blue triangles) with hybrid CdSe-Au nanodumbbells (black squares) in methylene blue dye. (c) and (d) are the schematic and SEM image of a single nanocrystal two-terminal device, respectively. *Courtesy: (a), (b) Nano Lett, 2008, 8, 637-641; (c), (d) Nano Lett, 2009, 9, 3676-3682.*

Unlike in the case of Pd, however, it was observed that a hollow shell of Fe_3O_4 develops around the Au core at the tip of the CdSe seeded CdS nanorod, as shown by the TEM image in Fig. 15 (a). This was comprehensively verified by HRTEM and HAADF-STEM techniques, as illustrated in Fig. 15(b)-(d). The use of light-induced techniques which exploit the transfer of electrons from the semiconductor to its metal tips in order to facilitate the deposition of other metals that are difficult to grow directly onto the semiconductor surface opens up new avenues for the hierarchical build-up of very complex metal-semiconductor nanostructures. Such structurally complex multicomponent nanoparticles can potentially exhibit unique physicochemical properties that address some of the current outstanding challenges in semiconductor NC-based applications, and perhaps form the basis for new ones.

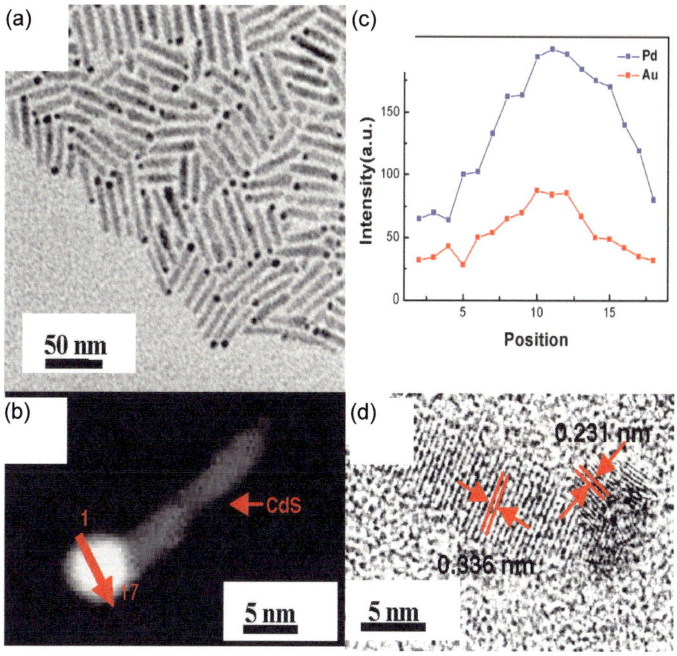

Fig. 14. (a) TEM image of Pd-Au-tipped CdSe-seeded CdS nanorod matchsticklike structures. (b) HAADF-STEM image of a typical rod, showing Au/Pd alloycomposition in it. (c) Corresponding EDX line scan across the spherical tip. (d) HRTEM image of a Pd-Au-CdSe-seeded CdS rod. The lattice spacing clearly related to the literature value here. *Courtesy: J. Am. Chem. Soc. 2011, 133, 672-675.*

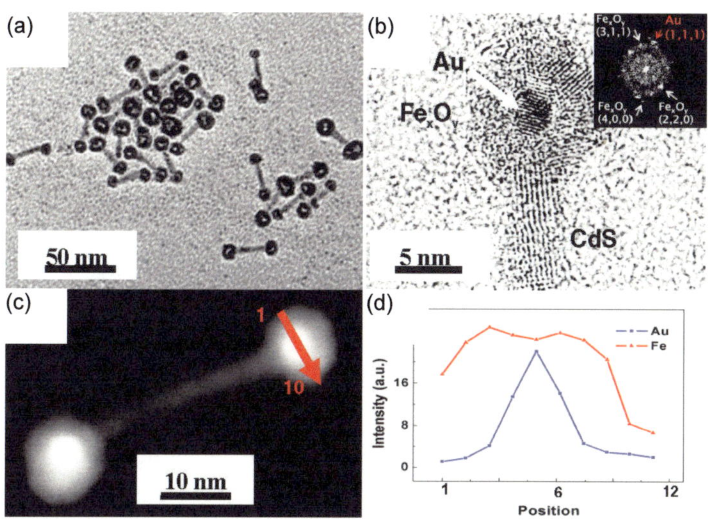

Fig. 15. (a) TEM image core Au nanoparticles with thick hollow shell of Fe_xO_y. (b) HRTEM image of part of the core-shell-tipped rod. The inset is an FFT image of the spherical tip region. (c) HAADF-STEM image showing three distinct materials composition within the same structure. (d) EDX line scan of the spherical tip shown in (C). *Courtesy: J. Am. Chem. Soc. 2011, 133, 672-675.*

5. Conclusion

In summary, this chapter describes a diverse set of wet-chemical synthesis routes to obtain monodisperse semiconductor nanorods of different materials composition. One of the most promising synthetic methods for obtaining semiconductor nanorods is that of the seeded growth approach, where very monodisperse rods with aspect ratios as large as 1:30 can be achieved. A number of strategies to achieve a wide range of metal-semiconductor heterostructures was presented, and the conditions used to achieve site-selective deposition were elaborated on. As the nature of the interface between the metal and the semiconductor is of paramount importance in the context of optoelectronic applications, a method to obtain epitaxial growth of metals on semiconductor nanorods was provided. The multicomponent architecture of the hybrid metal-semiconductor nanostructures potentially incorporates multiple functionalities within a single nanoparticle, and can demonstrate utility in a whole host of applications ranging from directed self-assembly to photocatalysis. Some of these salient applications were demonstrated in this chapter, though there is clearly still a lot of progress to be made. Lastly, we described a recently reported light-induced synthetic strategy which potentially expands the range of metals that can be deposited at the tips of semiconductor nanorods. These developments collectively suggest that hybrid metal-semiconductor nanostructures will continue to generate very active research interest from both the standpoints of fundamental science and impactful applications.

6. Acknowledgment

We wish to thank the Department of Chemistry, National University of Singapore for its continued support in our research efforts.

7. References

[1] R. Rossetti, R. Hull, J. M. Gibson and L. E. Brus, J. Chem. Phys. 1985, *82*, 552-559.
[2] A. I. Ekomov and A. A. Onushchenko, *JETP Lett.* 1982, *34*, 345–349.
[3] C. B. Murray, D. J. Norris and M. G. Bawendi, *J. Am. Chem. Soc.* 1993, *115*, 8706-8715.
[4] M. A. Hines and P. Guyot-Sionnest, *J. Phys Chem,* 1996, *100*, 468-471.
[5] H. Han, N. K. Devaraj, J. Lee, S. A. Hilderbrand, R. Weissleder and M. G. Bawendi, *J. Am. Chem. Soc.* 2010, *132*, 7838–7839.
[6] D. Battaglia and X. Peng, *Nano Lett.* 2002, *2,* 1027-1030.
[7] J. M. Pietryga, R. D. Schaller, D. Werder, M. H. Stewart, V. I. Klimov and J. A. Hollingsworth, *J. Am. Chem. Soc.* 2004, *126*, 11752–11753.
[8] S. Coe, W. K. Woo, M. G. Bawendi, and V. Bulovic, *Nature* 2002, *420*, 8706
[9] I. Gur, N. A. Fromer, M. L. Geier, and A. P. Alivisatos, *Science* 2005, *310*, 462
[10] X. Michalet, F. F. Pinaud, L. A. Bentolila, J. M. Tsay, S. Doose, J. J. Li, G. Sundaresan, A. M. Wu, S. S. Gambhir and S. Weiss, *Science* 2005, *307*, 538
[11] X. Peng, L. Manna, W. D. Yang, J. Wickham, E. Scher, A. Kadavanich and A. P. Alivisatos, *Nature* 2000, *404*, 59-61.
[12] T. Mokari, E. Rothenberg, I. Popov, R. Costi and U. Banin, *Science* 2004, *304*, 1787-1790.
[13] Z. A. Peng and X. G. Peng, *J. Am. Chem. Soc.* 2001, *123*, 1389-1395.
[14] M. P. Pileni, Nat. Mater. 2003, *2*, 145-150.
[15] S. M. Lee, S. N. Cho and J. Cheon, *Adv. Mater.* 2003, *15*, 441-444.

[16] C. D. Donega, P. Liljeroth and D. Vanmaekelbergh *Small*, 2005, *1*, 1152-1162.
[17] S. Kumar and T. Nann, *Small* 2006, *2*, 316-329.
[18] L. Manna, E. C. Scher, and A. P. Alivisatos, *J. Am. Chem. Soc.* 2000, *122*, 12700-12706.
[19] W. W. Yu, Y. A. Wang and X. Peng, *Chem. Mater.* 2003, *15*, 4300-4308.
[20] N. Zhao and L. M. Qi, *Adv. Mater.* 2006, *18*, 359
[21] E. Lifshitz,M. Bashouti, V. Kloper, A. Kigel, M. S. Eisen and S. Berger, *Nano Lett.* 2003, *3*, 857-862.
[22] L. Manna, D. J. Milliron, A. Meisel, E. C. Scher, and A. P. Alivisatos, *Nat. Mater.* 2003, *2*, 382-385.
[23] S. Deka, K. Miszta, D. Dorfs, A. Genovese, G. Bertoni and L. Manna, *Nano Lett.* 2010, *10*, 3770-3776.
[24] J. Hu, L. Li, W. Yang, L. Manna, L. Wang, A. P. Alivisatos, *Science* 2001, *292*, 2060-2063.
[25] W. U. Huynh, J. J. Dittmer, A. P. Alivisatos, *Science* 2002, *295*, 2425-2427.
[26] T. Mokari, C. G. Sztrum, A. Salant, E. Rabani and U. Banin, *Nat. Mater.* 2005, *4*, 855-863.
[27] Habas, S. E.; Yang, P.; Mokari, T. *J. Am. Chem. Soc.* 2008, *130*, 3294
[28] Deka, S.; Falqui, A.; Bertoni, G.; Sangregorio, C.; Poneti, G.; Morello, G.; Giorgi, M. D.; Giannini, C.; Cingolani, R.; Manna, L.; Cozzoli, P. D. *J. Am. Chem. Soc.* 2009, *131*, 12817.
[29] W. Koh, A. C. Bartnik, F. W. Wise and C. B. Murray, *J. Am. Chem. Soc.* 2010, *132*, 3909-3913.
[30] L. Carbone, C. Nobile, M. De Giorgi, F. D. Sala, G. Morello, P. Pompa, M. Hytch, E. Snoeck, A. Fiore, I. R. Franchini, M. Nadasan, A. F. Silvestre, L. Chiodo, S. Kudera, R. Cingolani, R. Krahne and L. Manna, *Nano Letters* 2007, *7*, 2942-2950.
[31] D. V. Talapin, R. Koeppe, S. Gotzinger, A. Kornowski, J. M. Lupton, A. L. Rogach, O. Benson, J. Feldmann and H. Weller, *Nano Letters* 2003, *3*, 1677-1681.
[32] D. V. Talapin, J. H. Nelson, E. V. Shevchenko, S. Aloni, B. Sadtler and A. P. Alivisatos, *Nano Letters* 2007, *7*, 2951-2959.
[33] G. Konstantatos and E. H. Sargent, *Nat Nanotechnol*, 2010, *5*, 391-400.
[34] R. K. Wahi, W. W. Yu, Y. Liu, M. L. Mejia, J. C. Falkner, W. Nolte and V. L. Colvin, *J. of Mol. Catal. A: Chemical*, 2005, *242*, 48–56.
[35] J. J. Shiang, A. V. Kadavanich, R. K. Grubbs and A. P. Alivisatos, *J. Phys. Chem.* 1995, *99*, 17417-17422.
[36] Z. Deng, H. Yan and Y. Liu, *Angew. Chem. Int. Ed.* 2010, *49*, 8695-8698.
[37] I. Gur, N. A. Fromer, M. L. Geier and A. P. Alivisatos, *Science* 2005, *310*, 462-465.
[38] F. Shieh, A. E. Saunders and B. A. Korgel, *J. Phys. Chem. B* 2005, *109*, 8538-8542.
[39] P. D. Cozzoli, T. Pellegrino and L. Manna, *Chem. Soc. Rev.* 2006, *35*, 1195.
[40] P. Peng, D. J. Milliron, S. M. Hughes, J. C. Johnson, A. P. Alivisatos and R. J. Saykally, *Nano Letters* 2005, *5*, 1809-1813.
[41] A. Figuerola, I. R. Franchini, A. Fiore, R. Mastria, A. Falqui, G. Bertoni, S. Bals, G. V. Tendeloo, S. Kudera, R. Cingolani and L. Manna, *Adv. Mater.* 2009, *21*, 550-554.
[42] M. T. Sheldon, P. E. Trudeau, T. Mokari, L. W. Wang and A. P. Alivisatos, *Nano Lett*, 2009, *9*, 3676-3682.
[43] J. Yang, H. I. Elim, Q. Zhang, J. Y. Lee and W. Ji, *J. Am. Chem. Soc.* 2006, *128*, 11921-11926.
[44] J. Maynadié, A. Salant, A. Falqui, M. Respaud, E. Shaviv, U. Banin, K. Soulantica and B. Chaudret, *Angew. Chem. Int. Ed.* 2009, *48*, 1814-1817.

[45] Chakrabortty, S.; Yang, J. A.; Tan, Y. M.; Mishra, Nimai; Chan Y. *Angew. Chem. Int. Ed.* 2010, *49*, 2888-2892.
[46] A. E. Saunders, I. Popov and U. Banin, *J. Phys. Chem. B* 2006, *110*, 25421-25429.
[47] C. Pacholski, A. Kornowski and H. Weller, *Angew. Chem. Int. Ed.* 2004, *43*, 4774 -4777.
[48] L. Carbone, A. Jakab, Y. Khalavka and C. Sonnichsen, *Nano Lett.* 2009, *9*, 3710-3714.
[49] P. D. Cozzoli and L. Manna, *Nat. Mater.* 2005, *4*, 801-802.
[50] M. Casavola, V. Grillo, E. Carlino, C. Giannini, F. Gozzo, E. F. Pinel, M. A. Garcia, L. Manna, R. Cingolani and P. D. Cozzoli, *Nano Lett.* 2007, *7*, 1386-1395.
[51] R. Costi, A. E. Saunders and U. Banin, *Angew. Chem. Int. Ed.* 2010, *49*, 4878.
[52] L. Carbone and P. D. Cozzoli, *Nano Today* 2011, *5*, 449.
[53] A. Figuerola, M. Huis, M. Zanella, A. Genovese, S. Marras, A. Falqui, H. W. Zandbergen, R. Cingolani and L. Manna, *Nano Lett.* 2010, *10*, 3028-3036.
[54] N. Zhao, K. Liu, J. Greener, Z. Nie and E. Kumacheva, *Nano Lett.* 2009, *9*, 3077-3081.
[55] S. Chakrabortty, G. Xing, Y. Xu, N. S.Wee, N. Mishra, T. C. Sum and Y. Chan, *Small,* 2011, 7, 2847-2852.
[56] D. R. Lide, *CRC Handbook of Chemistry and Physics,* CRC press, 2000, pp. 9-52.
[57] R. Costi, A. E. Saunders, E. Elmalem, A. Salant and U. Banin, *Nano Lett,* 2008, *8*, 637-641.
[58] E. Elmalem, A. E. Saunders, R. Costi, A. Salant and U. Banin, *Adv. Mater.* 2008, *20*, 4312-4317.
[59] X. Li, J. Lian, M. Lin and Y. Chan, *J. Am. Chem. Soc.* 2011, *133*, 672-675.